T5-AGA-254

THE ECONOMICS OF MANAGING CHLOROFLUOROCARBONS

THE ECONOMICS OF MANAGING

CHLOROFLUOROCARBONS

STRATOSPHERIC OZONE AND CLIMATE ISSUES

edited by
John H. Cumberland
James R. Hibbs
and Irving Hoch

RESOURCES FOR THE FUTURE / WASHINGTON, D.C.

Library of Congress Cataloging in Publication Data
Main entry under title:

The Economics of managing chlorofluorocarbons.

Bibliography: p.
1. Chlorofluorocarbons—Environmental aspects.
I. Cumberland, John H. II. Hibbs, James Russell,
1907- III. Hoch, Irving, 1926-
TD887.C47E26 1982 363.7′392 82-11279
ISBN 0-8018-2963-1

Distributed by The Johns Hopkins University Press, Baltimore, Maryland 21218
Manufactured in the United States of America

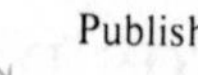

Published August 1982

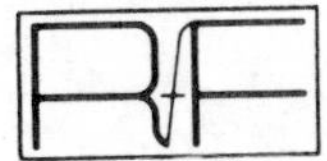

RESOURCES FOR THE FUTURE, INC.
1755 Massachusetts Avenue, N.W., Washington, D.C. 20036

Resources for the Future is a nonprofit organization for research and education in the development, conservation, and use of natural resources, including the quality of the environment. It was established in 1952 with the cooperation of the Ford Foundation. Grants for research are accepted from government and private sources only on the condition that RFF shall be solely responsible for the conduct of the research and free to make its results available to the public. Most of the work of Resources for the Future is carried out by its resident staff; part is supported by grants to universities and other nonprofit organizations. Unless otherwise stated, interpretations and conclusions in RFF publications are those of the authors; the organization takes responsibility for the selection of significant subjects for study, the competence of the researchers, and their freedom of inquiry.

Research Papers are studies and conference reports published by Resources for the Future from the authors' typescripts. The accuracy of the material is the responsibility of the authors and the text is edited at the authors' discretion. Research Papers are intended to provide prompt distribution of research that is likely to have somewhat more limited appeal than that of RFF books.

CONTENTS

LIST OF TABLES

LIST OF FIGURES

FOREWORD

New technologies frequently have unanticipated effects differing substantially from the principal original purpose for which the technology was created. Usually this is because the evaluation of a technology prior to introduction concentrates on those costs and benefits that are measured in the marketplace. Environmental effects may be neglected either because they do not enter marketplace economics directly or because these unintended consequences cannot be predicted accurately, or even be anticipated at all. When such unintended effects are identified or even suspected, it behooves us to evaluate them as best we can both for their positive and negative effects on present and future societies.

The several applications of chlorofluorocarbons present just such a case. They have proved very useful as spray can propellants, solvents, plastics, and, of special importance, as refrigerants. But the effects of this group of chemicals are not confined to those intended by the manufacturers of deodorants, refrigerators, and other consumer and industrial goods. Either directly or indirectly, chlorofluorocarbon emissions find their way to the atmosphere where they create complex reactions that potentially affect sunlight coming to the earth and the earth's climate in fundamental ways. It is most important that the social adjustments stimulated by these effects be studied and alternative forms of social behavior explored; indeed, that is what this volume is about. But it also is important that such social science analysis be based on appropriate natural science premises and this calls for a multidisciplinary effort. Happily, such evaluation is an important part of the work of Resources for the Future. We are so organized that interdisciplinary work comes to us easily and we have many links with the academic community. Both were vital to the evaluation reported upon in this research paper.

The chlorofluorocarbon problem is complex not only from a natural science standpoint, but also it involves some of the most difficult problems of group action and cooperative behavior. Moreover, the atmosphere is an open-access resource that is used jointly by all of the nations of the globe. It is hardly possible to imagine a more intractable problem, yet it is all too real and immediate. Given the complexity of the problem, it is not surprising that the results reported on often are tentative and speculative. But if some of the results are less than firm, other suggestions are practical and concrete. They do provide guidance about what we can do now to address this potentially serious problem.

We at Resources for the Future are pleased that we were able to work cooperatively with the University of Maryland on the administration of this project. But the scholars who participated in the conference reported on here were not limited to the University of Maryland and RFF. A dozen other universities and institutions also were involved. Thus, the job of coordinating and preparing a volume of this scope was even more difficult than is typical, and I would like to express my appreciation to the editors for their perserverence in making it possible. The conference papers required considerable work to make them suitable for publication by RFF, and the editors made the extra effort that was necessary to make the deliberations of the conference available to a broader audience. Resources for the Future is proud to add this volume to our list of publications and to have assisted financially in its production as well.

July 1982
Washington, D.C.

Emery N. Castle, President
Resources for the Future

PREFACE

As one of its activities in addressing the issue of the consequences of chlorofluorocarbon (CFC) emissions on ozone depletion and climate change, the U.S. Environmental Protection Agency made a grant in 1977 to the University of Maryland to consider conceptual, empirical, and policy issues involved in CFC management. James R. Hibbs was the EPA project officer for that grant, and John H. Cumberland served as principal investigator. Under the leadership of Cumberland and Hibbs, a group of scholars was assembled to investigate the economic aspects of ozone depletion and related changes in climate, and a set of papers was commissioned to report results. The group was international in its composition, though many of the scholars were affiliated with the University of Maryland or with Resources for the Future.

The participants in the project presented their initial papers at a conference at Port Deposit, Maryland, in July 1978, and those papers served as the basis of a report to the U.S. Environmental Protection Agency in November 1979. With the aim of disseminating the results to a wider audience, Resources for the Future funded the editing and revision of the initial set of papers for their appearance in this RFF Research Paper; primary responsibility at that stage was assumed by Irving Hoch of RFF. Besides considerable revision and some updating of the original papers, the RFF phase of the work included a greater emphasis on climate effects, given the explicit recognition that CFC climate impacts are expected to rival those of carbon dioxide in magnitude. In addition, Alphonse Forziati wrote an introductory chapter for this volume which presents a review of natural science information on the CFC emissions problem, and the editors wrote an introductory chapter which summarize each of the papers and presents conclusions. Cumberland was senior author of that chapter.

The purpose of this study has been to assemble the best available natural and social science information on the problem of managing CFC emissions and to assess the socioeconomic implications of various management strategies. The methodology has been to evaluate the current state of knowledge available in order to arrive at interim judgments on the benefits and costs of alternative levels of control. In so doing we have attempted to determine the advantages and disadvantages of utilizing alternative control instruments and to identify critical areas for further research.

In addressing these goals, this study explicitly accepts the need to live with the great uncertainties involved; to make the best use of existing information in order to arrive at reasonable estimates of the consequences of various courses of action; and to develop new methodologies and approaches to reduce the extent of the uncertainties.

Although there has been a time lag between the preparation of the initial papers and the appearance of their edited versions here, the research remains both current and relevant, given only modest change in the terms of the problem in the interim. The technical bases of all the papers are consistent with the central estimates presented by Forziati. Further, the papers generally have applicability to a wide range of issues and concerns, for analogous global environmental problems from unanticipated side effects are liable to be all too much with us in the years ahead.

Hence, an implicit goal which became explicit in the course of the work was to elucidate and address the general issues exemplified in, but not limited to, the problems of ozone depletion and temperature changes arising from the emission of CFCs.

The volume editors are grateful to the following persons who contributed to this study. In the University of Maryland phase of the work, Kwang Choi, James R. Kahn, and Derek Updegraff assisted in the preparation of the report to the EPA. Nancy Crawford supervised the production of that report and made the arrangements for the Port Deposit conference. Besides the authors of the chapters in this volume, conference participants included Robert D. Niehaus of the University of Maryland, and Daniel Golomb and Edward DeFabo of the EPA.

During the RFF phase of the work, the manuscript benefited from review comments received from John W. Firor, Kenneth D. Frederick, Helmut Landsberg, Walter O. Spofford, Jr., William D. Watson, and the members of the RFF Publications Committee. Jo Hinkel and Elsa Williams gave considerable help in the technical editing and production of the final volume, and John Mankin typed the several drafts of the manuscript.

Washington, D.C.
June 1982

Irving Hoch
Senior Fellow
Renewable Resources Division
Resources for the Future

LIST OF CONTRIBUTORS

Martin J. Bailey is Professor of Economics at the University of Maryland.

Peter Bohm is Professor of Economics at the University of Stockholm in Sweden. He was a visiting scholar at Resources for the Future at the time of the Port Deposit conference.

John H. Cumberland is Professor of Economics, and Director of the Bureau of Business and Economic Research at the University of Maryland.

Ralph C. d'Arge is John S. Bugas Professor of Economics at the University of Wyoming.

Alphonse Forziati was Director of Stratospheric Modification Research, U.S. Environmental Protection Agency from 1977 to 1979, retiring in 1980. He has a Ph.D. in chemistry.

Thomas N. Gladwin is Associate Professor of Management and International Business at the Graduate School of Business Administration of New York University.

James R. Hibbs is an economist with the U.S. Environmental Protection Agency, and has a Ph.D. in economics.

Irving Hoch is a Senior Fellow with Resources for the Future, and has a Ph.D. in economics.

Harry H. Kelejian is Professor of Economics at the University of Maryland.

Martin C. McGuire is Professor of Economics at the University of Maryland.

Ezra J. Mishan is a Visiting Professor of Economics at the University of Victoria, British Columbia, Canada. He was formerly Professor of Economics at The London School of Economics and was a Visiting Professor of Economics at the University of Maryland at the time of the Port Deposit conference.

Wallace E. Oates is Professor of Economics at the University of Maryland.

Mancur L. Olson is Professor of Economics at the University of Maryland, and is an Associate and Consultant to Resources for the Future.

Talbot Page is a Research Associate at the California Institute of Technology. He was a Senior Research Associate with Resources for the Future at the time of the Port Deposit conference. He has a Ph.D. in economics.

V. Kerry Smith is Professor of Economics at the University of North Carolina, Chapel Hill. He was a Fellow with Resources for the Future at the time of the Port Deposit conference.

Dennis J. Snower is an Assistant Professor of Economics at the University of Maryland. He is currently a Visiting Professor in the Department of Economics, University of London in England.

Ivar E. Strand, Jr., is an Associate Professor of Agricultural and Resource Economics at the University of Maryland.

Diana L. Strassmann is a doctoral candidate at the Department of Economics at Harvard University.

Judith L. Ugelow is a doctoral candidate in economics at the Graduate School of Business Administration of New York University. She has an M.S. in environmental engineering.

Bruce C. Vavrichek is an economist with the Congressional Budget Office. He was an Assistant Professor of Economics at the University of Maryland at the time of the Port Deposit conference.

Ingo Walter is Professor of Economics and Finance, and Chairman of International Business at the Graduate School of Business Administration at New York University.

Chapter 1

OVERVIEW

John H. Cumberland, James R. Hibbs, and Irving Hoch

One of the ironies of contemporary technological civilization is that some technological advances, which seem ideal for their immediate purposes, turn out to have unsuspected side effects with potentially serious consequences for society. A recently discovered example is the potentially serious damage to the earth's protective ozone shield from inert chlorofluorocarbons (CFCs) used as aerosol propellants, refrigerants, solvents, foam-blowing agents, plastics, and resins. It is also the case that the CFCs could have significant impacts on the global climate. The technological attraction of the CFCs lies in the combination of qualities which makes them so ideal for their immediate use by society--chemical stability, noncorrosiveness, noninflammability, nontoxicity, and low cost. However, the very stability which makes them so ideal for their purposes permits them to remain inert until they migrate from the troposphere (the lower atmosphere) to the stratosphere, where a complex set of photochemical reactions culminates in the destruction of stratospheric ozone. In addition, during the long period when the CFCs reside in the troposphere, they contribute to the "greenhouse effect" by absorbing the emissions of infrared radiation from the surface of the earth, thus contributing to a rise in temperature. This is in addition to a similar greenhouse effect expected from increased carbon dioxide (CO_2) emissions. Although the strength and magnitude of the complex physical linkages between these processes have not been precisely measured, the accumulating evidence indicates that the use of CFCs has damaged and is continuing to damage the earth's ozone shield. Additional, though less conclusive, evidence suggests that with damage to the ozone layer, increased amounts of biologically damaging ultraviolet radiation reach the earth's surface, and this is likely to increase the incidence of both melanoma and nonmelanotic skin cancer for human beings, and to injure crops, livestock, materials, and ecosystems. Further, on a molecular basis, it appears that CFCs are about 100,000 times more effective than CO_2 as contributors to the greenhouse effect. Thus, small amounts of CFCs in the troposphere correspond, in impact, to much greater amounts of CO_2. Consequently, it appears that by the year 2000, CFCs may well contribute as much to global warming as will CO_2. This prospective impact probably deserves more recognition than it has received.

In sum, improbable though it may seem, the growing use of an apparently harmless group of aerosol sprays, refrigerants, and related compounds on the earth's surface appears to have serious long-term effects in the stratosphere, direct effects on the earth's temperature, and the potential for irreversible damage to life on earth.

The issues involved in the response to that problem through the management of CFC emissions are addressed in this volume using the following three-part organization of chapters. Part I presents an overview and the setting of the problem. In addition to this chapter, which summarizes the remaining chapters, Part I includes a chapter by Forziati on natural science relationships and one by Gladwin, Ugelow, and Walter on the economic and institutional setting for CFC production, consumption, and trade. There is some focus in that chapter on international trade and the international regulatory process, a natural emphasis given the global dimensions of the CFC problem. Parts II and III apply economic analysis to the question of the management of CFCs. The chapters in part II focus on benefit-cost calculations, while those in part III are subsumed under the heading of "policy perspectives and analysis." A common concern in this last group is the economics of the regulatory process, ranging from practical advice on regulatory behavior to a conceptual framework for viewing the regulatory process.

We will summarize and review the individual chapters of the volume, in turn, and then carry out some integration of the findings in offering some conclusions and some suggestions for future research.

1.1 The Setting: Natural Science Relationships and CFC Production, Emissions, and Trade

Table 1-1 summarizes some of the key relationships reviewed by Forziati in his chapter, which summarizes the state of knowledge of the natural science relationships involved in the CFC emissions problem. The CFCs are the most important of the man-made compounds that deplete stratospheric ozone (O_3), a high-energy form of oxygen. Ozone is depleted or destroyed when it is transformed into other forms of oxygen. This can happen through many naturally occurring processes; however, additional ozone depletion now occurs by way of CFC interactions with ozone. The stratospheric ozone layer shields the earth's surface from destructive ultraviolet radiation (DUV); and with the depletion of the ozone layer, more DUV reaches the earth's surface. UV-B, a particular band of ultraviolet radiation, is involved in DUV. The amounts of UV-B, when weighted by the strength of biological impact of the wavelengths included, are referred to as DUV. The likely consequences of increased DUV include both higher rates of human skin cancer and nonhuman biological damage. The latter includes reduction in photosynthesis in plants; reduced crop yields; and adverse effects on fisheries, particularly on shellfish and the anchovy.

The CFCs are a subcategory of the halocarbons, which in turn are compounds of carbon and one or more halogens (fluorine, chlorine, bromine, or iodine). The CFCs are distinguished by containing both chlorine and fluorine and may also contain hydrogen. The major CFCs are CFC-11 and CFC-12, accounting for about 80 percent of CFC production; both have high emission rates and are long-lived in the stratosphere. The longer the stratospheric life, the greater the potential for ozone destruction. Less important CFCs include CFC-22, CFC-113, CFC-114, and others. CFC numerical designations indicate chemical composition; F is often used in place of CFC--for example, F-11. Some other chlorine-based halocarbons have ozone-depleting properties, but are relatively minor in impact, either because of their low emission rates (for example, carbon tetrachloride) or their short life span in the stratosphere (for example, methyl chloroform). Some nonchlorine halocarbons have ozone-depleting properties but are quite minor in impact.

Table 1-1. Key Relationships in the CFC Emissions Problem

Category	Description of relationship	Best estimate of parameter value
Major CFCs: F-11, F-12	1. Percentage of total CFC production	80%
	2. Lifetime in troposphere	100 years
	3. Lifetime in stratosphere	10 + years
Ozone depletion, in percentage, ultimate steady-state value	1. Central estimate, based on 1973 rate of CFC emissions continuing	11 - 16%
	2. Estimates assuming exponential increase in CFCs, paralleling growth in last decade	30 - 57%
Biological effects of ozone depletion	1. Percentage change in damaging ultraviolet radiation (DUV) relative to percentage change in ozone depletion	2:1
	2. Percentage of increase in incidence of skin cancer relative to percentage change in DUV	2:1
CFC-induced temperature increase, year 2000	1. Direct effect of CFCs: greenhouse effect	+ 0.7°C
	2. Indirect effect of CFCs: ozone depletion effect	- 0.1°C
	3. Total CFC effect	+ 0.6°C
	4. Comparative estimate, year 2000 CO_2 greenhouse effect	+ 0.5°C

Levels of CFC production and emissions are summarized in table 1-2, drawing on information developed by Gladwin, Ugelow and Walter in chapter 3.

Besides their use as propellants for aerosols (now banned in the United States and in a few other countries), the major uses of CFCs are as refrigerants (in refrigerators and air conditioners), as solvents, and in urethane foam plastics. The latter group includes rigid foams (insula-

Table 1-2. Distributions of Production and Emissions of CFCs and Related Halocarbons

Chlorine halocarbons	Approximate share of CFC production	Estimated world emissions (1,000 metric tons)	Lifetime in stratosphere
CFC-11	30	700	Greater than 10 yr
CFC-12	50		
CFC-22	12	60	1 to 10 yr
All other CFCs	8	40	Various
Carbon tetrachloride	--	40	Greater than 10 yr
Methyl chloroform	--	320	1 to 10 yr

tion, packaging) and flexible foams (furniture, bedding). CFC emissions are released immediately in many of the uses of those chemicals, but in a number of cases, emissions occur some time after production, for example, when refrigerators are junked. Because CFCs are gases at room temperature, all production eventually will translate into emissions.

The most likely range of ultimate ozone reduction from CFCs was estimated as ranging from 11 to 16 percent in a 1979 National Academy of Sciences Report,[1] based on emissions continuing at the rate attained in the early 1970s. However, if the rate of emissions accelerates in the same fashion as it has in the last decade, ultimate ozone depletion is estimated to range from at least 30 percent to as much as 57 percent. A 1 percent decrease in stratospheric ozone implies roughly a 2 percent increase in DUV, which in turn implies a 4 percent increase in human skin cancer rates.

Besides their ozone-depleting properties, CFCs are strong absorbers of the infrared radiation emitted by the earth's surface, thus contributing to the greenhouse effect, which increases temperatures. Whereas the greenhouse effect is a direct climate impact, there is also an indirect climate effect, because ozone depletion, per se, may lower temperatures a small amount, somewhat offsetting the greenhouse effect. The combined direct and indirect effects yield an estimated temperature increase on the order of 0.6° Centigrade (C) or 1.1° Fahrenheit, by the year 2000. This contrasts with a year 2000 increase of around 0.5°C projected for the effects of CO_2, indicating that in terms of relative concentrations in the atmosphere (parts per million), CFCs are much more effective than CO_2 in raising temperature. Specifically, Forziati notes that the CFCs are more than 100,000 times as effective as CO_2, in terms of the greenhouse effect relative to atmospheric concentration.

There has been considerable scientific uncertainty and controversy over the physical relationships involved in CFC emissions, but the broad outlines of the problem have now become clear. Hence, the economics of managing CFC emissions can be addressed in meaningful fashion, a task carried out by the remaining chapters in the volume.

In chapter 3, Gladwin, Ugelow, and Walter carry out a detailed empirical and institutional review of the industrial structure and economic organization of CFC production and use, with an extensive review of products, quantities and firms involved in international trade. The U.S. share of world production of F-11 and F-12 has fallen considerably over time, dropping from around 70 percent in 1960 to less than 40 percent in the late 1970s. The authors evaluate the likelihood that unilateral U.S. regulation could shut down U.S. production and encourage the movement of capacity to nonregulating nations by looking at the specific industrial organization of U.S. firms in terms of their product lines, technology, international linkages, and potential adjustment capabilities in response to various alternative U.S. regulatory policies. They find that, among CFC users, producers of refrigerators and air conditioners potentially are especially sensitive to U.S. regulatory action. Much would depend upon the nature of the regulation. The most devastating type would be unilateral, near-term, nonselective, comprehensive prohibition of the use of CFCs. In contrast, a multinational, gradually applied, selective discontinuation of the most damaging CFCs in low priority uses would be unlikely to cause serious economic damage to U.S. producers. Gladwin, Ugelow, and Walter do not expect widespread international relocation of CFC producers, even in the event of considerably increased regulation of U.S. production. Institutional reasons for this view include expected convergence by OECD nations on any reasonable U.S. standards and the small scale of foreign plants, designed to meet local needs rather than to capture large-scale international markets. It is also suggested that foreign nation protectionism and risk factors would limit large-scale international relocation of CFC production facilities. The authors conclude that even the most severe CFC regulation in the United States would have a mild effect on international relocation relative to the amount of domestic dislocation, especially for U.S. producers of refrigeration and air conditioning equipment, who would probably face plant closings, and unemployment.

The real problem appears to be CFC emissions in the rest of the world, and the authors therefore emphasize the desirability of a multinational approach. They draw attention to the possible emergence of international consensus on environmental protection, largely through the efforts of the United Nations Environmental Programme (UNEP),[2] although other multinational approaches are considered. They also examine a range of measures available to the United States for unilateral efforts to reduce CFC emissions in other nations, including import controls through the Toxic Substances Control Act. Unilateral national policies to protect the global commons may become necessary as emerging technologies pose a threat to the global environment.

1.2 Benefit-Cost Analysis

Part II focuses on the benefits and costs of CFC management. The first chapter in the grouping deals with methodological and conceptual issues in benefit-cost analysis, the next two carry out alternative benefit-cost assessments of the control of CFC emissions, and the remaining three undertake analyses of specific sector impacts of CFC emissions.

Mishan and Page (in chapter 4) begin the discussion by asking whether the concepts and methodology of benefit-cost analysis allow its use in evaluating the control of CFC emissions, which are characterized by long latency periods, global effects, the involvement of multiple generations, and possible irreversibilities. They focus on the following five issues:

1. Implications of the concept of economic efficiency
2. Conceptual problems of valuation
3. The legitimacy of using discounted present value (DPV) in ranking public projects
4. Questions of intergenerational equity
5. Treatment of risk and uncertainty.

In considering economic efficiency, Mishan and Page emphasize the unique assumption of economists in basing value on the subjective preference of individual members of society rather than on some noneconomic concept of the general good. Consequently, they insist, for rigorous proof of superiority, upon the principle of Pareto improvement, or of making at least one person better off without making anyone worse off. However, in practice they recognize that the criterion of a potential Pareto improvement (generation of net benefit after compensating losers out of gains) is acceptable.

Mishan and Page eschew political weighting schemes, perceiving a unique distinction between such systems and an ethical-moral weighting scheme, arguing that the benefit-cost analysis be based upon the ethical consensus of the society. They also find cause for concern that prices may bear no relationship to long-run net utility where new goods and services are based on new technologies with possible massive externalities.

In considering discounting to value goods and services which extend over long time periods, the authors distinguish intragenerational from intergenerational discounting. They conclude that discounting is legitimate, provided that the lives of the individuals involved in an intragenerational problem overlap in time, but the legitimacy of using discounted present value is considered much more questionable in ranking intergenerational public projects. If there are no common points overlapping in time between the generations involved, then they see no basis for an individual in the current generation to compare his valuations of preference with those of future generations. Government procedures which purport to make intergenerational judgments on behalf of nonoverlapping generations do not meet the test for Pareto improvements based upon individual valuations by individual members of society.

Technical, as well as ethical, questions about discounting are posed. The traditional view of discounting in environmental management is that without environmental protective measures, there will be an added output of goods and services, resulting in savings and investment which will enrich future generations to a greater extent than any environmental damage which they inherit. Mishan and Page argue that this is not valid unless the incremental benefits from nonregulation are actually invested and transferred to future generations in a way that at least compensates them for the environmental damage which they inherit.

In contrast to Bailey (whose position is discussed below), Mishan and Page question the validity of "hyperrationality." If individuals are hyperrational, they perceive increased environmental quality as an increase in their total wealth and permanent income, replacing income lost because of expenditures on environmental improvements. Therefore, the costs of improved environmental quality are paid for by reductions in savings, which reduces the welfare of future generations, since less savings are passed on to those future generations. Mishan and Page expect, however, that an increase in environmental quality leading to a decrease in income would come primarily out of present consumption and not out of savings and investments, thus weakening the argument for discounting.

Given these arguments, Mishan and Page reject discounting for very long-run intergenerational problems, and instead recommend making behavioral assumptions explicit by tracing out the intertemporal distribution of benefits and costs with and without regulation and then making a choice. Since the unregulated use of CFCs clearly has immediate benefits and potentially large future costs, it presents a true intergenerational conflict of interests. To address this problem they believe there is need for an ethical consensus involving a presumption in favor of equality of incomes between generations, in the sense of equality of average incomes.

The Mishan and Page analysis of the role of discounting in benefit-cost analysis leads to a set of positions on present environmental regulatory policies. Currently the government is required to show "a reasonable basis for concluding that there may be an unreasonable risk" before taking regulatory action. Mishan and Page reject this requirement in the case of possible major irreversible environmental damage, even if the risk cannot be calculated and may be small. They prefer, if necessary, bearing heavy cost for insurance against even low-probability, major irreversible events, such as ozone shield damage or global climate changes imposed by CFC emissions. They recommend careful assessment of the consequences of excess caution over excess risk. In this regard they postulate two regulatory guidelines. Rule A, currently in effect, would permit continuing any economic activity until evidence of damage beyond reasonable doubt could be provided. Rule A has been followed since the industrial revolution and is still appropriate for individual nations struggling for economic development. However, Mishan and Page believe that Rule A is no longer appropriate for large affluent industrial nations, since even very large discounted present values of the use of materials like CFCs cannot justify even the small risk of future catastrophe. They prefer Rule B, which would ban potentially catastrophic practices until plausible evidence of safety is provided. In implementing Rule B they are even willing to depart from the economist's general preference for taxes rather than bans in order to dramatize the importance of environmental protection. They believe that this need is sufficiently compelling that, if necessary, the United States should act alone in banning CFCs on the basis that, as the largest user in the world, we could both greatly reduce the risk and exert moral influence to encourage other nations to follow suit.

In contrast to Mishan and Page, two investigations integrate relevant data in overall evaluations of net costs and benefits of alternative management options. The importance of uncertainty and the inadequacy of current data is recognized, but agencies and officials responsible for decisionmaking in environmental management have a need for assembling in coherent form what information is currently available. Consequently, the comprehensive benefit-cost studies by Smith and d'Arge (chapter 5) and Bailey (chapter 6) can be of help in meeting that need.

In chapter 5, Smith and d'Arge calculate benefit-cost measures and apply a number of sensitivity tests, employing such data and methodology as are currently available. The limiting costs of controlling CFC emissions are taken to be the costs of doing without these chemicals, or finding substitutes for them. The appropriate procedure for measuring these costs is to estimate the consumer surplus lost from foregoing their uses.

There are two major approaches to measuring consumer surplus. The first is derived from the demand for products embodying CFCs. Since CFCs are not purchased by consumers for their direct utility, but are embodied in refrigerants, insulating foams, and other products, the CFCs respond to a demand derived from demand for the end products into which they are inputs. Smith and d'Arge investigate the area under the derived curve for these products and compute the consumer surplus for F-11 and F-12 (that is, the derived demand for inputs) as approximately $3 billion (in 1971 dollars) for all future years discounted back to present value in 1974.

If products such as CFCs are essential inputs into consumer goods and there are no acceptable currently available substitutes for them in their end uses, then an alternative measure is the consumer surplus under the demand curve for the end products. Using this end-product concept of consumer surplus, Smith and d'Arge find an upper bound of approximately $107 billion on the present value of CFCs. To the extent that alternative inputs and technologies can be developed, the upper limit of this consumer surplus would fall toward the $3 billion figure for the derived demand, depending upon how prices of CFCs would change as substitutes were introduced.

To be useful for decision making on regulation these estimates must be confronted with estimates of damages from CFC emissions. Smith and d'Arge recognize that the estimation of damages is extremely difficult since it could include not only the cost of biological damage from increased DUV, and effects of temperature change on crops and energy use, but also far-reaching damage from possible melting of the polar ice caps and extensive ecological changes. They do not attempt to estimate all of these costs but do derive estimates for costs of increased skin cancer, materials weathering, fisheries damage from ozone depletion, and the probable effects upon agriculture, forests, and energy use from temperature changes. These damages are estimated at about $7 billion in present value for all future damages discounted to 1974 in 1971 dollars. Related temperature increases would result in benefits to fisheries (in contrast to Strand's study (chapter 9), which shows costs based on DUV damages), forestry, some agriculture (cotton), and households from reduced housing and clothing expenses. However, these benefits would be far outweighed by increased electricity use for cooling purposes. [The electricity estimates are based on results from a CIAP energy study, and as noted below, Hoch's results (chapter 8) diverge from the CIAP results, and suggest lower estimates for electricity use.]

Smith and d'Arge then develop a benefit-cost analysis with various sensitivity tests. With control costs based on derived consumer surplus of approximately $3 billion dollars, and benefit estimates of $16 billion dollars, they find a benefit-cost ratio of banning all uses of F-11 and F-12 of 5.86. However, this depends on the assumption that substitute chemicals become available. To the extent that substitute materials and processes are not available and final product consumer surplus must then be used as the measure of control costs, which now rise to $107 billion, the benefit ratio falls to 0.16 for a total ban on F-11s and F-12s. A partial ban on these two products still appears to be economically feasible. However, if it turns out that the use of CFCs does not result in a climatic change, then even if substitutes become available, the cost of banning CFCs outweighs the benefits, leaving a benefit-cost ratio of 0.30.

Recognizing the global nature of the impact of CFC emissions, and the dominance of the international aspects of the problem, Smith and d'Arge also look at the sensitivity of benefits and costs to the U.S. economy of a wide range of international strategies. At one extreme, the assumption is made that the United States takes unilateral action to reduce the use of F-11 and F-12, and no other nations take such steps. At the other extreme, the assumption is made that all countries reduce the production of F-11 and F-12 to 1973 levels. Regardless of the extent of international control, the results are found to be very sensitive to the discount rate. At a 3 percent discount rate, all control strategies are economically beneficial to the United States acting either unilaterally or multilaterally. At a 5 percent discount rate, all strategies would be economically beneficial for the United States using the derived surplus measure. Since many of the benefits from regulating the use of CFCs involve preventing damage occurring far in the future, a high discount rate wipes out much of the value of such benefits, whereas a low discount rate results in accumulation of very large damages from unregulated use of CFCs.

Bailey, in chapter 6, also addresses the discounting issue in assessing the benefits and costs of CFC regulation. He separates the efficiency argument from the equity argument by emphasizing the use of pretax, private market rate of return on investment as the proper opportunity cost in discounting, which he estimates to be 11 percent.

Bailey stresses the importance of discounting future benefits from environmental quality at this rate, not to downgrade the benefits, but to measure alternative actions having different time profiles and different time streams of benefits in common units of present wealth. He argues that application of a low discount rate for future benefits from environmental protection, no matter how well intentioned, could result in a sacrifice of tangible welfare by present generations through costly environmental benefits less valuable to them than the value to present generations of their current sacrifices. In addition, some sacrifice of material benefits by future generations is entailed. As noted earlier in the discussion of Mishan and Page, Bailey applies the concept of hyperrationality in support of this position.

Using a discount rate of 11 percent, Bailey derives dollar estimates of benefits for two levels of control on use of CFCs, providing estimates of uncertainty factors and a range of upper and lower limits as well as central values for the benefits of controlling CFC emissions, discounted to the present and valued in 1978 dollars. Present values of benefit streams over the years are estimated for the alternatives of stopping the growth of emissions at 1974 levels versus totally prohibiting emissions, beginning in 1978. Limiting CFC emissions to the 1974 level would result in an annual reduction in skin cancer cases of thousands of cases per year, with a peak reduction of one million cases in the year 2050. If all CFC emissions had stopped in 1978, many more skin cancer cases could be prevented annually, rising from 14,000 around the turn of the century to a peak of 1.4 million in the year 2050, and decreasing to 1 million cases at the beginning of the twenty-second century. These reductions translate to "central estimate" discounted present values of $820 million from controlling growth of CFC emissions and $1.1 billion from the complete cessation of emissions. In addition to benefits from reducing skin cancer, there are even greater benefits from reducing damage to materials, with central estimate values of $1.5 billion and $2.3 billion, respectively, for the alternatives of no-growth versus zero emissions. Total benefits for the two cases then were $2.3 billion and $3.4 billion, respectively. The lower and upper limits for the latter case ranged from $530 million to $10 billion.

These estimated benefits are then compared to the corresponding costs of cutting back CFC emissions. Cost measures are obtained as alternative estimates of consumer surplus; they range from a low value consisting of the area under the demand curve for F-11 and F-12, the most widely used CFCs, to a high value consisting of the consumer surplus for all products in which CFCs appear as input. Central estimates for present values of costs were $4.9 billion for the no-growth case and $8 billion for the complete cessation case. Lower and upper limits for the latter ranged from $3 to $22 billion.

Bailey concludes that on the basis of the specific cases examined, only a modest cutback of the least essential uses of CFCs would be justified. His findings are consistent with the EPA policy of prohibiting the use of CFCs for "inessential" purposes in aerosol sprays. However, Bailey would support the application of taxes on use of CFCs as a least-cost, more efficient regulatory measure which would allow the market, rather than regulators to determine which were essential and inessential uses of CFCs.

In chapter 7, Kelejian and Vavrichek find that the impacts upon U.S. agriculture from potential climate changes related to CFC emissions are likely to be quite different from those felt in the rest of the world. If the release of CFCs continues to grow at the rate of 10 percent annually until the year 2000, corn and wheat production will not be much affected in the United States, but will decline in the rest of the world.

Kelejian and Vavrichek develop a linear climatological model to examine the effects of three pollutants on global mean temperature: CO_2, CFCs, and nitrous oxides, thus embedding the CFC effects within a general set of temperature effects caused by pollutants. The use of a two-dimensional meteorological model allows them to examine the effect on crop change of temperature modifications by latitude, since latitude is crucial to changes in crop patterns. The model is based on a survey which provides percentage changes in both temperature and precipitation for four latitude zones. The climatological model is then linked to an econometric model that includes equations for the supply of land, demand for agricultural products, supply of agricultural products and international trade as affected by climatic changes. In contrast to the usual one-crop model, climatic changes result not only in direct output changes, but in substitutions as well. Climatic change in one region also has economic consequences in other regions because of changes in trading patterns. A bumping process can occur so that climatic change causing one crop to contract will cause another crop to expand, as the respective crops exhibit different responses to temperature changes. The effect of changes in climate on benefits can be evaluated directly in terms of producer surplus and consumer surplus. Also, priorities can be derived for the optimal distribution of research funding since the model indicates which parameter changes have large economic effects, thus justifying their further investigation.

The model yields extensive empirical estimates even in a limited initial application. The starting point is a total global temperature increase of 1° to 2°C from the joint effect of all three pollutants of concern with about a 5 percent increase in precipitation. Translation of these impacts into crop changes suggests relatively small effects for the United States, as compared with the rest of the world. Slight temperature increases appear to aid U.S. output of wheat and corn, suggesting that present U.S. temperatures are slightly suboptimal. However, the model indicates that larger temperature increases could result in serious economic losses in agriculture.

The situation is even more serious for the rest of the world since temperature increases of 1° to 2°C could yield up to 22 percent decreases in wheat and corn crops. This is probably an overestimate because each effect is considered individually, rather than synergistically. If emissions of all three pollutants continue to grow until the year 2000, potentially serious fluctuations in food production could occur globally. Kelejian and Vavrichek give much emphasis to the importance of the uncertainties involved in the parameters, and they make a number of sensitivity tests which confirm the importance of further research to reduce uncertainties.

Consumers would be affected in complex ways by any extensive climatic changes. Impacts would be felt not only through changes in energy costs for heating and cooling, but also through eventual wage adjustments, interpreted as reflecting changes in consumption of clothing, housing, and medical care, as well as in amenity values attributable to climate.

In chapter 8, Hoch investigates this problem first by examining residential energy use and then wage rates as a function of climate and other explanatory variables. The energy study consisted of individual analyses of the use of petroleum products, natural gas, and electricity, employing state data, and an analysis of electricity consumption, employing metropolitan area data, with good agreement between the two sets of results for electricity. In similar fashion, metropolitan area samples for a number of specific occupations were used to relate wage rates to climate. It was hypothesized that workers require wage differentials as compensation for higher costs of living or for the psychic costs of lower amenities, and that such costs increase with population size, and can be related to climate, so that places perceived as having unpleasant or cost-imposing climate have to pay more for the same work than do places having pleasant or cost-reducing climate. The equations used in estimation employed both a number of climate and a variety of other economic variables; results obtained supported the basic hypotheses that both energy use and wage rates were significantly related to climate.

For present purposes, predicted temperature changes then implied corresponding predicted impacts on energy use and on wage rates. For the United States as a whole, a decrease of 2°F (or 1.1°C) implied somewhat more than 2 percent increase in energy costs (summing heating and cooling costs), and a 0.6 percent increase in wage rates. An increase of 2°F implied a 1 percent decline in residential energy costs and a 0.4 percent decrease in money wage rates, interpreted as corresponding to an equivalent reduction in costs, including psychic costs. The case of a 2°F increase corresponds to the estimate presented by Forziati (chapter 2) of a year 2000 temperature rise, with roughly half attributable to CFCs and half to CO_2.

Using a 1975 base, the 2°F temperature increase would reduce energy costs by about $400 million and total costs by roughly $5 billion. It is hypothesized that the difference between energy costs and total costs is attributable to other money costs (including clothing, medical care, and housing) plus the value consumers place on climate amenities.

Hoch's results thus indicate that net economic benefits should result from the temperature changes attributed to CFCs. However, there is evidence that the cost relationship is U-shaped, so that beyond a certain range of temperature increase, with rising costs for cooling processes, net economic gains should become negative. Hence, if CFC- or CO_2-induced temperature increases turn out to be larger than expected, or if year 2000 effects are not steady state (or ultimate), but rather are part of a longer-run upward trend, then the impact of temperature change could well become negative.

Some additional points seem of consequence. Hoch's results for energy use show some disagreement with the energy studies carried out for the earlier Climatic Impact Assessment Program (CIAP). For the cases considered, the relation of economic behavior and climate is complex and nonlinear, with interactions of the climate variables. Finally, costs and benefits of climate change can vary considerably by locale. Thus, a small temperature decrease is beneficial for the South, and a temperature increase is costly, reversing the U.S. results. Again, divergent cost impacts can occur between metropolitan areas in the same region.

While climatic change may have long-term impacts on agriculture and consumer welfare, damage to the ozone shield and consequent increases in DUV radiation can have direct impacts on some sectors such as fisheries. Until recently, few scientists were seriously concerned about such effects because DUV penetrates only a few meters below the water's surface and was thought to affect very few species of marine life. Further, previous investigators had assumed that future distribution of fish landings would continue to be similar to those of the current period so there would be no major international impacts from ozone-induced changes in fishery resources. However, more recent research indicates that ozone depletion may have extensive adverse effects on world fisheries. Those marine microorganisms which inhabit the upper layer of water, as well as the eggs and larvae of some species which spend crucial periods of their life near the surface, can be highly vulnerable, especially those which grow rapidly.

Laboratory research also suggests that a 15 percent reduction in ozone protection, which is consistent with the 1979 central estimate cited in chapter 2,[3] could seriously damage some fisheries species which are already near their tolerance level for UV-B. Consequently, the world distribution of fish landings could change significantly, with extensive impacts on some aspects of the international economy.

Strand applies these revised perceptions and estimates in chapter 9. Species which appear particularly vulnerable to ozone depletion are mackerel, the anchovy, crab, and shrimp. Damage to them could have differential effects throughout the world, probably falling most heavily upon the developing nations. Since protein is a major dietary need in developing nations, with fisheries providing one of the most promising sources, extensive damage to fisheries could result in aggravated protein deficiencies in those nations. To examine the national aspects of economic costs to the United States of CFC-induced damage to fisheries, Strand develops an optimizing model with which he first calculates the ideal economic time path of development for U.S. fisheries. He then estimates the worst-case impact on the fisheries from increased DUV, based on the case of the highly vulnerable shrimp. He finds that the maximum gain from control of ozone damage, using a 5 percent discount rate for producer and consumer surplus through the year 2025, would be $43 million or approximately $1 million per year for shrimp alone. Extrapolation of shrimp damage to all fisheries establishes that under the worst of conditions, damage through the year 2025 would cost the United States $300 million at a 5 percent discount rate or $70 million at a 10 percent discount rate.

These results should be contrasted with earlier findings of little damage to fisheries from ozone depletion and an earlier estimate by d'Arge of gains to fisheries of $661 million from the effects of CFC emissions.

1.3 Policy Perspectives and Analysis

The chapters in part III address a range of policy issues that bear on the management of CFC emissions. A common concern is the economics of the regulatory process. Thus, chapter 10 develops a general conceptual framework for CFC regulation in the context of uncertainty and irreversibility.

Chapter 11 carries out an analysis of a deposit-refund system applicable to refrigeration systems employing CFCs. Chapters 12 and 13 extend the consideration of the international dimension of CFC regulation, while chapter 14 examines impacts of variations in market structure on the effectiveness of effluent charges. Finally, chapter 15 considers the relation of environmental standards and CFC regulation from two vantages, first providing a set of practical guidelines for short run regulation through application of production standards, and then deriving implications for long-run regulatory policy in a dynamic macroeconomic framework.

In chapter 10, Mancur Olson proposes a conceptual framework for dealing with the uncertainty and irreversibility inherent in the CFC problem. He identifies three unique features of the problem. First, it involves "scant sets"; since there is only one ozone layer, and one global climate system, knowledge about them is difficult to obtain. Second, the cost of unsuccessful experimentation may be the value of the entire system: catastrophe is possible. Third, there may be considerable inertia in both natural and institutional systems, causing delays in corrective responses to changed conditions.

Olson concludes that benefit-cost analysis should not be ruled out simply because of large uncertainties, though adjustments must be made for uncertainty. More serious is the existence of very large distributional consequences between regions, classes, nations, income groups, and genetic groups, as well as between intergenerational groups. Net benefit-cost outcomes are necessarily inconclusive if they involve unacceptably large distributional changes. Decision tree analysis, pursuing many paths and establishing many branches, is the recommended response to uncertainty. Olson also emphasizes the value of deriving additional information. For example, permitting some uses of CFCs to continue generates additional information. Obviously difficult problems remain in deciding how many alternative research and development paths to follow and how long to let possibly dangerous but productive information-creating processes run on before irreversibility occurs. Olson applies the concept of hysteresis (the retardation of effect when forces are changed) to warn that possibly dangerous practices may be permitted to run too far and too long because inertia and uncertainty are barriers to prompt regulatory action. But regulatory action may also generate inertia by creating a vested interest in regulation and in the creation of substitute materials and processes, even if the dangers of the original state appear to have been exaggerated.

In a final set of policy hints, Olson provides ten useful guidelines:

1. The readily estimable costs and benefits are not a sufficient basis for policy decisions on regulating CFC emissions.

2. There is a need for better monitoring, or "social and environmental indicators," to reduce the extent of the lags and the seriousness of the hysteresis loops.

3. There is a need for "quick-response" research capabilities, so that unexpected results of the monitoring system can be analyzed promptly.

4. Beware of "monism" in sources of scientific and policy advice; even the best scientists may be wrong, sometimes even about some matters about which they are nearly certain.

5. Beware of "monism" in policies, since a single policy that is followed consistently over a large area has a greater chance of causing a catastrophe than varied policies.

6. Beware of even the small chance of the really catastrophic outcome.

7. There is no such thing as a safe lunch, either.

8. Two risky technologies may be better than one.

9. "Irreversibility" needs to be emphasized in policymaking and in further research, particularly when irreversibility is tied to indivisibility.

10. The research process and the policymaking process cannot be altogether separated where CFC emissions are concerned.

Expanding on the last hint, Olson derives the unconventional conclusion that, where there is one indivisible element in the scant set, and where, therefore, we must take into account not only information but the process by which information is gained, research and policy cannot be sharply separated in matters of global environmental issues such as the ozone layer and climate.

In chapter 11, Peter Bohm analyzes the workings of a deposit-refund system for the recapture of CFCs used in refrigeration equipment, including refrigerators, chillers, and both stationary and automobile airconditioningers. Acknowledging that current limitations in computing marginal social costs make it impossible to fine-tune control instruments, Bohm demonstrates the applicability of a deposit-refund system in these circumstances. He points to the lack of incentives as the major disadvantage of regulation; taxes seem unlikely to help much because labor is the most important input in salvaging refrigerators; and charges and fines to control disposal entail problems of supervision and control. Because subsidies on CFC recovery and taxes on disposal, that is, a deposit-refund system, could overcome those disadvantages, a detailed examination is made of the full set of linkages in the production and consumption of refrigeration equipment, with specific suggestions on how the deposit-refund system would function at each stage. Bohm estimates that a refund credit of $2.50 per pound for CFCs, assuming a 5 percent discount rate, an average fifteen-year lifetime for equipment, and a required deposit rate of $1.25 per pound on F-11 and F-12, would result in a maximum price increase for refrigerators of 5 percent. The effect might be mildly negative on production and consumption of refrigerating equipment, but this could be offset by more rapid turn-ins.

After developing a set of specific proposals for the structure of the deposit-refund system, Bohm identifies a number of advantages for the deposit-refund system over conventional regulation for emissions control. By contrast to a regulatory system, which would require equipment returns in a specific way without incentives, the deposit-refund system would generate reduced emissions at each level of manufacture and use. The rates established could be determined by social willingness to pay for the reduction in risk from ozone depletion. The deposit-refund system would not interfere with efficiency or with the development of improved technology. The proposed system would allow costs of operating the system to be recovered and would not have adverse distributional effects. It could also generate net revenue. Furthermore, the deposit-refund system could be combined with other regulatory instruments, including taxes on CFCs, thus permitting regulatory measures to achieve highly specific objectives, and to be modified as new information became available.

It was earlier indicated that a key characteristic of controlling CFC emissions is the global, and hence, international nature of the problem.

In examining the prospects for global management of CFC emissions in chapter 12, Bohm finds little justification for optimism either from the global public good paradigm or from the experience with conventional international cooperation. The analogy of building up from the national/local public goods problem to a global/national resolution of the public goods problem is not an acceptable model because no global constitution or consensus exists for an international body to promulgate judgments which would be optimal for all of the national units involved. The primary difficulty is that of gaining widespread international acceptance of the parameters for a binding international agreement, especially on income distribution, the social rate of discount and the existing power structure which influences other countries' responses.

In the absence of an accepted global vision of the problem, a second-best substitute is a set of national visions which can serve as the basis for bargaining. An initial step could be each nation's internally acceptable solution which would control CFC emissions to the point where internal marginal costs equal internal marginal benefits of control. The next step could then be international bargaining for further control measures which might move towards a global improvement.

Bohm argues that the United States should play a leadership role in regulating CFC emissions and in negotiating for global management. In this effort, the United States could consider two alternatives: the push strategy and the pull strategy. The push strategy involves offering a quid pro quo to other nations under which the United States would agree to undertake further reductions of CFC emissions in return for regulatory efforts by other nations, with all bargaining based upon efficiency criteria and mutual self-interest. Bohm, however, prefers the alternative of a "pull" policy based on example, set by the United States which, as he points out, largely initiated the use of CFCs and until recently produced and consumed approximately 50 percent of the global total (and a much larger percentage of the historical cumulative total). It is sobering to recognize that the United States has benefited over the decades from protecting the health of its citizens through refrigeration and their comfort through air conditioning, at the possible long-run cost to the health and safety not only of the United States, but also of much of the rest of the world. This situation may also prevail for other environmental impacts, such as those of radioactivity, agricultural chemicals, and carbon dioxide emissions from burning fossil fuels.

Beyond the moral and ethical aspects of major health and environmental impacts, Bohm points out important operational advantages of the United States playing a pathfinder role in global protection. Aside from increasing the willingness of other nations to follow a constructive example, serious U.S. efforts to reduce CFC emissions could improve the ability of other nations to follow suit in several important aspects. U.S. development of technological advances and cost reductions for control technologies is a first and obvious possibility. But Bohm attaches equal or even greater weight to the development and testing of new policy options which could be added to the arsenal of control possibilities. He gives particular emphasis to deposit-refund systems which he analyzed in the preceding chapter, and he notes the desirability of examining other combinations of regulatory and incentive systems. An important point is that optimal control of CFC emissions may require not a single instrument, but a combination of instruments including both regulation and incentives, and, as noted by Snower in the last chapter, some variation of instruments over time.

In chapter 13, McGuire points out that the ozone shield and the global climate are relatively pure examples of global public goods, in the sense that citizens of all nations benefit from their services (though not necessarily equally), and that the enjoyment of those services by one nation does not reduce the supply of the services to other nations. Therefore, there is a good prima facie basis for seeking international coordination of policies to control CFC emissions.

Applying modern foreign trade theory to the case of goods produced internationally by firms in which there is some international mobility of factors of production (as presumably would be true in the case of multinational corporations), McGuire finds that unilateral regulation could be highly self-defeating. The application of an effluent charge or comparable environmental regulations to CFC emissions would be analogous to a partial factor tax. Under conditions of factor mobility, McGuire finds that unilateral regulation could possibly, but not necessarily, drive producers of CFCs completely out of the unilaterally regulating nation.

McGuire concludes that there are a number of policy implications of this analysis. In contrast to unilateral regulatory efforts, which could cause a complete shutdown of the firms producing the CFCs in the regulating nation, international coordination of emissions charges or regulations could permit all nations to share in reducing the output of the pollution-related substance. Conceivably, however, unilateral regulation could generate net environmental benefits for the United States, even if all other nations refused to participate. McGuire emphasizes that such unilateral regulation by the United States would amount to a grant-in-aid to all other nations, with the greatest benefits going to the affluent, fair-skinned citizens of nations of the Northern Hemisphere. Whether U.S. unilateral regulation would actually result in the shutdown of domestic CFC producers clearly depends upon a number of specific factors relating to industrial structure and associated elasticities. McGuire recommends that priority be given to empirical research on this problem, and notes that if such research indicates net benefits for the United States through more stringent regulation, it should consider accompanying any such regulation with possible offset measures which could cushion the impact of a U.S. shutdown on producers and workers. Therefore, while efficiency conditions could justify stringent regulation of CFC production, equity considerations could warrant ameliorative actions to avoid severe hardships to individuals and activities upon which the burden of adjustment would fall.

In earlier work, Baumol and Oates concluded that the ideal instrument for achieving efficiency in resource allocation under conditions of adverse pollution externalities was imposition of Pigouvian charges or taxes on emission equal to the difference between marginal private costs and marginal social costs at the optimal level of output.[4] In practice, where problems of measuring marginal social damage prevent identification of the intersection between this function and marginal treatment costs, a second-best alternative is to set an environmental standard based on the best information available, and then to impose the Pigouvian charge, varying this experimentally if necessary, in order to bring environmental quality to the desired standard.

In chapter 14, Oates and Strassmann extend the analysis to imperfect market situations including monopolistic firms, public agencies, and regulated utilities, in order to determine the robustness of policy conclusions drawn for the competitive case. They first examine the case of monopolistic firms and find that failure to conform to the competitive norm interposes no new complications on the earlier choice of standards and charges as the most efficient instrument available. For monopolistic firms that are profit maximizers, or more generally, that maximize utility, the solution of adopting the effluent charge which is consistent with all firms meeting the environmental standard is the least-cost method for society of implementing that environmental standard.

For example, if monopolistic firms emit CFCs, and society establishes a preferred ambient concentration of CFCs, then setting the effluent fee at levels which will induce each firm to adjust its production to the point where marginal treatment cost equals the effluent fee will result in the lowest cost method of achieving that specified concentration of CFCs. Therefore, the standards and charges approach remains intact as the preferred regulatory instrument.

Oates and Strassmann then go on to the case of the government agency as a polluter, drawing on a behavioral model in which the public agency neither maximizes profits nor minimizes costs but seeks to maximize its discretionary budget. For this objective, the public agency would still need to maximize perquisites rather than profits. Therefore, as long as public agencies seek to minimize total costs including the cost of treatment emissions, the standards and charges control instrument is still the preferred regulatory technique.

One possible limitation on this prescription arises in the case where a Pigouvian charge on emissions by a monopolistic firm may induce that firm to cut its production below the optimal level, thus resulting in too much of a good thing (an excessive cut in output). Empirically, however, the gains from abatement appear in most cases to exceed the losses from monopolistic restriction of output. In the case of public agencies, the imposition of effluent fees can be beneficial in three ways: by reducing pollution, by reducing excess production of output, and by reducing the cost per unit of output.

In the case of regulated firms, such as public utilities, Oates and Strassmann do find possible significant limitations on the role of standards and charges as the preferred regulatory instrument. This is because regulated firms are not necessarily cost-minimizers. The reason for this is the well-known Averch-Johnson problem of possible excessive installation of capital equipment by utilities in order to expand their rate base. Therefore they may not select a total cost level where marginal total cost is equal to the effluent fee. However, Oates and Strassman find no overwhelming empirical evidence of excess capital construction by utilities sufficient to warrant dropping the standards and fees approach as the appropriate control instrument even for regulated companies.

In sum, utilizing economic incentives in the form of effluent fees is likely to generate important savings for society in attempting to reach any particular standard for ambient quality, regardless of industry structure. Since the production and use of CFCs involve a wide variety of enterprises that hardly fit the model of the competitive firm, these results are directly relevant to the problem of managing CFCs.

In the concluding paper (chapter 15), Dennis Snower relates environmental standards to the management of CFC emissions, viewed from two distinct perspectives. In the first, short-term environmental or production standards are given, and criteria for optimal production responses are developed. In the second, desirable dynamic relations between CFC environmental standards and the state of the economy are formulated.

Under the first perspective, Snower notes that environmental policymakers frequently use production standards, as well as environmental standards, to fight pollution. He develops the point that under some circumstances, production standards may be convenient and efficient. Corresponding to this case, he develops a model that makes use of data on CFC production and emissions but is unaffected, in terms of regulatory behavior, by a number of variables of interest, including levels of CFC concentration in the atmosphere, health and welfare effects, and the valuation of those effects. A highly simplified artificial example is used in an application of the model, in which a variable production standard consists of an upper limit on the tonnage of F-11 and F-12, set for a one-year period. Three forms of output responsible for CFC emissions are identified as roughly corresponding to three sectors in an input-output table, including plastic materials and resins (foams), cleaning and toilet products (aerosols), and motor vehicles (automobile airconditioners). Emission coefficients (emissions per unit of output) are estimated in each case, and then the model is applied to derive the optimal order for the withdrawal of the products as the production limit is lowered. Aerosols comprise the first product withdrawn, followed by foams and then by vehicle airconditioners. Though highly simplified, the example illustrates that the policymaker need not decide on a precise value of pollution prior to setting standards. The model can be applied to the practical problem of relating production to environmental standards.

Under the second perspective, Snower develops the general point that setting environmental standards independently of production and consumption activities is generally a suboptimal procedure. As part of his development, he presents these major corollaries as "contentions":

1. Environmental standards for CFCs should be pegged to the levels of consumption and production activities.

2. Environmental standards for CFCs should be permitted to change through time. The optimal amount of CFC pollution cannot be expected to remain constant.

3. The optimal intertemporal relation between environmental standards for CFCs, on the one hand, and consumption and production activities, on the other, should depend on factor supplies and international transfers of CFC emissions.

Snower constructs an environmental macroeconomic model which describes the structure of the economy in terms of the production of output; relates CFC emissions and treatments to production and consumption; contains a list of CFC standards; and defines the relative social valuations of CFC concentrations and consumption levels. He applies the model and develops results which support his contentions. In particular, the model implies that a fall in resource supplies, as in the OPEC constraints on trade in oil, calls for less stringent environmental standards. Again, an increased flow of CFC emissions from abroad requires the pursuit of less ambitious environmental policies at home.

1.4 Conclusions

This volume, while focusing upon control of CFC emissions, has addressed a number of issues with broader significance for the future of the global environment. It is now generally recognized that technological advance, which has brought enormous gains in real welfare to the inhabitants of the earth, often has unanticipated, but far-reaching side effects. The adoption of CFCs for refrigerants and related uses provides compelling evidence of this fact.

Further, short-run decisions on environmental management made at the national level are seen as essential for meeting market forces, contributing to economic development, responding to political realities and assuring national survival. However, these actions, which seem expedient and rational in the short run, can conflict with long-run international goals of global environmental protection, which all nations would support, given the choice and given the availability of institutions for achieving these goals.

The papers in this volume help illuminate the broad reaches of the problem not only by their many areas of concordance, but also by their often sharp conflicts. We note both in considering some conclusions.

There are gaps in our knowledge of many of the key variables in the CFC control problem, so that accounting for and dealing with uncertainty becomes a central problem. There have been several recent major reversals in the physical science modeling of ozone-depletion processes, but the physical scientists are now sanguine that further major surprises will not be forthcoming. However, much more biological information is needed. Our understanding of the interactions between CFC emissions, ozone depletion, climate, and economic activities remains highly imperfect, and subject to major revision. Witness Strand's estimates of fisheries impacts which reversed earlier conclusions reached by d'Arge. And Hoch's energy-use results diverged from the earlier CIAP estimates, which in turn affected estimates by Smith and d'Arge who had drawn on the CIAP figures.

Again, limitation on resources and time necessarily required us to examine individually the important sectors of agriculture, fisheries, and households. Actually, all of these sectors are more highly interrelated than our methodology suggests. Consequently, more attention should be given to the impact of CFC-induced change in ozone and climate on interrelated ecologies in the biosphere, including the human ecology. The importance of these interrelationships is suggested by Kelejian and Vavrichek, who, by considering relationships between corn and wheat, revealed that these interdependencies are complex and cannot adequately be studied one by one.

In addition to the physical science questions and relationships, many social science issues remain unresolved. In the case of discounting, Mishan and Page agree with Bailey that discounted present value (or its equivalent, cumulative terminal value) is necessary for making valid measures of opportunity costs within a generation. Bailey and Mishan and Page object to artificial lowering of the discount rate on future benefits, and to raising the discount rate for environmental damage in an effort to deal with uncertainty. However, there is a major lack of agreement on intergenerational discounting. While it is possible to distinguish between some of the efficiency aspects and some of the equity aspects of intergenerational transfers, it no longer appears possible or desirable to avoid a direct inquiry into social values concerning what present generations owe to future generations. We have found no scientific or noncontroversial way of determining how much present cost to accept (in terms of reduced material wealth) in return for endowing future generations with reduced danger of environmental damage. Failure to undertake environmental safeguards today may be regarded as imposing upon future generations a form of taxation without representation. However, it must also be recognized that future generations will probably, if the past is still a reliable guide, enjoy higher levels of material wealth from which they may be better able to offset or correct possible environmental damage left to them as part of the legacy from current generations. Again, even to the extent that the role of discounting can be established, there still remains a debate over the appropriate rate of discount to be employed.

Another set of unresolved issues lies in the area of international trade, regulation, and cooperation. The fact that both climate and the ozone layer are global public goods raises to new levels the free rider problem as a result of which no nation individually has an incentive to deal unilaterally with emissions of CFCs, recognizing that such action will impose upon it heavy costs with benefits going to the rest of the world from which no compensation can be collected. There is potential for dramatic benefits from international cooperation, but the methods for achieving this cooperation remain unclear. Views of those investigating that topic for this volume vary between pessimism concerning the likelihood of achieving international agreement, to optimism about the benefits of both unilateral U.S. regulation of CFCs, and the potential payoff of vigorous, generous U.S. leadership in efforts at managing the global atmosphere. It is unclear whether U.S. efforts to date in limiting the production and use of CFCs unilaterally will serve as a guide to other nations or whether we have merely given away some bargaining chips. Some reconciliation of those views emerges by drawing out some commonly held positions. If regulation is pursued in the United States, a conciliatory internationally approved approach is much more likely to be successful and cost effective than unilateral actions. However, it is conceivable that, under some circumstances, benefit-cost calculations would make even unilateral action worthwhile in the United States. Technical questions still remain about the effects upon specific U.S. producers of unilateral U.S. regulatory action. The possible consequences have been estimated to range from negligible to the displacement abroad of all U.S. production of CFCs.

Another unresolved issue is the optimal choice of control instruments once the appropriate level of control has been determined. Most economists writing in this volume, as well as in the profession generally, have a strong preference for the use of control instruments which are based upon economic incentives in order to achieve maximum efficiency and lowest cost of regulation. However, some investigators here have found possible reasons in economic analysis for the widespread preference by noneconomists for using regulatory controls rather than incentive-based taxes and subsidies. Therefore, economists may still benefit from an open mind on the issue of optimal control instruments, especially with respect to tailoring the instrument to the need and to using more than one instrument when appropriate, such as combinations of regulations, standards, emissions charges, and deposit-refund systems.

Given conditions of uncertainty as to the ultimate extent of damage from ozone depletion, the high cost of imposing additional limitations on use, the wide extent of uncertainty, and the high value of information, it is clear that the evidence is not conclusive in support of further control of CFC emissions at this time. Much valuable information on the physical linkages between CFCs, ozone damage, and climate can be obtained simply by letting important uses continue, while additional information is collected. Thus, much valuable information becomes available as the result of taking no action beyond very limited control over nonessential uses. However, the potential cost of inaction rises sharply if inaction begins to lead into a case of irreversible catastrophe where the potential damage becomes infinite. At this point, we have not been able to resolve the issue of how far to continue in the assembly of valuable information and what guidelines may be derived concerning the point at which action should be taken before the situation becomes irreversibly damaging. Clearly, further research is needed on this key issue, and we consider those research needs in more detail below.

Despite the imposing set of unanswered and unresolved issues, there is a general consensus among the authors that it is worthwhile to undertake benefit-cost analysis of the management of CFC emissions, since decisions concerning the global commons must be made before irreversible damage has occurred. The models we have been able to implement and operate thus far do not yield the ultimate answers, but have shown where the critical areas are, where the research priorities exist, and what characteristics should be incorporated in the models which will be needed to organize new scientific information as it becomes available.

Even given the incompleteness and uncertainty of the information with which we have had to work, and given the early stages of knowledge about the effects of CFC emissions, some important findings have emerged. This volume demonstrates that current prices of CFCs do not necessarily reflect their full social cost. Uses of these products, beneficial as they are, in the past have set in motion a process which will be very costly to society in terms of increased morbidity and mortality from skin cancer. However, in economic terms alone, the ozone-induced costs of thousands of additional skin cancer cases per year will probably be exceeded by other costs due to damage from materials weathering. Further, even the adverse health and material damage issues may be far outweighed by the economic consequences of changes in climate resulting from the addition of CFC-induced climate changes to the greenhouse effects of continued CO_2 releases.

We have confirmed the importance of disaggregating the investigation of the CFC emissions problem by time period, by geographic region, and by type of economic activity. Clearly, the effects of both ozone depletion and of climate change impinge upon different economic units with different intensities, even with possible benefits to some of those units over a range of values. Light-skinned populations of North America, Europe, and Oceania are more vulnerable to DUV damage than are darker-skinned populations, who tend to be located near the equator. On the other hand, Kelejian and Vavrichek (see chapter 7) show that relatively mild temperature increases of 1°C to 2°C will aid U.S. agriculture, and only greater increases will cause serious economic loss. In contrast, even the mild increases have seriously adverse effects on agriculture in the rest of the world. Again, Hoch's results (see chapter 8) show that a temperature increase of 2°F will benefit most consumers in the United States, although those in the South will be adversely affected. By extension, it can be inferred that populations living closer to the equator will be adversely affected and those living further from the equator will be benefited. Again, there is evidence that, with a high enough temperature increase, the United States as a whole will be adversely affected.

On the methodological level, the models studied emphasize the importance of incorporating interdependence among the variables since changes in one sector can influence other sectors. Models for managing CFC emissions must also be dynamic in nature because management targets and objectives change over time as a result of the long lead times and lag times inherent in the total system. These models also clearly demonstrate the necessity for building uncertainty explicitly into the model.

Our experience has also identified the need for an interdisciplinary approach to the CFC problem. We have learned that economists' state of the art can be advanced significantly by undertaking the effort necessary to understand some of the principles of chemistry, physics, and climatology involved. We also would like to think that some of our colleagues in the physical and life sciences will benefit from our emphasis upon scarcity, opportunity costs, substitution, incentives, and optimal control strategies.

Aside from these specific findings, this volume has broad implications for the future. Technological advances in design and production processes, which seem ideal for their limited specific applications, may have profound long-run, adverse side effects. Therefore, technology assessment should take place within the context of the sensitivity of the entire biosphere. The case of ozone damage and climate change from release of CFCs reveals that nations acting individually have strong incentives to adopt products and technologies which may be harmful to the entire globe, and which all nations acting together would find in their advantage to regulate and control. However, in the absence of effective international mechanisms, individual nations will lack the incentives to undertake those management activities which all nations together would prefer. The understandable pursuit by developing nations of improved health and welfare through the use of refrigeration and air conditioning, to reduce the pressing social costs of disease and famine despite possible long-run damage to the environment, dramatically illustrates why and how impoverished nations discount the future.

Despite modest cutbacks in their growth rates in the United States, CFC production and use are growing rapidly in the rest of the world, where refrigeration and air conditioning are urgently sought for improved health and comfort. As the affluent nation which invented and introduced their use on a large scale, the United States is not in a strong position to urge other nations to cut back on refrigeration and air conditioning based on CFCs, particularly in view of the fact that light-skinned populations in the affluent Northern Hemisphere are more vulnerable to DUV damage than are darker-skinned populations. (In contrast, however, projected temperature increases, at least to the year 2000, should benefit lighter-skinned populations, who tend to be located far from the equator, and impose costs on darker-skinned populations, who tend to be located near the equator.)

1.5 What Can Be Done?

Since the studies of this volume have revealed that the continued release of CFCs will be associated with heavy real costs from resulting deaths, illnesses, and materials damage, with possible greater effects on climate and ecology, the economic issue is how much cost is justified to prevent or mitigate these CFC-induced damages. We find that additional control measures could be undertaken beyond the inexpensive controls now in effect, but the cost of further action at this time would be heavy in terms of losses in consumer satisfaction, and increases in prices for CFCs and for products using CFCs, as well as some reductions in performance where substitutes were available. Given the heavy economic cost of further action, and the wide range of uncertainties in the basic science involved, further control measures are difficult to justify at this time on economic grounds, except for the purpose of providing insurance against what now appears to be a small but finite risk of serious biological and economic damage occurring some time in the next century. The cost of insurance against this catastrophe would be very high for the United States, especially if other nations do not yet feel the evidence is strong enough to justify their cooperation.

However, just as stronger efforts to control CFC emissions at this time would be costly, making the mistake of not instituting strong controls could turn out to be even more costly if the damage proves to have consequences ranging toward the upper bounds which now appear possible. Since making a mistake in either direction would be very costly, any time gained from delays in taking action should be used with the greatest dispatch and efficiency for collecting the most important information needed to reduce the ignorance and uncertainty about critical parameters in both the natural and social science areas. Even if this essential information can be acquired, the chance of implementing appropriate actions and policies is low, given the present institutions available for international action.

The findings about the inadequacy of present approaches and institutions suggest that our chances of developing proper policies for managing CFC emissions would be greatly enhanced by undertaking a comprehensive interdisciplinary, international study of CFC emissions as they affect the earth's stratosphere, climatic change, and alternative human responses to these forces.

This study must of necessity be interdisciplinary because of the need for designing optimal responses and optimal management instruments over time. The effective bases for understanding and acting to control CFC emissions must be international since the present national structure of decision making provides every nation with strong positive incentives to undertake short-run actions which are expedient from the standpoint of market and political forces, but which, in the long run, are at best suboptimal, and, at worst, potentially disastrous for all nations. The success of such an undertaking could certainly not be guaranteed since it would involve not only immediate economic costs, but perhaps more seriously, the prospect of yielding some national sovereignty in return for global environmental protection. However, many possible formats could be explored, including ultimate self-financing of such global environmental protection efforts through the imposition of international charges on the most hazardous emissions with the proceeds to be used for further research and management. Obviously, such an effort would best be staffed by distinguished scientists whose credibility could be perceived as lying beyond political considerations. The United States as the inventor, leading producer, and consumer of CFCs to date, could appropriately initiate such a study. Such an interdisciplinary effort to protect the earth's atmosphere could help provide the experience and the institutions that are necessary to protect the biosystem. The CFC emissions problem clearly demonstrates that international efforts have become one of the inescapable costs which must be paid for out of the gains from economic growth in order to protect the earth's ecosystem and population from the environmental consequences of that economic growth.

References

1. National Research Council, National Academy of Sciences, Panel on Stratospheric Chemistry and Transport, Stratospheric Ozone Depletion by Halocarbons: Chemistry and Transport (Washington, D.C., NAS, 1979).

2. United Nations Environmental Programme (UNEP), UNEP Ozone Layer Bulletin, no. 1 (Nairobi, UNEP, January 1978).

3. National Academy of Sciences, Stratospheric Ozone Depletion.

4. W. Baumol and W. Oates, The Theory of Environmental Policy (Englewood Cliffs, N.J., Prentice-Hall, 1975).

Chapter 2

THE CHLOROFLUOROCARBON PROBLEM

Alphonse Forziati

2.1 Introduction

This chapter briefly reviews what is known in the physical, biological, and health sciences about chlorofluorocarbon (CFC) emissions and the attendant issues of stratospheric ozone depletion and climatic change. It is intended that this review will serve as background information for the other chapters in this volume.

CFCs are synthetic compounds whose release into the atmosphere has the potential for reducing the earth's stratospheric levels of ozone (O_3), a high energy form of oxygen. A decrease in ozone concentration would permit more ultraviolet radiation to reach the earth's surface, with likely adverse consequences, including greater incidence of skin cancer and biological damage to a number of plant and animal species. In addition, CFCs could have considerable impact on the climate, indirectly through the effects of ozone depletion per se, but also by way of the CFC contribution to the greenhouse effect.

CFCs are compounds that contain chlorine and fluorine (and sometimes hydrogen) directly bonded to carbon atoms. When CFC emissions reach the stratosphere, solar radiation can release the chlorine, which then acts as a catalyst in breaking down ozone. In turn, the reduction in ozone allows more of the harmful wavelengths of ultraviolet radiation to reach the earth's surface. Ozone can be destroyed by a number of chemical reactions besides those involving CFCs, but CFC emissions are long-lived, rapidly increasing, and potentially of major importance in ozone depletion. Further, CFCs are also very strong absorbers of the infrared radiation emitted by the surface of the earth. On a molecular basis, CFCs are about 100,000 times more effective than CO_2 as contributors to the greenhouse effect.

These topics will be considered in detail in the following order. Section 2.2 discusses the CFCs, in terms of their chemical structure and nomenclature, and then of their use and disposition. Section 2.3 describes some of the key characteristics of the atmosphere, solar radiation, and stratospheric ozone that underlie the CFC problem, while section 2.4 considers the destruction of ozone, reviewing several chemical reactions involved in the destruction, but focusing on the CFC reaction. Section 2.5 considers both direct and indirect climatic impacts of CFCs, and section 2.6 assesses the biological effects of ozone depletion, in terms of nonhuman biological damage and human health effects. A short concluding section presents some perspective on the importance of the CFC problem.

2.2 The Chlorofluorocarbons (CFCs)

2.2.1 Chemical Structure of CFCs and Their Nomenclature

A series of chemical compounds that necessarily contain chlorine, fluorine, and carbon, and sometimes contain hydrogen, are called chlorofluorocarbons (CFCs). Those having only one carbon atom, the chlorofluoromethanes (CFMs), are the most common. Specific CFCs are identified by a system of code numbers involving two or three digits, as developed by E.I. du Pont de Nemours & Co. in 1956,[1] and adopted by the American Society of Heating, Refrigerating, and Air-Conditioning Engineers in 1957.[2] The system, although unnecessarily complicated, is in commercial use in this country and abroad. Briefly, the identifying digits are determined as follows:

1. The number of fluorine atoms in the compound is represented by the first digit on the right. Thus, CFC-12 has two fluorine atoms.
2. The number of hydrogen atoms _plus one_ determines the second digit from the right. Thus, CFC-12 has no hydrogen atoms.

3. The number of carbon atoms <u>minus one</u> is indicated by the third digit from the right. If the resulting digit is zero (that is, when there is only one carbon atom), the zero is omitted from the code number (as in the case of CFC-12).

The system may be clarified by referring to figure 2-1, which shows the molecular structure of several CFC compounds; note the attachment (bonding) to the carbon atom or atoms at the center of each diagrammed molecule. The number of chlorine atoms is not designated in the code number but is determined by subtracting the sum of the number of fluorine and hydrogen atoms from the total number of atoms that could be attached to the indicated number of carbon atoms to satisfy the requirement that each carbon atom has four atoms attached to it. One of these four may be another carbon atom, for example, see figure 2-1, diagram 4.

The CFC terminology is analogous to that employed for the hydrocarbons. In that usage methane is CH_4, with one carbon and four hydrogen atoms, with an arrangement paralleling diagram 1; and ethane is C_2H_6, with two carbon and six hydrogen atoms, paralleling diagram 4. Dichlorotetrafluoroethane is often written $CClF_2CClF_2$ (rather than as the simple formula, $C_2Cl_2F_4$) to clearly indicate that it is a "derivative" of ethane, where fluorine and chlorine atoms have taken the place of the hydrogen atoms appearing in ethane. The hydrocarbon series has the general formula C_nH_{2n+2}; beyond methane and ethane, for example, propane has n=3, butane has n=4, and pentane has n=5. The general formula for CFCs is $C_nCl_xF_yH_{2n+2-x-y}$, with x and y greater than zero.

In common usage, the code numbers appear with one capital letter preceding them, with F and R the most commonly employed; for example, F-11, F-22, or R-22. The letter F indicates that the product was manufactured by E.I. du Pont de Nemours & Co., and is an abbreviation for their trademarked name Freon. The letter R simply indicates that the product has the chemical purity and physical properties suitable for use as a refrigerant. Other letters are used by other manufacturers in this country and abroad. For simplicity of exposition, the other chapters in this volume usually refer to specific CFCs employing the F designation.

If the refrigerant contains bromine, the same code number as the corresponding chlorine-containing molecule is used, but the code number is

Figure 2-1. Structural formulas and code numbers of typical CFCs

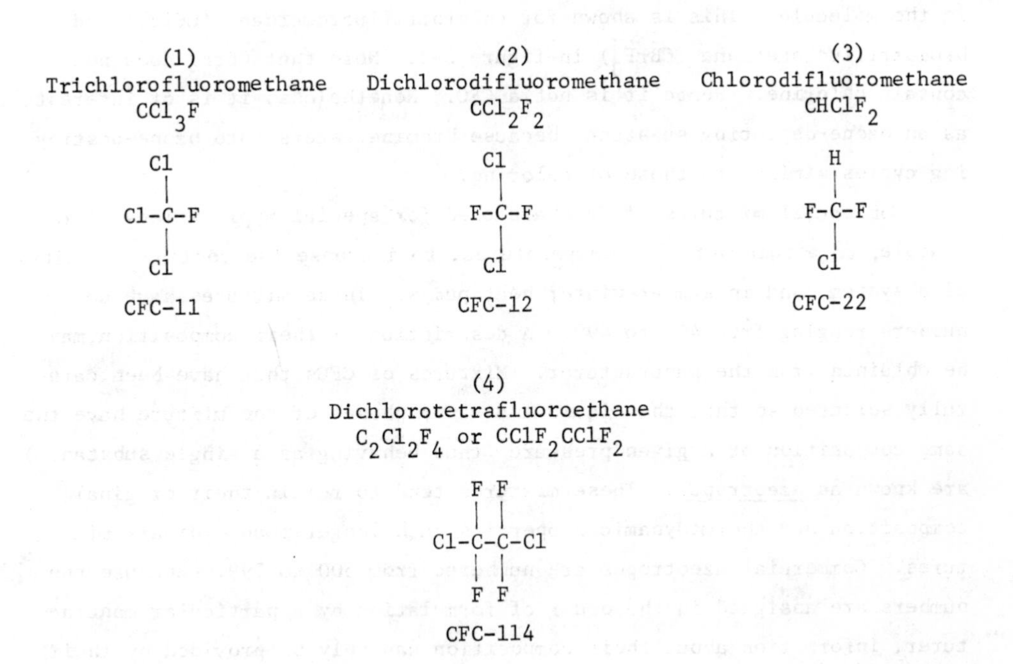

Figure 2-2. Comparison of formulas and code numbers for $CClF_3$ and $CBrF_3$

Chlorotrifluoromethane
$CClF_3$

```
      F
      |
Cl -  C  - F
      |
      F
```

CFC-13

Bromotrifluoromethane
$CBrF_3$

```
      F
      |
Br -  C  - F
      |
      F
```

R-13B1

followed by a B and a number corresponding to the number of bromine atoms in the molecule. This is shown for chlorotrifluoromethane ($CClF_3$) and bromotrifluoromethane ($CBrF_3$) in figure 2-2. Note that $CBrF_3$ does not contain chlorine. Hence it is not a CFC. Nonetheless, it is of interest as an ozone-depleting substance because bromine enters into ozone-destroying cycles similar to those of chlorine.

Commercial mixtures of CFCs are used for special applications, for example, to attain very low temperatures, to increase the cooling capacity of a system, and in summer-winter heat pumps. These mixtures have code numbers ranging from 400 to 499. A description of their composition may be obtained from the manufacturer. Mixtures of CFCs that have been carefully selected so that the liquid and vapor phases of the mixture have the same composition at a given pressure (thus behaving as a single substance) are known as azeotropes. These mixtures tend to retain their original composition and thermodynamic properties much longer than ordinary mixtures. Commercial azeotropes are numbered from 500 to 599. Because the numbers are assigned in the order of formulation by a particular manufacturer, information about their composition can only be provided by their manufacturer.

Other organic and inorganic substances are used as refrigerants, and are listed in ASHRAE Standard 34-78.[3] Because these substances are not CFCs and do not appear to enter into stratospheric ozone-depleting cycles, they will not be considered in this chapter.

To complete the review of nomenclature, the term halocarbon refers to the resultant of the direct bonding of any halogen and carbon (as in the examples above), where the halogens include bromine and iodine, as well as fluorine and chlorine. Hence, CFCs are a subcategory of the halocarbons. Some halocarbons other than the CFCs also deplete stratospheric ozone; these include carbon tetrachloride (CCl_4) and methyl chloroform (CH_3CCl_3). Because of that ozone-depleting characteristic, these compounds are often included with the CFCs in discussions of ozone depletion.

2.2.2 Use and Disposition of CFCs

CFCs were developed around 1930 to overcome the flammability, toxicity, corrosiveness, and instability of ammonia, methyl chloride, ethyl chloride, sulfur dioxide, and carbon dioxide then in use for commercial and home refrigeration. The CFCs were not only nontoxic, very stable over long periods of time, and inert toward the electrical and mechanical components of the refrigeration system, but also had superior thermodynamic properties enabling engineers to design more efficient refrigerators. As new CFCs with different boiling points, critical temperatures and pressures, and solvent characteristics were synthesized, additional uses were developed by industry. Thus the fraction of the total CFC production used for refrigeration dropped steadily. It was estimated that only 25 percent of the 1974 worldwide production went into refrigeration systems and that only 14 percent of the CFCs released into the atmosphere came from such systems. By 1978, about 50 percent of the CFCs released in the United States and 81 percent worldwide resulted from their use as propellants for aerosols (hair sprays, deodorants, perfumes, polishes, oven cleaners, air fresheners, and insecticides). An increasing amount was released during the manufacture of polyurethane, polystyrene, and polyolefin foams. (CFCs-11, -12, and -114 were the most common foaming agents.) CFCs are now also used as solvents for cleaning printed circuits and other components of high-precision electrical devices; for the extraction of edible oils from agricultural products and of essential oils from fish, coffee, and spices; and for the concentration of active ingredients for perfumes. In addition, CFCs are used in very large quantities for degreasing machined parts before assembly or plating. In chapter 3 (table 3-4), Gladwin, Ugelow, and Walter note that CFC-11 and CFC-12 jointly account for about 80 percent and CFC-22 for another 10 percent of worldwide CFC production.

The rapid growth in the amount of CFCs released into the atmosphere may be appreciated by noting that releases during 1975 (about 1.5 billion lb) were equal to one-eighth of the total worldwide production over the sixteen-year span 1958 to 1973 (12.4 billion lb). All of this production will eventually be discharged into the atmosphere. In view of the long lifetimes of CFCs (fifty to one hundred years), continual discharges of this magnitude will lead to high concentrations in the troposphere (the lower portion of the atmosphere) and ultimately damage the stratosphere, as will be discussed later in this chapter.

Several tropospheric "sinks," or CFC-destruction mechanisms, have been suggested. These include (1) solubility in and removal by microorganisms in the oceans; (2) entrapment in ice; (3) photolytic processes; (4) chemical reactions with neutral molecules; (5) absorption and microbial action in soil and vegetation; (6) ionic processes; (7) heterogeneous processes; (8) chemical reaction with aerosols; (9) enhanced removal in urban areas; (10) removal by lightning-thermal decomposition and photolysis; and (11) thermal decomposition in combustion processes. Only dissolution in the oceans followed by microbial degradation and adsorption on desert sands followed by photolysis appear to offer significant "sinks" for tropospheric CFCs. Under the most favorable circumstances, removal of CFCs by the "ocean sinks" would decrease the calculated steady-state ozone reduction by 20 percent. The photolysis of CFCs, adsorbed on desert sands, by intense solar radiation has been reported recently. This and the ocean sinks should be investigated further. Nevertheless, it seems clear that the major process for removing CFCs from the troposphere appears to involve transport into the stratosphere where the CFCs are decomposed by solar radiation. That process will be considered in detail in the following sections. Section 2.3 sets out the terms of the problem by discussing underlying key components and processes, while section 2.4 discusses the destruction of ozone.

2.3 The Atmosphere, Solar Radiation, and Stratospheric Ozone

2.3.1 The Atmosphere

The atmosphere may be conveniently divided into three regions--the troposphere, the stratosphere, and the mesophere. The troposphere contains about 87 percent of the air mass, the stratosphere slightly less than 13 percent, and the mesophere only about 0.1 percent. Starting in the troposphere at the surface of the earth, the average temperature of the air decreases steadily with increasing altitude to about -80°C at 16 kilometers (km) at the equator and about -50°C at 8 km at the poles. Above these heights, the temperature slowly begins to rise, attaining 0°C at a height of about 50 km. The temperature then decreases again with increasing altitude. The first region of temperature reversal, known as the tropopause, forms a diffuse but effective boundary between the troposphere and the less dense air of the stratosphere. The troposphere is a region of rapid vertical mixing because warmer air from near the ground tends to rise and is replaced by cooler air from above. Thus any material released into the troposphere soon becomes evenly distributed throughout it. Water vapor evaporated from surfaces of the earth condenses as it rises into cooler air and is eventually precipitated as rain or snow. Solid particles, condensable vapors, and soluble materials in the troposphere are then removed by the rain and snow. Only a few parts per million of water vapor pass into the stratosphere because of the very low temperature of the tropopause. Transfer of gases between the troposphere and the stratosphere is also very limited because of the boundary effect of the high-density air in the tropopause.

In the stratosphere, the rise in temperature with altitude is accompanied by a decrease in air density in the upward direction, resulting in a very stable air mass with high-density air at the bottom and lowest-density air at the top. As there is no convective drive, mixing occurs very slowly in the vertical direction. Gases are transported and mixed by a large-scale, nearly horizontal movement of air masses.

The second region of temperature reversal forms the stratopause. As noted above, the temperature decreases with altitude in this region. However, air density at this altitude is so low that temperature-density effects can be neglected for present purposes.

2.3.2 Solar Radiation

The human sensation of light is stimulated by radiation with wavelengths between 400 and 750 nanometers (1 nm = 10^{-9} meters). Radiation in that range is referred to simply as "light." Radiation with wavelengths less than 400 nm is sometimes referred to as ultraviolet light but, as it is not visible, it should more precisely be named ultraviolet radiation. Several ranges of wavelengths have been defined for convenience in discussing stratospheric ozone phenomena. The 320-to-400-nm range is designated as UV-A. This radiation is not involved in ozone formation or in destructive reactions. The 290-to-320-nm range, known as UV-B, is largely responsible for sunburn, skin cancer, and other biological effects. Radiation in this range is also effective in altering deoxyribonucleic acid (DNA), the material carrying genetic information of all living cells. The DNA-altering effectiveness increases by a factor of about 5,000 from 320 nm to 290 nm. Thus, even a very small increase in 290 nm radiation may be more significant biologically than a large increase at 320 nm. The total amounts of UV-B, weighted by the relative DNA-altering effectiveness of the wavelengths included, are referred to as _damaging_ UV or DUV. Radiation in this range of wavelengths is partially absorbed by the stratospheric ozone shield, thus limiting the amount reaching the earth.

Radiation of 240-to-290-nm wavelengths is almost completely absorbed by the stratospheric ozone shield and does not reach the earth's surface. Radiation between 190-and-240-nm wavelengths is absorbed by molecular oxygen in the upper stratosphere to form ozone. Most of the radiation with wavelengths less than 190 nm is absorbed before it reaches even the high stratosphere. Thus, practically no radiation with a wavelength less than 290 nm reaches the surface of the earth.

The energy (E) associated with radiation is calculated from the relation $E=hC/\lambda$, where h is a universal constant (Planck's constant), C is the velocity of light, and λ is the wavelength of the radiation. Thus, as the wavelength of radiation _decreases_, the energy _increases_. As C/λ is equal to the frequency (v) of the radiation, the energy equation is simplified to $E=hv$. The product hv represents the energy contained in one unit of radiation of the frequency (and related wavelength) indicated. The unit is known as a "quantum"; several units are "quanta." Quanta are included in photochemical reactions to indicate the frequency (and wavelength) and the amount of radiation entering into the reaction.

2.3.3 Stratospheric Ozone

Molecular oxygen in the stratosphere is dissociated by 190-to-240-nm radiation into atomic oxygen in accordance with Equation (1):

$$O_2 + hv\ (190\ \text{nm} - 240\ \text{nm}) \longrightarrow O + O \qquad (1)$$

In the presence of another molecule M (other than oxygen), the oxygen atoms rapidly combine with molecular oxygen to form ozone:

$$O + O_2 + M \longrightarrow O_3 + M \quad (2)$$

The other molecule M serves to stabilize the ozone molecule by removing the excess energy released by the combination of an oxygen atom and an oxygen molecule. (The excess energy would tend to dissociate the ozone molecule, if not removed.) Thus, although solar ultraviolet radiation is more intense at the upper stratosphere, most ozone formation takes place at lower, denser stratospheric levels where third molecule (M) collisions are more frequent. The most favorable ozone-forming region, considering radiation intensity, oxygen concentration, and overall gas density, appears to be between 25 km and 35 km at tropical latitudes. However, the greatest ozone concentration (in molecules of O_3 per cubic centimeter of air) is found in the polar regions at altitudes of about 15 km. This is the result of compression of the atmosphere at the polar regions and motions of the stratospheric air masses.[4]

Ozone absorbs 240-to 320-nm radiation and is dissociated into atomic and molecular oxygen:

$$O_3 + h\nu \text{ (240 nm - 320 nm)} \longrightarrow O_2 + O \quad (3)$$

Although this is a reversal of Equation (2), it does not result in a decrease in ozone concentration because the dissociation products rapidly recombine upon colliding with a third molecule, liberating the absorbed energy as heat. Thus absorption at 240 nm to 320 nm serves not only to protect the earth's surface from biologically damaging ultraviolet radiation but also warms the stratosphere causing the temperature inversion above the tropopause that stabilizes and isolates the stratosphere from the troposphere. Another heat-liberating reaction is the combination of an ozone molecule with an oxygen atom to form two molecules of oxygen:

$$O_3 + O \longrightarrow 2(O_2) \quad (4)$$

From 1930 to about 1950, it was believed that this reaction, competing with Equations (1) and (2), controlled the amount of ozone at equilibrium at any time. However, careful studies have found less ozone in the stratosphere than was predicted from the rates of reactions shown in Equations (1), (2), and (4). A search for other reactions that could destroy ozone showed that species containing hydrogen, nitrogen, and chlorine or bromine could become involved in catalytic cycles, destroying many molecules of ozone per individual molecule or radical released into the stratosphere. These reactions are discussed in the following section.

2.4 The Destruction of Ozone

At present, of the stratospheric ozone destroyed, about 20 percent is destroyed by reaction with atomic oxygen; about 10 percent is destroyed by reactions with hydrogen (in the HO_x cycle); about 60 percent is removed by several cycles involving nitrogen (in sequences of the NO_x cycle); perhaps 4 to 7 percent is destroyed by reaction with chlorine (the ClO_x cycle, involving CFCs); and the small remainder is accounted for by various other reactions, including reactions with bromine (the BrO_x cycle), thus involving other halocarbons. We focus here on the CFCs because of their importance relative to the other halocarbons, and because they may become the major source of ozone destruction. The various mechanisms of ozone destruction will now be considered in some detail.

2.4.1 The HO_x Cycle

HO_x species are produced by the reaction of naturally occurring water vapor or methane with excited oxygen atoms formed when ozone is decomposed by ultraviolet radiation:

$$O_3 + h\nu \text{ (less than 310 nm)} \longrightarrow O_2 + O('D) \quad (5)$$

$$O('D) + H_2O \longrightarrow 2HO \quad (6)$$

$$O('D) + CH_4 \longrightarrow CH_3 + HO \quad (7)$$

O('D) is the spectroscopic notation for oxygen atoms excited to a higher energy state.

The HO radicals destroy ozone by means of the cycle:

$$HO + O_3 \longrightarrow HO_2 + O_2 \quad (8)$$

$$HO_2 + O \longrightarrow HO + O_2 \quad (9)$$

The net effect of Equation (8) plus (9):

$$O_3 + O \longrightarrow 2O_2$$

The HO radicals are regenerated to enter into the cycle many times until removed by some other reaction such as:

$$HO + HO_2 \dashrightarrow H_2O + O_2 \qquad (10)$$

It is estimated that about 10 percent of the stratospheric ozone is removed by variations of the HO_x cycle.

2.4.2 The NO_x Cycle

Nitrous oxide (N_2O) is produced by the action of bacteria on nitrogenous material in soil and water. As N_2O is only slightly soluble in water, it accumulates in the troposphere and eventually some passes into the stratosphere. Most of it is converted to nitrogen and oxygen by ultraviolet radiation but about 1 percent reacts with excited oxygen atoms to form nitric oxide (NO):

$$O('D) + N_2O \dashrightarrow 2NO \qquad (11)$$

and start the NO_x cycle destroying ozone:

$$NO + O_3 \dashrightarrow NO_2 + O_2 \qquad (12)$$

$$O + NO_2 \dashrightarrow NO + O_2 \qquad (13)$$

The net effect of Equation (12) plus Equation (13):

$$O + O_3 \dashrightarrow 2O_2$$

This cycle is repeated until NO_2 is removed by reacting with HO in the presence of a third body, M, to form nitric acid:

$$HO + NO_2 + M \dashrightarrow HNO_3 + M \qquad (14)$$

The nitric acid gradually passes downward into the troposphere where it is removed by rain. The NO_x cycle may also be interrupted by the reaction:

$$HO_2 + NO \dashrightarrow HO + NO_2 \qquad (15)$$

The products ($HO + NO_2$) then react as in Equation (14) to produce HNO_3, which again is removed by being rained out. The NO_x cycle accounts for about 60 percent of the _natural_ ozone destruction rate. Large fleets of supersonic aircraft would release significant amounts of NO_x directly into the stratosphere. It was initially believed this would increase the destruction rate of ozone by the NO_x cycle above the natural 60 percent of the total rate, but more recently it has been concluded that supersonic transport emissions probably will not have much impact on stratospheric ozone.

2.4.3 CFCs and the ClO_x Cycle

Although chlorine-containing compounds have been discharged into the atmosphere for some fifty years, it was not until 1974 that the potential catalytic destruction of ozone by chlorine and oxides of chlorine was recognized by Molina and Rowland,[5] Wofsy, McElroy and Sze,[6] and Stolarski and Cicerone,[7] essentially independently. Chlorine and its oxide, ClO, catalyze the destruction of ozone in this fashion:

$$Cl + O_3 \longrightarrow ClO + O_2 \qquad (16)$$

$$O + ClO \longrightarrow Cl + O_2 \qquad (17)$$

The net effect of Equations (16) and (17):

$$O + O_3 \longrightarrow 2O_2$$

Thus, one chlorine atom can destroy thousands of ozone molecules before it is removed by reacting with naturally occurring methane to form hydrogen chloride, a very water-soluble gas:

$$Cl + CH_4 \longrightarrow HCl + CH_3 \qquad (18)$$

The HCl serves as a temporary reservoir as it can react with HO to release the Cl atoms:

$$HO + HCl \longrightarrow H_2O + Cl \qquad (19)$$

Some of the HCl gradually moves down into the troposphere where it is rained out. Recently, Rowland, Spencer, and Molina proposed chlorine nitrate ($ClONO_2$) as another temporary reservoir for active chlorine.[8] It is formed in the presence of a third body, M, by the reaction:

$$ClO + NO_2 + M \longrightarrow ClONO_2 + M \qquad (20)$$

Chlorine nitrate is destroyed by photolysis:

$$ClONO_2 + h\nu \longrightarrow ClO + NO_2 \qquad (21)$$

The importance of these reservoirs depends on the proportion of potentially active chlorine tied up in the reservoir, unavailable for participation in ozone-destroying cycles. This proportion is determined by the rates of formation and destruction of the reservoir species. For $ClONO_2$, this proportion is such that inclusion of Equations (20) and (21) in the ozone destruction models reduced the calculated ozone destruction by a factor of 1.85. That is, ozone reduction calculations that include the $ClONO_2$ reactions yield values that are about one-half those calculated without $ClONO_2$ formation. This marked effect is due to the ability of the $ClONO_2$ reactions to interfere with the ClO_x and the NO_x cycles, both major ozone-destroying mechanisms. In fact, the ClO_x cycle is about three times more efficient in destroying ozone than the NO_x cycle, on a molecule per molecule basis. At present it is believed that the ClO_x cycle accounts for only a small percentage of the current ozone reduction because of the few chlorine-containing molecules in the stratosphere. However, tropospheric lifetimes and release rates imply a rapidly increasing effect. (The lifetime of a molecule in the atmosphere is the time required for its concentration to become reduced to about one-third of its original value, or more precisely, to 1/e, where e is the base of the natural logarithm, e = 2.71828...). Given the very long tropospheric lifetimes of some chlorine-containing compounds (ranging from three months for chloroform to fifty or more years for trichlorofluoromethane) and the rapid increase in release rates (roughly doubling every five to seven years), the ClO_x cycle may become the predominant ozone-destroying mechanism in the not too distant future unless rates of release are substantially reduced.

In the stratosphere, CFCs are decomposed by solar radiation with wavelengths of 190 nm to 210 nm, as follows:

$$CCl_3F + h\nu \text{ (190-210 nm)} \longrightarrow CCl_2F + Cl \quad (22)$$

$$CCl_2F_2 + h\nu \text{ (190-210 nm)} \longrightarrow CClF_2 + Cl \quad (23)$$

Sometimes an additional chlorine atom is split off during the primary photolysis but the dominant reactions of the CFC radicals in the presence of oxygen are:

$$CCl_2F + O_2 \longrightarrow FClCO + ClO \quad (24)$$

$$CClF_2 + O_2 \longrightarrow F_2CO + ClO \quad (25)$$

The Cl from Equations (22) and (23) and the ClO from Equations (24) and (25) then destroy ozone in accordance with the ClO_x cycle, Equations (16) and (17), until the cycle is interrupted by the formation of HCl, Equation (18).

As transport into the stratosphere is slow, it is estimated that if the releases of CFCs were terminated at any point in time, the tropospheric CFC burden would be reduced to 1/e (about one-third) of its value at the time of termination, after approximately one hundred years. (A report by E.I. du Pont de Nemours & Co. suggests that CFC tropospheric lifetimes are about fifty years.[9]) Thus, CFCs accumulate and persist for many years.

The 1976 NAS report estimated that continued release of CFCs at the 1973 production rate would cause the ozone to decrease steadily until a probable ultimate reduction of about 6 to 7.5 percent was reached,[10] with half that reduction attained in about forty years. (The 6 percent figure occurs if CFCs are indeed removed from the troposphere by ocean sinks.) It was noted that less importance for $ClONO_2$ would imply increased ozone depletion by a factor approaching 1.85, in turn implying ozone reduction of 11.1 to 13.9 percent, that is, $6 \times 1.85 = 11.1$ and $7.5 \times 1.85 = 13.9$. It was also estimated that if CFCs were released at the 1973 rate until 1978 and then all release halted, the reduction in ozone would continue to increase for another ten years and then recover one-half of its maximum loss (about 2 percent) in an additional sixty-five years.

The persistence of CFCs in the troposphere has been confirmed by measurements of CFC-11. As of September 1, 1975, the average value for the tropospheric burden of CFC-11 was 110 parts per trillion (ppt) $\pm$ 40 percent. As the troposphere contains about 90 percent of the total atmosphere, 110 ppt corresponds to 1.09×10^{34} molecules of CFC ($110 \times 10^{-12} \times 0.90 \times 1.1 \times 10^{44}$, where 1.1×10^{44} represents the total number of molecules in the atmosphere). The Manufacturing Chemists Association (now the Chemical Manufacturer's Association) estimated the total world production of CFC-11 through August 1975 as 3.32 million metric tons (about 7 billion lb).[11] The association estimated that 85 percent was released into the atmosphere, which corresponds to 1.24×10^{34} molecules.[12] Thus, about 88 percent of the total amount of CFC-11 introduced was still in the atmosphere after at least two decades. (Note: $1.09 \times 10^{34} \times 100 \div 1.24 \times 10^{34} = 87.9$ percent.)

The importance of the ClO_x cycle as an ozone-destroying mechanism was greatly enhanced by the discovery of an error in the rate constant for the reaction $HO_2 + NO \longrightarrow HO + NO_2$. It was found to be much greater than previously thought. Because this reaction is capable of interrupting the NO_x cycle (see NO_x cycle), a large increase in its rate reduced the effectiveness of the NO_x cycle in destroying ozone. The change in the rate of reaction seen in Equation (15) was a major factor that led to the conclusion that NO_x emissions from currently projected fleets of supersonic transports and present use of fertilizers would not have a significant impact on stratospheric ozone.

As the reaction shown in Equation (15) produces HO radicals, an increase in the rate of Equation (15) increases the concentration of HO. The HO radicals in turn release more Cl atoms from HCl (Equation 19) to feed back into the ClO_x cycle. It now appears that the anticipated ozone reduction caused by CFCs is about twice as large as previously calculated in the 1976 NAS report.[13] The most likely range of ultimate ozone reduction is now estimated as 11 to 16 percent, on the basis of calculations given in the 1977 NASA report.[14] That estimated central range was subsequently reported and applied in the 1977 and 1979 NAS reports.[15] The revised estimate of an 11 to 16 percent reduction adheres to the original assumption that CFC emissions will continue at the 1973 rate. If the rates of emissions accelerate as they have during the past decade, ultimate ozone depletion is estimated to range from at least 30 percent to as much as 57 percent, based on the increasing emissions scenario.[16]

Parenthetically, it is interesting to note that the two changes--the introduction of the $ClONO_2$ formation reaction and the increase in the rate constant (or removal) of HO_2 and NO from NO_x cycles--have effects that are almost equal and opposite, so that the currently accepted ultimate ozone reduction estimate is now about the same as had been calculated with 1975 models. The 1975 results, as reported in the 1976 NAS report,[17] were scaled down to account for the presumed full $ClONO_2$ effect. Thus, this sequence of central range estimates of ozone depletion has appeared in the literature: 1975 NAS model, 11.1 to 13.9 percent; 1976 NAS report, 6.0 to 7.5 percent; 1977 NASA report, 11 to 16 percent; 1977 and 1979 NAS reports, 11 to 16 percent.

As noted earlier, a more recent report by E.I. du Pont de Nemours & Co.[18] suggests that the trophospheric lifetimes of CFCs are only half the values previously calculated. If this suggestion is supported by further research results, CFCs entering the stratosphere would be reduced proportionately and estimated central level of ozone depletion would be 5.5 to 8 percent. These conflicting research results are excellent indicators of the complexity and interdependence of the processes involved.

2.5 Climatic Effects

The major expected climatic effects of CFCs involve temperature changes, although other induced climatic changes may also occur. Available evidence suggests that the direct effect of CFCs far outweighs the indirect effect by way of ozone depletion. By the year 2000, the direct effect of increased CFC emissions would be a surface temperature increase on the order of 0.7°C (or 1.25°F); in contrast, the indirect effect might be a decrease in temperature of 0.1°C. This section reviews the evidence on climate impacts, considering direct and indirect effects, in turn. (Some additional discussion of both direct and indirect climatic effects appears in chapter 7. See table 7-1, in particular.)

2.5.1 Direct Climatic Effects

The direct climatic effects of CFCs occur because CFCs absorb and emit infrared (thermal) radiation with 7- to 14- μm (1 μm = 10^{-6} meters) wavelengths, normally not absorbed by the atmosphere. The CFCs in the atmosphere trap the infrared radiation emitted by the earth's surface and warm the atmosphere. The increased infrared radiation from the warmer troposphere heats the surface of the earth, giving rise to the well-known greenhouse effect.

If releases of CFC-11 and CFC-12 were continued at the 1973 rate, the atmospheric steady-state concentration would be 0.7 ppb (parts per billion) and 1.9 ppb for the two compounds, respectively. The terrestrial radiation flux passing through such an atmosphere into space would be reduced by 0.3 percent. The absorbed energy would be radiated back to the earth, raising the average surface temperature by 0.5°C. If CFC releases increased at 10 percent per year, the calculated concentrations are 0.5 ppb for CFC-11 and 0.8 ppb for CFC-12 by the year 1990 and 1.4 and 2.1 ppb, for the two CFCs, respectively, by the year 2000. The average surface temperature increases would be 0.3°C by 1990 and 0.7°C by 2000.

As an indication of the greenhouse effectiveness of CFCs, it may be noted that the projected concentration of CO_2, in the atmosphere, for the year 2000, is 390 ppm (parts per million) and the accompanying average surface temperature rise about 0.5°C. Thus, the CFCs are more than 100,000 times as effective as CO_2 ($0.7 \times 390 \times 10^{-6} \div 0.5 \times 3.5 \times 10^{-9} = 1.5 \times 10^{5}$).

2.5.2 Indirect Climatic Effects

Indirect climatic effects arise from the destruction and redistribution of ozone in the stratosphere and consequent changes in the ultraviolet flux reaching the troposphere and the surface of the earth.

Changes in the average surface temperature due to ozone reduction have been estimated in several one-dimensional studies, using models that retain variation only in the vertical dimension, averaging over the other dimensions of latitude and longitude. (The lack of a satisfactory three-dimensional model of the climate system makes it very difficult to estimate the temperature distribution within the atmosphere and at the surface of the earth.) The one-dimensional studies also assume uniform ozone reduction throughout the vertical column. Within those constraints, the one-dimensional studies indicate that removing ozone cools the surface. The cooling is small and proportional to the ozone reduction up to about 40 percent. The calculated decrease in surface temperature is 0.1°C for a 10 percent reduction in ozone. That estimate is quite uncertain, however, because the cooling is the result of two opposing and much larger effects. Removal of stratospheric ozone allows more radiation, particularly ultraviolet radiation, to pass through to the troposphere and the earth's surface, thereby warming the earth's surface. At the same time, the decrease in absorption by stratospheric ozone tends to cool the stratosphere, reducing the infrared radiation emitted toward the ground. This tends to cool the troposphere and the ground.

Ozone absorbs solar radiation in the ultraviolet, visible, and infrared regions of the spectrum, to varying degrees. Absorption of solar ultraviolet is the major source of heating within the stratosphere. Hence, reductions of ozone also leads to a lowering of stratospheric temperatures, though there are differences of opinion as to the distribution of the lowering. Cooling of the stratosphere near the tropopause could lead to increased ice-crystal formation in this region. The consequent increase in solar energy reflected into space would tend to further cool the troposphere and the earth's surface. These effects need further study. However, a more commonly held view is that the reduction of stratospheric ozone would lead to enhanced solar heating of the lower stratosphere and tropopause, increasing the flux of water vapor into the stratosphere. Photolysis of the added water vapor would increase the concentration of the very reactive HO radicals that destroy ozone by participating in the HO_x cycle and by releasing Cl atoms from HCl. This latter effect alone is such that doubling stratospheric water vapor doubles the amount of ozone reduction produced by a given amount of stratospheric chlorine. This would lead to further warming of the tropopause in a reinforcing process.

Changes in the temperature profile of the stratosphere likely would also lead to significant changes in the dynamics and transport processes within the stratosphere that might affect its stability. More rigorous models are being developed to explore these effects.

Insofar as climatic effects at the surface of the earth are concerned, some of the most significant impacts might result from changes in regional temperatures, wind patterns, precipitation, and other weather elements caused by changes in the radiative properties of the troposphere and stratosphere.[19]

2.6 Biological Effects of Ozone Depletion

It is generally accepted that a 1 percent decrease in stratospheric ozone is accompanied by about a 2 percent increase in biologically damaging ultraviolet radiation, DUV, received at the surface of the earth. Thus, the 11 to 16 percent ozone reduction, predicted in the 1979 NAS report,[20] would result in a 22 to 32 percent increase in DUV. (The earlier 6 to 7.5 percent reduction predicted in the 1976 NAS report would correspond to a 12 to 15 percent increase in DUV.)[21] This section reviews the evidence on the consequences of increased DUV, first considering nonhuman biological effects and then human health effects.

2.6.1 Nonhuman Biological Effects

Exploratory studies have been undertaken on the effects of increased DUV on agriculturally important crops, insect pests, nitrogen fixation, gas exchange rates in crops under field conditions, eye cancer in cattle, growth impairment in forest tree seedlings, plant-pathogen interactions, and commercial fisheries. Preliminary results indicated adverse impacts in all cases, but it was not possible to determine reliable magnitudes of adverse impacts. Considerable variations among species and among cultivars (cultivated varieties) within species were found. The most common responses by crop plants to increases in DUV were reductions in leaf area, height, and weight, and delays in germination.[22]

Apparently, increases in DUV altered the growth and productivity of higher plants by reducing photosynthetic rates, increasing respiratory carbon consumption in the dark, and impairing leaf-cell division.[23] In sensitive plant species, the amount of photosynthesis occurring under standard light conditions decreased with total DUV radiation dose. Even doses equal or less than those currently received in the natural environment caused some reduction in photosynthesis compared with plants given no DUV. The effect of a given total dose was the same over a wide range of dose rates. Thus, even small increases in natural DUV radiation might have significant effects on the production of plant leaves. It has been reported that only plants possessing the "C_3" pathway of photosynthesis, the most common pathway, were sensitive to DUV. Plants native to those areas of the earth receiving more intense UV radiation possess a physiologically more elaborate photosynthetic system known as "C_4" and were not sensitive to DUV. As only a very limited number of crop species were included in this study, this observation must be explored further. Nonetheless, if this hypothesis ultimately proves to be true, plants may be able to adapt to increases in DUV by converting to "C_4" photosynthesis.[24]

Impairment of photosynthesis did not appear to involve cellular nucleic acids, nor did the radiation repair mechanism appear to be operative in the case of DUV damage, as was believed previously.[25] Reduced foliage after DUV exposure seemed to be due to inhibition of cell division rather than reduced cell expansion. The plant phytochrome system did not appear to be directly involved. The reduced foliage area combined with the reduced photosynthetic rate to decrease the plant's ability to fix carbon, resulting in the observed impairment of growth after exposure to DUV.

Generally, the results of studies and theoretical considerations indicated that exposure to increased DUV levels would reduce crop yields, increase disease susceptibility of some crops, and increase or decrease pesticide usage depending on the crop.[26] In addition, some nitrogen fixing bacteria in algae have been shown to be impaired.

Preliminary results on animals indicated that increases in DUV radiation caused reductions in insect life span and in the number of offspring per surviving adult, and delay in development; eye damage in cattle;[27] and the formation of brain and retinal lesions in fishes and aquatic invertebrates. Embryonic forms of mackeral and anchovy are sensitive even at levels very near those naturally experienced close to the water's surface.[28] Other observed adverse impacts on aquatic ecosystems were depressed chlorophyll-a concentrations, reduced biomass production, decreased community species diversity, and altered community structures.[29] Some of the aquatic species (and some land species too) appeared to be very near their limits of tolerance of DUV. These species would be adversely affected by even small increases in DUV.

In sum, the variation in sensitivity among species and the extreme sensitivity of some species "imply that shifts in composition of natural plant and animal communities will occur if DUV irradiation increases, and perhaps changes in agricultural practices may be expected. Quantitative predictions of the size and significance of these changes are not feasible with current knowledge."[30]

2.6.2 Human Health Effects

Even before the rise of public concern about the effects of continued releases of CFCs, it was commonly accepted that increased exposure to sunlight increased the risk of skin cancer. Three types of skin cancers are involved: (1) basal cell, (2) squamous cell, and (3) melanoma. Most skin cancers are of the first two types. They are easily detected and successfully treated in practically all cases. A few of these cancers are caused by exposure of individuals to arsenic, pitch, and X-rays in the course of their work or during treatments; for example, skin disorders are sometimes treated with X-rays. The third type of skin cancer, melanoma, is a serious life-threatening hazard; about one-third of melanomas cause death. The incidence of melanoma has been rising very rapidly in recent years.

Fortunately, most human exposures to sunlight do not result in skin cancers. The skin of about two-thirds of the white population, after short, repeated exposures, develops a tan that absorbs ultraviolet radiation, preventing it from penetrating down to sensitive cells. The remaining third experience little or no tanning. Within this group, sensitivity to sunlight continues or even increases, leading to severe burns and often skin cancer. This group consists mainly of inhabitants of high northern latitudes, very light-skinned people, and those of Celtic origin--the Celts are particularly prone to develop skin cancer. Most blacks are not susceptible to solar-induced skin cancer. Genetic factors other than pigmentation are not responsible for this immunity as albino Negroes fall into the high-risk category. Solar-induced skin cancers appear to result more frequently from exposures interrupted by long periods of nonexposure than from continual repeated exposures. Thus, consideration of life-style is very important in studies of the etiology of skin cancer.[31]

In summary, skin cancer associated with solar UV exposure is a disease of less-pigmented people and thus of major concern in the United States, Europe, and Australia. Despite the large number of inhabitants of Celtic origin, it does not appear to be a problem in the United Kingdom, probably because of the limited periods of bright sunshine. Some Britons believe they would actually benefit from increased ultraviolet radiation because of the known conversion, in the skin, of 7-dehydrocholesterol to vitamin D_3, preventing rickets. However, large doses of ultraviolet radiation are not required for this conversion. Even exposures experienced at high latitudes in winter suffice to meet the needs of most people and consequently, any increases in ultraviolet radiation would usually yield no benefits on this score.

A key step in cancer production appears to be changes in the body's deoxyribonucleic acid (DNA) molecules. It has been suggested that sunlight-induced skin cancers probably result from the action of sunlight on DNA or compounds with similar "action spectra." This hypothesis has been confirmed in animals for nonmelanotic skin cancer.

The graph resulting from plotting the reciprocal of the amount of radiation required to produce one unit of an effect under study against the wavelength of the radiation is known as the "action spectrum" of that particular effect. More often, action spectra are displayed as plots of the effectiveness of equal amounts of radiation as a function of the wavelength of the radiation. The action spectrum for skin cancer is not known but it is believed to resemble the erythema (sunburn)-producing and the DNA-damaging action spectra. The action spectrum of melanoma also is not known. Again, the erythemal action spectrum is used to estimate potential hazard from melanomas as DUV radiation increases.

Five lines of evidence have been suggested for sunlight being the major or contributing cause of skin cancer:[32]

1. Increase in incidence (or mortality in the case of melanomas) with decrease in latitude, corresponding generally to increases in UV exposure;
2. Location of skin cancers on the more commonly exposed areas of the body;
3. Increased incidence or mortality among workers exposed to bright sunshine for very long periods;
4. Changes in incidence or mortality over time, consistent with changes in exposure behavior;
5. Experimental confirmation in animals--for nonmelanotic cancers only.

Data collected from the Surveillance, Epidemiology and End Results (SEER) program, the Health and Nutrition Examination Survey (HANES), the National Ambulatory Medical Care Survey, the Survey of Discharges from Short Stay Hospitals, and from the eighth revision of the International Classification of Diseases in 1968, support the lines of evidence cited above. Several models have been proposed to predict the effect of increases of DUV on the incidence of skin cancers other than melanomas.

Generally, these models predict that a 1 percent increase in DUV will result in a 2 percent increase in skin cancer. Recalling that a 1 percent decrease in ozone is accompanied by a 2 percent increase in DUV, the recently predicted 11-to 16-percent ozone reduction, at a steady state, would mean a 44-to 64-percent increase in skin cancers. In the 1975 report of the Federal Task Force on Inadvertent Modification of the Stratosphere,[33] it was estimated that there were approximately 300,000 new cases of nonmelanotic skin cancers in the United States annually. Assuming a 54 percent increase (from the range 44 to 64 percent), 162,000 additional cases of skin cancer would probably occur from the predicted 11-to 16-percent decrease in ozone. This would raise the annual new skin cancer incidence to 462,000. The average annual incidence rate for melanomas was 4.2 per 100,000 for the period 1969-71. Assuming this incidence rate remains constant, and using the value of 250 million for the U.S. population, the calculated number of new melanomas per year is 10,500. About one-third (3,500) of these patients die within five years. If the same percentage increase of skin cancers applied to melanomas (and it appears that it does), the postulated 11-to 16-percent decrease in ozone might increase the number of melanomas by 5,770 and the number of deaths by 1,890 annually.

There are other less serious undesirable health effects of increased ultraviolet radiation. Among them are increased atrophy, wrinkling, and areas of hyper- and hypopigmentation of the skin (aging); conjunctivitis, photokeratitis of the cornea, blepharospasm (excessive winking from involuntary contraction of the eyelid muscle), and sometimes cataracts of the eye; potentially excessive vitamin D formation in individuals who ingest large quantities of vitamin D supplements, and photosensitivity to various chemicals.

2.7 Concluding Remarks

The detection of man-made influence on ozone is difficult because of large, natural variations in the amount of total ozone. For example, a station in Switzerland had daily variations in ozone of almost 10 percent in winter and 5 percent in summer; the variations are much greater at higher latitudes. In addition, there is an approximately 25 percent annual variation and 2 percent quasi-biennial variation. The Federal Task Force on Inadvertent Modification of the Stratosphere reported that there are also natural long-term variations (some possibly connected to the eleven-year solar cycle) on the order of 5 percent in a decade.[34] Given such variations, it might be argued that ozone depletion caused by CFCs is tolerable.

However, long-term global average reductions in ozone attributable to CFCs are not small but now appear to be 11 to 16 percent and could become much greater (at least 30 percent and possibly 57 percent) if CFC releases continue to rise as in the past decade.[35] The large day-to-day changes are caused by winds transporting ozone-rich or ozone-poor air from one area to another. These are short-term changes easily tolerated by ecosystems. However, altering the average ozone concentration profile might cause even larger daily changes and significant average changes. Furthermore, reductions in ozone concentration could eventually exceed even the large seasonal variations and interrupt or enhance those effects that now constitute the normal environmental patterns important to ecosystems. Any damage to the stratospheric ozone shield could take hundreds of years to repair!

References

1. E.I. du Pont de Nemours & Co., Freon Products Division, Bulletin R-29, Refrigerant Numbering System, Wilmington, Delaware, no date.
2. American Society of Heating, Refrigerating, and Air-Conditioning Engineers, Inc., ASHRAE Standards 34-57 and 34-78, New York, N.Y., 1957.
3. See ASHRAE, ASHRAE Standards 34-57 and 34-78.
4. National Research Council, National Academy of Sciences, Panel on Atmospheric Chemistry, Halocarbons: Effects on Stratospheric Ozone (Washington, D.C. NAS, 1976) p. 23.
5. M.J. Molina and F.S. Rowland, "Stratospheric Sink for Chlorofluoromethanes - Chlorine Atom - Catalyzed Destruction of Ozone," Nature (London) vol. 249 (1974).
6. S.C. Wofsy, M.B. McElroy, and N.D. Sze, "Freon Consumption: Implications for Atmospheric Ozone," Science vol. 187 (1975).
7. R.S. Stolarski and R.J. Cicerone, "Stratospheric Chlorine: A Possible Sink for Ozone," Canadian Journal of Chemistry vol 52 (1974).
8. F.S. Rowland, J.E. Spencer, and M.J. Molina, "Stratospheric Formation and Photolysis of Chlorine Nitrate, $C\ell ONO_2$," Journal of Physical Chemistry vol. 80 (1976), p. 2711.
9. E.I. du Pont de Nemours & Co., Freon Products Division, Wilmington, Delaware, Fluorocarbon/Ozone Update: The Chlorofluorocarbon/Ozone Depletion Theory--A Status Report, October 1980.
10. National Research Council, NAS, Halocarbons: Effects on Stratospheric Ozone, p. 161.
11. Manufacturing Chemists Association (now Chemical Manufacturers Association), Worldwide Annual Fluorocarbon Production (Washington, D.C.), 1976.
12. R.L. McCarthy, "World Production and Release of Chlorofluorocarbons 11 and 12 Through 1976 (E.I. du Pont de Nemours & Co.), Wilmington, Del., 1977.
13. National Research Council, NAS, Halocarbons: Effects on Stratospheric Ozone, p. 161.
14. R.D. Hudson, ed., Chlorofluoromethanes and the Stratosphere (Washington, D.C., NASA), 1977.
15. National Research Council, National Academy of Sciences, Committee on the Impacts of Stratospheric Change, Response to the Ozone Protection Sections of the Clean Air Act Amendments of 1977: An Interim Report (Washington, D.C., NAS, 1977; and National Research Council, National Academy of Sciences, Panel on Stratospheric Chemistry and Transport, Stratospheric Ozone Depletion by Halocarbons: Chemistry and Transport (Washington, D.C., NAS, 1979), p. 4.
16. National Research Council, Stratospheric Ozone Depletion by Halocarbons, p. 176.

17. National Research Council, NAS, Halocarbons: Effects on Stratospheric Ozone, p. 161.

18. E.I. du Pont de Nemours & Co., Fluorocarbon/Ozone Update, 1980.

19. V. Ramanathan and R.E Dickinson, "The Role of Stratospheric Ozone in the Zonal and Seasonal Radiative Energy Balance of the Earth-Troposphere System," Journal of Atmospheric Science, vol. 36 (1979), pp. 1084-1104.

20. National Research Council, Stratospheric Ozone Depletion by Halocarbons, p. 4.

21. National Research Council, NAS, Halocarbons: Effects on Stratospheric Ozone, p. 161.

22. R.H. Briggs and S.V. Kossuth, "UV-B Biological and Climate Effects Research, Terrestrial; Impact of Solar UV-B Radiation on Crop Productivity," Final Report to the U.S. Department of Agriculture (Washington, D.C.), 1978.

23. M.R. Kaufman, "The Effect of Ultraviolet (UVB) Radiation on Pine Seedlings," Final Report EPA-IAG-D6-0168 (Washington, D.C., EPA), 1978.

24. National Research Council, NAS, Response to the Ozone Protection Sections of the Clean Air Act Amendments of 1977, p. 50.

25. National Research Council, National Academy of Sciences, Committee on Impacts of Stratospheric Change, Halocarbons: Environmental Effects of Chlorofluoromethane Release (Washington, D.C., NAS) 1976, p. 67-76.

26. H.R. Carns, J.H. Graham, and S.J. Rairtz, "Effects of UVB Radiation on Selected Leaf Pathogenic Fungi and on Disease Severity," Final Report, Interagency Agreement EPA-IAG-D6-0168 (Washington, D.C., EPA), 1979.

27. K.E. Kopecky, G.W. Pugh, and D.E. Hughes, "Biological Effect of Ultraviolet Radiation on Cattle: Bovine Ocular Squamous Cell Carcinoma," Final Report, USDA and EPA Interagency Agreement, EPA-IAG, D6-0168 (Washington, D.C.), 1978.

28. J.R. Hunter, J.H. Taylor, and H.G. Moser, "Effect of Ultraviolet Irradiation on Eggs and Larvae of the Northern Anchovy, Engraulis Mordax, and the Pacific Mackerel, Scomber Japonicus, During the Embryonic Stage," Photochemistry and Photobiology, vol. 27, 1978.

29. H. Van Dyke and R.C. Worrest, "Assessment of the Impact of Increased Solar Ultraviolet Radiation Upon Marine Ecosystems," NAS 9-14860, Mod. 78 (Houston, Texas, NASA Lyndon B. Johnson Space Center), 1977; J. Calkins, "Effects of Real and Simulated Solar UVB in a Variety of Aquatic Microorganisms--Possible Implication for Aquatic Ecosystems," CIAP Monograph 5 (Washington, D.C., Department of Transportation), pp. 33-69, 1975; D.M. Damkaer, G.A. Heron, D.G. Deay, and E.F. Prentice, "Effects of UVB Radiation on Near-Surface Zooplankton of Puget Sound," Technical Report (Seattle, Wash., Pacific Marine Environmental Laboratory/NOAA), 1978;

30. U.S. Environmental Protection Agency, Results of Research Related to Stratospheric Ozone Protection--A Report to Congress, EPA 600/8-78-002 (Washington, D.C. EPA), 1978.

31. E.L. Scott and M.L. Straf, "Ultraviolet Radiation as a Cause of Cancer," Origins of Human Cancer (H.H. Hiatt, ed.), Cold Springs Harbor Laboratory, pp. 529-546, 1977; T.R. Fears, J. Scotto, and M.A. Schneiderman, "Mathematical Models of Age and Ultraviolet Effects on the Incidence of Skin Cancer Among Whites in the United States," American Journal of Epidemiology, vol. 105, pp. 420-427; J. Scotto, A.W. Kopf, and F. Urbach, "Nonmelanoma Skin Cancer Among Caucasians in Four Areas of the United States," Cancer, vol. 34, p. 1333; T.J. Mason and F.W. McKay, "U.S. Cancer Mortality by County: 1950-1969," NIH Publication 74-615 (Bethesda, Md., National Institutes of Health), 1974; J.E. Dunn, E.A. Levin, G. Linden, and L. Harzfeld, "Skin Cancer as a Cause of Death," California Medicine, vol. 102, pp. 361-363; H.O. Lancaster and J. Nelson, "Sunlight as a Cause of Melanoma: A Clinical Survey," Medical Journal of Australia, vol. 1, pp. 452-456; and M. Movshovitz and B. Modan, "Role of Sun Exposure in the Etiology of Malignant Melanoma Epidemiology Inference," Journal of the National Cancer Institute, vol. 51, pp. 777-779, 1973.

32. National Research Council, NAS, Halocarbons: Environmental Effects of Chlorofluoromethane Release (Washington, D.C. NAS), p. 84, 1976.

33. Federal Task Force on Inadvertent Modification of the Stratosphere (IMOS), Fluorocarbons and the Environment (Washington, D.C., Council on Environmental Quality and Federal Council for Science and Technology, June 1975).

34. Ibid.

35. National Research Council, NAS, Stratospheric Ozone Depletion by Halocarbons: Chemistry and Transport (Washington, D.C., NAS), p. 176, 1979.

Chapter 3

A GLOBAL VIEW OF CFC SOURCES AND POLICIES TO REDUCE EMISSIONS

Thomas N. Gladwin, Judith L. Ugelow, and Ingo Walter

With the exception of work by the Organisation for Economic Co-operation and Development (OECD) and d'Arge and coworkers,[1] consideration of the potential international aspects and ramifications of CFC regulation has been virtually absent from most of the economic and technical analyses of CFC restrictions.[2] The inclusion of relevant international costs and benefits should improve benefit-cost analyses and regulatory policy for CFC emissions.

This chapter provides a preliminary assessment of the international economic implications of existing, planned, and possible CFC regulation in the United States. The paper points out apparent cause-and-effect linkages and tentatively evaluates the significance of international trade and factor movement shifts that could result from the exogenous "shock" of CFC regulations. It provides a survey of the status and probable trend of CFC regulations around the world, and analyzes the international commerce of five CFC-related industries, and considers how CFC regulation can affect that commerce. (The five industries are chlorofluorocarbons, precursor chemicals, aerosols, refrigeration and air conditioning, and urethane foams.) The industrial findings are then integrated to assess adverse effects of U.S. restrictions on the nation's competitive position in world trade, and to develop insight into the transboundary nature of the CFC-pollution problem, in effect involving a "global commons." Finally, there is a brief consideration of a range of unilateral and multilateral policy tools which the United States might employ to improve the effectiveness of its own policy and to foster ozone management safeguards abroad.

This report employed field interviews as well as secondary sources and a literature review in developing its findings. Despite this coverage, the preliminary nature of the analysis must be stressed. In many cases the necessary data are not available, and a number of assumptions can be subjected to serious question. These caveats, along with the contingency logic used in drawing policy conclusions, point up the need for careful and balanced additional research on the issue.

3.1 CFC Magnitudes and Regulation Around the World

3.1.1 Worldwide CFC Production

Table 3-1 presents 1974 statistics on the shares of production and consumption on the major CFCs, F-11 and F-12, for each of the member countries of the OECD. The United States produced about half the OECD total of 755,000 metric tons, and that total, in turn, represented about 87 percent of world production of 872,000 metric tons, as estimated in a survey by the Manufacturing Chemists Association (MCA).[3] This grand total is composed of worldwide production estimates supplied by twenty OECD-based producers, plus estimates for Eastern bloc countries, and four producers in India and Argentina. The twenty OECD producers accounted for 93 percent of world production in 1973 and 89 percent in 1975.

Figure 3-1 and 3-2 show trends in F-11 and F-12 production, respectively, drawing on MCA survey data[4]. The average annual growth rate in production from 1961 through 1976 was 6.9 percent for the United States and 16.9 percent outside of the United States. The EPA estimates that, through 1975, U.S. production accounted for about 50 percent of world F-11 production and about 60 percent of world F-12 production. For 1976-77, however, the estimated shares dropped to about 35 percent and 50 percent, respectively. Figures 3-1 and 3-2 show that worldwide production of the compounds began to fall in 1975, with most of the decline occurring in the United States (a drop of about 20 percent from 1974 to 1975).

3.1.2 Worldwide CFC Consumption

As compared to production, the consumption of CFCs is much more widespread, with most countries relying on imports for their supplies. As shown in table 3-1, the United States accounted for roughly half of OECD

Table 3-1. Distribution of 1974 OECD Production and Domestic Consumption of F-11 and F-12 Chlorofluorocarbons

Country	Production (metric tons)	Share of total OECD production (%)	Net Export (-)/Import (+) (metric tons)	Domestic consumption (metric tons)	Share of total OECD consumption (%)
Germany	88,300	11.7	-27,600	60,000	8.7
Australia	14,000	1.9	+ 768	14,768	2.1
Austria	-	-	+ 5,000	5,000	0.7
Belgium	-	-	+ 3,163	3,163	0.5
Canada	23,500	3.1	- 1,410	22,090	3.2
Denmark	-	-	+ 4,246	4,246	0.6
Spain	8,700	1.2	- 1,513	7,187	1.0
U.S.A.	376,000	49.8	-14,796	361,204	51.5
Finland	-	-	+ 4,752	4,752	0.7
France	72,000	9.5	-27,000	45,000	6.4
Greece	-	-	+ 3,000	3,000	0.4
Ireland	-	-	+ 2,900	2,900	0.4
Iceland	-	-	+ 200	200	0.01
Italy	38,000	5.0	+ 7,000	45,000	6.4
Japan	34,200	4.5	- 2,200	32,000	4.6
Luxembourg	-	-	+ 300	300	0.04
Norway	-	-	+ 2,900	2,900	0.4
New Zealand	-	-	+ 5,000	5,000	0.7
Netherlands	29,000	3.8	-15,000	14,000	2.0
Portugal	-	-	+ 3,409	3,409	0.5
U.K.	72,000	9.5	-24,000	48,000	6.8
Sweden	-	-	+ 5,630	5,630	0.8
Switzerland	-	-	+ 8,370	8,370	1.2
Turkey	-	-	+ 1,400	1,400	0.2
Yugoslavia	-	-	+ 700	700	0.1
Total	755,700	100.0	-54,781	700,919	100.0

Source: OECD, Economic Impact of Restrictions on the Use of Pluorocarbons (Paris, OECD document ENV/Chem/77.2, 24 January 1978).

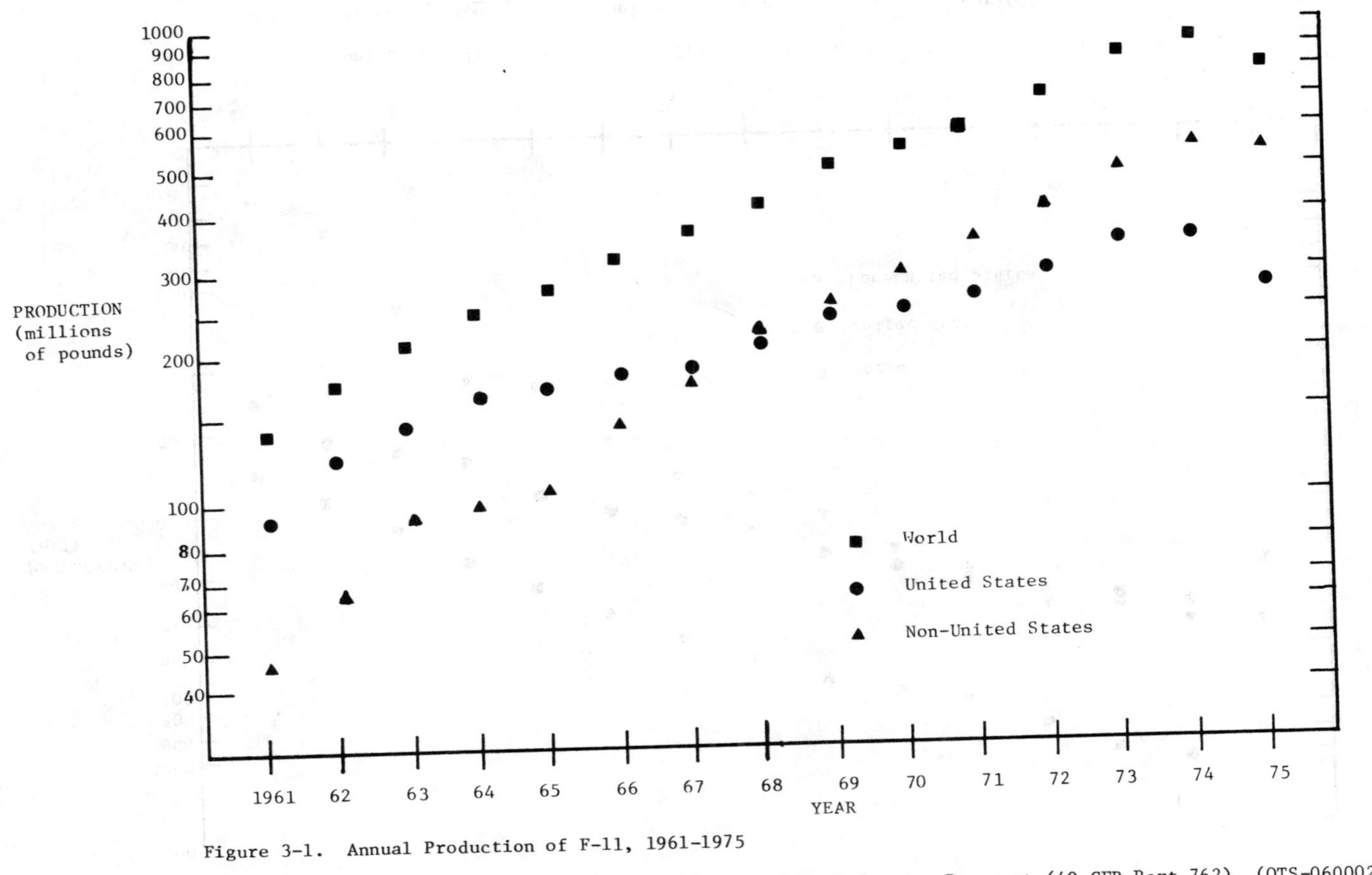

Figure 3-1. Annual Production of F-11, 1961-1975

Source: , EPA, Chlorofluorocarbon Problem Assessment: Support Document (40 CFR Part 762), (OTS-060002), Toxic Substances Control, Fully Halogenated Chlorofluoroalkanes, May 1977, p. 13.

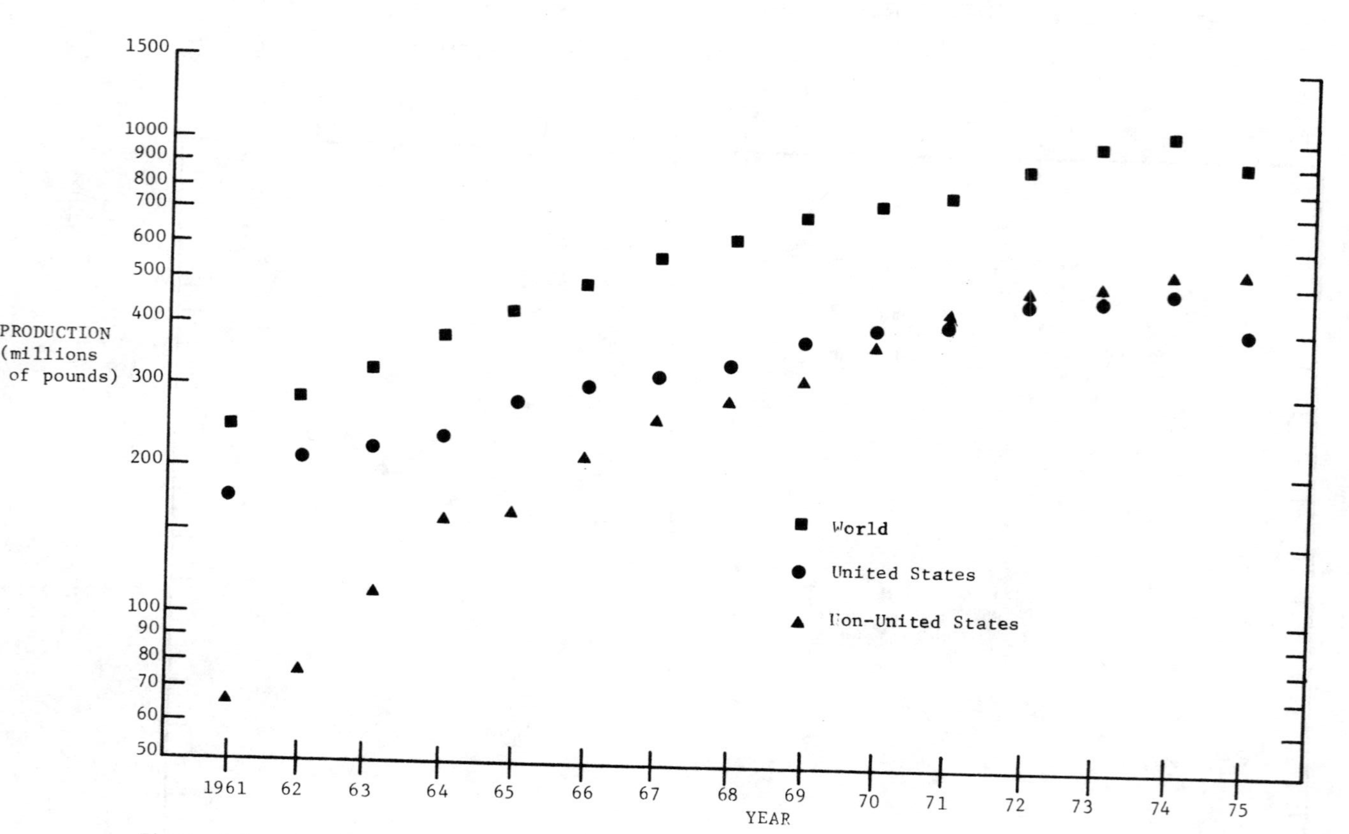

Figure 3-2. Annual production of F-12, 1961-1975

Source: EPA, Chlorofluorocarbon Problem Assessment: Support Document (40 CFR Part 762), (OTS-060002), Toxic Substances Control, Fully Halogenated Chlorofluoroalkanes, May 1977, p. 14.

consumption, as well as half the OECD production. Table 3-2 provides a breakdown of 1974 end-use consumption for these compounds for fourteen nations, including weighted averages for all those nations, and for the OECD countries other than the United States. From that table it can be seen that personal aerosols accounted for about half the consumption in both the United States and the other countries. In contrast, the percentages employed in domestic aerosols and plastic foam-blowing agents were considerably higher for the other OECD nations than for the United States, while the reverse held for refrigeration and air conditioning.

3.1.3 Worldwide CFC Emissions

World emissions of F-11 in 1975 were estimated at 340.6 kt (thousand metric tons), equivalent to 96 percent of 1975 production, while world emissions of F-12 were estimated at 413.0 kt, equivalent to 99 percent of 1975 production.[6] Table 3-3 provides estimates of worldwide releases for 1975, broken down by use. Aerosols accounted for 74.5 percent of the F-11 and F-12 releases, whereas air conditioning and refrigeration accounted for 13.7 percent and plastic foams for 11.8 percent.

Table 3-4 shows estimated world and U.S. production and atmospheric emissions of five principal halocarbons for 1973. The U.S. share of total world emissions of F-11, F-12, and F-22 was estimated to be 45 percent. The volume of world CFC emissions in 1973 was estimated to be about 1,730 million pounds, with propellant emissions accounting for about 62 percent, refrigerants for 27 percent, and other uses (including solvents and foaming agents) for 12 percent. Tables 3-5 and 3-6 provide estimates of the world aerosol propellant and refrigerant emissions for 1973, showing that U.S. emissions accounted for just under one-half the emissions in both categories, and those from Europe for just over one-third.

With the U.S. ban on CFC use in aerosols, it is estimated that the U.S. share of world CFC emissions dropped to less than 20 percent by 1979. The EPA Administrator has found that "chlorofluorocarbon emissions anywhere in the world deplete the ozone layer and adversely affect the health and the environment of the United States."[7] This raises the important issue of environmental policies and CFC regulations overseas and the problem of the expected 80 percent share of CFC emissions occurring outside the United States.

Table 3-2. Breakdown of F-11 and F-12 Consumption by End Use, 1974 (metric tons and percentages)

Country	Total consumption	Aerosols personal	Aerosols domestic	Aerosols industrial	Aerosols medical	Refrigeration and air-conditioning	Plastics	Other
Germany	60,700	35,100 (57.8%)	8,500 (14.0%)	2,400 (4.0%)	1,200 (2.0%)	4,000 (6.6%)	9,000 (14.8%)	500 (0.8%)
Australia	14,760	5,316 (36.0%)	5,169 (35.0%)	591 (4.0%)	148 (1.0%)	2,068 (14.0%)	- 1,476 - (10.0%)	
Austria	5,000	2,800 (56.0%)	600 (12.0%)	70 (1.4%)	30 (0.6%)	400 (8.0%)	1,100 22.0	- -
Canada	22,090	9,765 (44.2%)	3,400 (15.4%)	725 (3.3%)	455 (2.1%)	3,900 (17.6%)	3,765 (17.0%)	90 (0.4%)
Denmark	4,246	1,350 (31.8%)	375 (8.8%)	63 (1.5%)	9.0 (0.2%)	848 (20.0%)	1,552 (36.6%)	49 (1.1%)
U.S.A.	361,204	180,000 (49.8%)	20,600 (5.7%)	12,300 (3.4%)	10,600 (2.9%)	84,200 (23.3%)	31,300 (8.7%)	22,204 (6.2%)
Finland	4,752	2,683 (56.5%)	769 (16.2%)	262 (5.5%)	2 -	230 (4.8%)	801 (16.9%)	5 (0.1%)
Italy	45,000	24,000 (53.3%)	6,400 (14.2%)	1,200 (2.7%)	400 (0.9%)	2,800 (6.2%)	10,000 (22.2%)	200 (0.5%)
Japan	32,000	10,000 (33.3%)	1,700 (5.3%)	1,440 (4.5%)	600 (1.9%)	9,300 (29.1%)	8,200 (25.6%)	100 (0.3%)
Norway	2,900	1,134 (39.1%)	324 (11.2%)	107 (3.7%)	33 (1.2%)	600 (20.7%)	700 (24.1%)	- -

Country	Total consumption	Aerosols personal	Aerosols domestic	Aerosols industrial	Aerosols medical	Refrigeration and air-conditioning	Plastics	Other
Netherlands	14,000	6,400 (45.7%)	3,700 (26.4%)	700 (5.0%)	50 (0.4%)	1,200 (8.6%)	1,800 (12.9%)	150 (1.0%)
U.K.	48,000	29,140 (60.7%)	5,520 (11.5%)	4,140 (8.6%)	460 (1.0%)	3,220 (6.7%)	4,600 (6.9%)	920 (1.9%)
Sweden	5,630	2,100 (37.3%)	470 (8.4%)	690 (12.2%)	95 (1.7%)	725 (12.9%)	1,550 (27.5%)	- -
Switzerland	8,370	5,110 (61.1%)	1,895 (22.6%)	430 (5.1%)	65 (0.8%)	400 (4.8%)	300 (3.6%)	170 (2.0%)
Turkey	1,400	461	133	44	12	400	- -	350 - (25.0%) -
OECD total	700,919	351,861 (50.2%)	65,887 (9.4%)	28,037 (4.0%)	16,121 (2.3%)	126,866 (18.1%)	84,811 (12.1%)	27,336 (3.9%)
OECD excluding United States	339,715	171,861 (50.8%)	45,287 (14.6%)	15,737 (4.8%)	5,521 (1.4%)	42,666 (11.0%)	53,611 (16.7%)	5,132 (0.7%)

Source: OECD, Economic Impact of Restrictions on the Use of Fluorocarbons (Paris, OECD document ENV/Chem/77.2, 24 January 1978).

Table 3-3. Worldwide Releases of F-11 and F-12 (millions of pounds)[a]

Source	Million pounds
Aerosols 1115.1 (74.5%)	
Personal care	
Antiperspirants/deodorants	458.4
Hair care	401.5
Medicinal	37.3
Fragrances	2.3
Shave lathers	0.9
Others	34.4
Household	
Room deodorants	17.7
Cleaners	9.6
Laundry products	23.4
Waxes and polishes	9.2
Others	9.2
Miscellaneous	
Insecticides	33.3
Coatings	22.9
Industrial	39.0
Automotive	8.0
Vet. and pet.	2.3
Others	5.7
Air conditioning refrigeration 204.7 (13.7%)	
Mobile A/C	89.8
Chillers	42.9
Food store	33.1
Beverage coolers	5.8
Home refrigerators and freezers	5.3
Miscellaneous	27.3
Plastic foams 176.5 (11.8%)	
Open cell	100.0
Closed cell	76.5
	1496.3

Source: Committee on Impacts of Stratospheric Change, Assembly of Mathematical and Physical Sciences, National Research Council, Halocarbons: Environmental Effects of Chlorofluoromethane Release (Washington, D.C., National Academy of Sciences, 1976), p. 117.

[a]Based on the annual incremental releases of F-11 and F-12 indicated for 1975 in the MCA report and the detailed percentage analysis by uses in the United States for 1973 in the Arthur D. Little report. Russia and Eastern Europe, which have not reported production or release data (about 11 percent of total) not included.

Table 3-4. Estimated World and U.S. Production and Atmospheric Emissions of Five Principal Halocarbons, 1973

(in thousands of metric tons)

Halocarbons	World production	U.S. production	World emissions	U.S. emissions	Lifetime in the atmosphere
CFCs[a]					
F-11	930	150	700	140	greater than 10 years
F-12		220		170	
F-22	120	60	60	28	1 to 10 years
Subtotal	1,050	430	760	340	
Other halocarbons[b]					
Carbon tetrachloride	950	470	40	20	greater than 10 years
Methyl chloroform	420	250	320	200	1 to 10 years
Subtotal	1,370	720	360	220	
Total[c]	2,420	1,150	1,120	560	

Sources: Arthur D. Little, Inc., production and emission estimates. Atmospheric lifetime estimates by Cicerone, et al., "Stratospheric Ozone Destruction by Man-Made Chlorofluoromethanes," Science, Vol. 185 (27 September 1974), p. 1165 and references contained therein, and A.P. Altshuller, EPA (personal communication).

[a]Production of F-11, F-12, and F-22 accounts for approximately 90 percent of total CFC production.

[b]Production of carbon tetrachloride and methyl chloroform accounts for 8 percent of total U.S. production of the commercially important other halocarbons. Vinyl chloride represents 28 percent. Emissions of other chemicals in this group are not presented here because of their shorter atmospheric life.

[c]Values may not add due to rounding.

Table 3-5. World Emissions of CFC Aerosol Propellants, 1973

Region	Estimated emissions		
	Thousand metric tons	Millions of pounds	Percentage of total
United States	232	511	48
Europe	170	375	34
Other	86	190	18
Totals	488	1076	100

Source: Arthur D. Little, Inc., estimates

Table 3-6. World Emissions of CFC Refrigerants, 1973

Region	Estimated emissions		
	Thousand metric tons	Millions of pounds	Percentage of total
United States	96	212	47
Europe	81	179	39
Other	30	66	14
Totals	207	457	100

Source: Arthur D. Little, Inc., estimates

3.1.4 Determinants of National Environmental Policies

Substantial variation in environmental policy occurs among countries.[8] Generally, the developing countries appear to have less rigorous environmental policies than the developed countries. This is to be expected if environmental quality is, in part, viewed as a consumption good that is sensitive to income, so that poorer countries are more likely to accept lower environmental quality levels whenever this involves sacrificing other social or economic objectives. However, a number of forces are at work that may significantly reduce the diversity of environmental policy over time. There is evidence, for example, that the demand for improved environmental quality, which normally costs something, is a positive and elastic function of income. If this holds generally, some convergence in environmental standards may well occur as levels of real per capita income rise.

In certain areas, such as the European Community, attempts at regional harmonization of member-nation environmental policies currently under way are, indeed, moving ahead, albeit at a rather slow pace. Other international organizations, such as the United Nations Environment Programme and the Organisation for Economic Co-operation and Development, are beginning to exert an influence on the setting of national environmental policy agendas. And environmental policy innovations--for example, environmental impact statements, toxic substances control legislation, and coastal zoning acts--which emerge in one country tend to be diffused and adopted by other nations as well. Hence, there are mechanisms which begin to imply greater homogeneity in national environmental programs, at least among the advanced countries.[9] They have not yet exerted a great deal of influence, however, and the environmental setting of the international economy remains diverse. Let us examine CFC regulation as a specific case in point.

3.1.5 International Survey of CFC Regulation

From the evidence, it is clear that there is a significant divergence in approach between the United States and the European Community, with the former having moved rapidly to eliminate CFCs in nonessential aerosol applications and now considering regulation of non-aerosol CFC uses, and the latter taking a much more cautious and conservative approach.[10] Among other countries, only Canada, Norway, Denmark, and Sweden have largely moved with the United States; most of the others have looked to the European Community for policy guidance.

Following the determination of U.S. policy with regard to chlorofluro-carbon use in aerosols, the principal focus of the effort to achieve a worldwide consensus on the CFC controversy shifted to Europe, where a particular combination of facts gave rise to a variety of responses.

The European response to the agitation over the CFC controversy in the United States was rather uniform in the mid-1970s. It appears that both public and official opinion in Europe did not rate the problem as highly as did American opinion. This lack of concern seems to have been derived from the following factors:

1. There was a lack of awareness and concern of the population as a whole. Either Europeans were not aware of the existence of a potential environmental hazard, or they downgraded such an ill-defined, not readily recognizable hazard.

2. The consensus within European scientific circles was that there was not enough evidence to warrant the imposition of restrictions.

3. The various European governments assigned a low ranking, within their priorities, to such a unique environmental problem. This low ranking is presumably also related to the direct economic consequences, usually limited to the locality in which related production facilities are located, of any immediate change in the regulation of CFC production and use.

Since 1975, several approaches for dealing with the problem have been developed by the various European countries that produce or consume CFCs. Through 1981, definitive regulatory actions prohibiting the manufacturing or use of certain CFCs for aerosol application had been taken only in Sweden, Norway, and Denmark, all countries with rather strong environmental ethics and small economic stakes in CFC production or consumption. The government of the Netherlands, after a three-year review of the scientific evidence and potential economic impacts of regulation, merely issued a requirement for a warning label on CFC-using aerosols in 1978. It also stated its intention, however, subsequently to ban CFC use in these products if the hypothesis about the damaging effects of CFCs was not disproved. In the Federal Republic of Germany, a large producer and consumer of CFCs, the government reached an agreement with industry to voluntarily reduce CFC use of aerosols by a third using 1976 as a benchmark. (As of late 1981 an actual reduction of more than 40 percent had been achieved.) The most laggard of the major CFC-producing nations have been France and the United Kingdom, with the former prior to 1981 essentially only encouraging and entering into research agreements, and the latter merely encouraging CFC-related firms to intensify their search for alternatives and to endeavor to minimize CFC releases from industrial equipment. Thereafter, however, both nations adopted the recommendations made by the European Community to reduce by 30 percent the use of two major CFCs (F-11 and F-12) as propellants by the end of 1981, using 1976 as the base year. Despite West German demands for further Community-endorsed restrictions on CFC use, it appeared in late 1981 that most environmental ministers in the Community were adhering to the established line that continued use of CFCs at present levels would not cause undue environmental harm, that further restrictions would hurt industry, and that it was therefore best to wait until the scientific uncertainties were resolved.

Table 3-7 presents an overview of the national policies toward CFC regulation. It is clear that as of 1981 only the United States, Sweden, Norway, and Denmark had taken concrete, mandatory steps to ban CFCs in non-essential aerosols, with the Netherlands, West Germany, France, and Canada moving in a similar direction, but proceeding in a much more cautious and voluntary way. Other CFC producing and -consuming nations such as Italy, Japan, Switzerland, and Australia were generally relying upon the conservative approach of the European Community for guidance in formulating their own policies.

3.2 An Assessment of the International Economic Impact of Regulation

3.2.1 Survey of CFC Related Industries

A variety of industrial sectors are involved in CFC manufacture and in the production, distribution, and servicing of products using CFCs. Each is likely to be affected economically in various ways and degrees--both domestically and internationally--by regulations regarding CFC use. This section, therefore, separately examines the five most important CFC-related sectors: CFC production, precursor chemicals, aerosols,

Table 3-7. Summary of National Chlorofluorocarbon Policies (as of 1981)

Country	Nonessential aerosol ban	Voluntary phase-out	Production leakage	Labeling require-ments	Foam blowing agents	Refrig-erants	Search for al-ternatives	Further study	Legal con-trol frame-work exists	Production limits
Australia										
Austria									x	
Canada		x							x	
EEC		x	p		p		p	x		p
Belgium									E	
Denmark	x								E	
France		x						x	E	
Germany		x							E	
Ireland									E	
Italy									E	
Luxembourg									E	
Netherlands	p			x					E	
U.K.		x	x				x	x	E	
Finland									x	
Japan										
New Zealand									x	
Norway	x								x	
Portugal									x	
Spain									e	
Sweden	x									
Switzerland									e	
U.S.	x			x				x		x
Yugoslavia										

x = implemented; p = proposed; E = EEC policy will apply; e = EEC policy likely to apply refrigeration/air conditioning, and urethane foams. For each, we will (1) provide background data on production, consumption, and market trends; (2) highlight special features associated with particular segments of the sector, (3) profile the sector's foreign economic involvement, and (4) identify actual or potential economic impacts of CFC regulation.

Information was solicited by personal interviews (most in person and a few by phone or letter) with twenty-six executives from twelve firms in the five CFC-related industries. By agreement, the cooperating firms must remain anonymous, and our field interview findings are presented in aggregate form only. Contacts were also made with representatives of a number of industry trade associations, trade journals, consulting firms and governmental agencies. Data from a range of secondary sources (governmental trade statistics, regulatory agency hearings, commissioned studies by consultants, industry reports, and industry and environmental magazines and journals) were also collected, reviewed, and integrated. Our review does not consider the CFC end-use sectors of solvents, fluoroplastic materials, and miscellaneous applications, which, in total, accounted for only about 15 percent of total CFC consumption in the United States in 1974.

It is helpful to establish an aggregate perspective before delving into the specific sectors. Raymond McCarthy of DuPont, for example, estimated the total retail value of the U.S. industry involved in the production, packaging, and distribution of CFCs to be $8 billion in 1974, and the total related employment to be approximately one million workers.[11] The distribution of these workers among major CFC-related sectors, as compiled by the U.S. Department of Commerce, is shown in table 3-8. Upon closer analysis, the Commerce Department concluded that only about 594,000 of these workers truly represented CFC-dependent employment. Of these workers, 83.3 percent were in the refrigeration/air conditioning industry, 11.4 percent were in foams and plastics, 4.6 were in aerosols, and 0.7 percent were in precursor chemical and CFCs production.[12] When the CFC-dependent employment in each sector is related to the annual amount of CFC emissions from each sector, the following results are obtained. In 1974, approximately 4.85 jobs per ton of emissions existed in the refrigeration/air conditioning sector. The figures for the urethane foam and aerosol foam and aerosol sectors were 0.649 and 0.056, respectively. These figures for the urethane foam and aerosol reflect the economic sensitivity

Table 3-8. Estimate of U.S. Jobs Related to CFC Production and Use, 1974

Industry category	Jobs
Refrigeration	
Packaged air conditioners, refrigerators, freezers, manufacturing[a]	150,000
Packaged air conditioners, refrigerators, freezers: sales and service[a]	225,000
Contractors: refrigeration and air conditioning[b]	400,000
Fluorocarbon manufacturers and related suppliers	2,500
Total refrigeration	777,500
Aerosols[c]	
Propellant manufacturers and related suppliers	5,000
Can, valve, and cap manufacturing	9,000
Other ingredients and loading machinery manufacturing	5,000
Can loading	25,000
Total aerosol manufacturing	44,000
Selling and supporting activity	9,000
Total aerosol manufacturing	53,000
Foaming agents[c]	
Foamed plastic manufacturing	15,000
Raw materials manufacturing	10,000
Foamed product end-use manufacturing	150,000
Total foaming agents	175,000
Solvents	
Employment in plants of major users of fluorocarbon solvent[d]	160,000
Fluorocarbon manufacturing and related suppliers	1,000
Total solvents	161,000
Total jobs related to fluorocarbon manufacture and end use	1,166,500

Source: Compiled by U.S. Department of Commerce from Council on Environmental Quality, Federal Council for Science and Technology, Fluorocarbons and the Environment (June 1975), p. 98.

[a]ARI data for 1972.

[b]ARI estimate of 460,000 people in 1972 adjusted to eliminate heating specialists.

[c]duPont estimates based on industry contacts.

[d]Bureau of Census data.

(measured from the standpoint of potential job losses) of each of the sectors to regulations aimed at reducing levels of CFC emissions. In this perspective, the refrigeration/air conditioning sector was about eighty-six times as "sensitive" as the aerosol sector in 1974.

The refrigeration/air conditioning industry is also the most sensitive sector in regard to the potential impact of U.S. CFC regulations on foreign trade. Table 3-9 presents available U.S. international trade data for 1977 for CFC-related products. One major omission is aerosol cans, which could not be obtained as a separate category. However, other evidence indicates relatively little international trade in empty cans, as these tend to be manufacturered in reasonable proximity to the fillers, for most reasons.

As shown in table 3-9, the United States appears to be a major net exporter of organofluorine compounds, with imports of slightly under $1 million (1.4 million pounds) and exports of over $23.5 million (34 million pounds). The unit value of exports is slightly higher than the unit value of imports (69 cents versus 63 cents per pound), implying that exports are relatively more heavily concentrated in the higher-value categories of organofluorine compounds. Similarly, chloroform exports of 20 million pounds greatly exceed imports of slightly over 2 million pounds. The picture on the import side is quite different, with the U.S. appearing on balance as a major net importer of hydrofluoric acid and fluorspar--$104 million of imports versus $8 million of exports. With respect to the production of CFCs themselves, the United States thus appears to be an importer of the inputs and an exporter of the products. According to the Arthur D. Little study,[13] approximately 7 percent of the CFCs produced in the United States in 1973 were exported.

As noted, CFCs find their primary uses in propellants, refrigerants, solvents, and foam-blowing agents, all of which enter the channels of international trade in one form or another. Propellants are used in personal and household products, paints and finishes, insect sprays and food products, among others. All such products are internationally traded, but only a small proportion is in aerosol-packaged form, and this is not broken out separately in the trade statistics. Nevertheless, in personal products, the product group which incorporates the largest proportion of aerosols, the United States, is, in fact, a major net exporter. In 1977, almost $37 million in personal products were imported into the

Table 3-9. U.S. Trade in CFCs and Related Products, 1977

	Imports		Exports	
	Quantity	Value[a]	Quantity	Value
F11 and F12	161,955 lb.	150		
F22	1,202,392 lb.	706	34,222,664 lb.[b]	23,470
Chloroform	2,136,486 lb.	333	19,992,674 lb.	2,970
Hydroflouric acid	180,147,582 lb.	44,990	52,080,146 lb.	7,186
Fluorspar	867,281 tons	59,888	6,642 tons	975
Perfumes	1,179,891 lb.	15,494	n.a.	22,467
Cosmetics, etc.	n.a.	21,206	n.a.	65,372
Air conditioners	70,699 units	58,656	1,081,291 units	399,547
Refrigerators, compressors	2,191,298 units	233,284	896,060 units	858,330
Buses, trucks, etc.	276,931 units	1,262,708	218,162 units	1,989,312
Automobiles	2,795,650 units	11,100,574	716,227 units	3,461,248
Mattresses/cushions	n.a.	1,303	n.a.	17,902
Polyurethane	1,847,161 lb.	1,380	-	-
Polystyrene	15,661,669 lb.	8,715	13,515,885 lb.	7,436
Resins foamed	n.a.	n.a.	16,279,263 lb.	19,633
Polymers foamed	n.a.	n.a.	13,860,819 lb.	11,635
Fire extinguishers	n.a.	n.a.	n.a.	24,407

Source: U.S. Department of Commerce.

[a]Value in thousands of dollars. Imports classified by Trade Statistics-United States of America. Exports classified by Schedule B.

[b]Organofluorine compounds.

United States, but there were over $87 million in exports, indicating a strong U.S. export-competitive interest in this sector. Little evidence, on the other hand, is available for international trade in solvents, plastics and resins, or foam-blowing agents, as indicated in table 3-9.

In the area of refrigerant use, table 3-9 shows that the United States is a major net exporter of air conditioners, selling over one million units abroad ($400 million) in 1977, as compared with imports of 70,699 units ($58 million). With respect to refrigerators and compressors, however, the United States had imports of over two million units worth over $233 million, with exports of about 900,000 units worth over $858 million. Most of the imports appear to consist of small refrigerators, while exports were predominantly large refrigerators and commercial freezers. Another route of refrigerant exports is through automobile and truck air conditioners. This is an area where the United States has a virtual monopoly in international trade. Despite the fact that the United States imports about four times as many automobiles as it exports, a sizable proportion of exported cars are air conditioned, but few of the imported vehicles reportedly carry foreign-made air conditioners--these are either installed after arrival in the United States or are previously exported from the United States for installation abroad.

We now present additional detail on each of our five industries, in turn.

3.2.1.1 Chlorofluorocarbons. Table 3-10 exhibits levels of U.S. CFC production for selected years from 1960 through 1975. The major products, F-11 and F-12, accounted for the bulk of that production (78 percent in 1975). The stabilization and decline of production in 1974-75 can perhaps be attributed primarily to the economic recession and only in a minor way to the CFC ozone-depletion controversy.

Table 3-10. U.S. Chlorofluorocarbon Production and Plant Capacity (millions of pounds)

Year	F-11	F-12	F-22	F-113	F-114	Others[a]	Total production	Plant capacity
1960	72	165	40	b/	10	b/	287	400
1965	171	271	64	14	22	b/	461	690[c]
1970	245	375	100	36	22	b/	778	970
1973	334	489	136	59	26	12	1,056	1,145[d]
1974	347	509	141	64	27	14	1,102	1,145
1975[c]	266	390	108	49	21	11	845	1,195

Source: International Research and Technology Corporation, The Economic Impact of Potential Regulation of Chlorofluorocarbon-Propelled Aerosols, prepared for EPA, FDA, and CPSC (McLean, Virginia, April 1977).

a/ Includes F-13, F-14, F-21, F-115, and F-116.

b/ Less than 10 million pounds.

c/ Average of 1964 and 1966 amounts.

d/ Set at 1145 since both 1972 and 1974 amounts equaled that figure.

e/ Preliminary.

The U.S. share of world production of F-11 and F-12 has declined considerably over time, as shown by these data on production in million pounds:[14]

Year	World	U.S.	U.S./World
1960	328.8	237	.721
1970	1,232.9	620	.503
1974	1,762.4	856	.486
1975	1,509.9	656	.434
1976	1,636.1	624	.381

However, some reversal of the obvious trend may occur given the greatly increased markets for CFCs in energy conservation, for example, in rigid foam insulation.

Ranked in order of sales, the six U.S. producers of CFCs, as of 1976, were:

Company	Sales in million dollars	CFC sales as percentage of company sales
DuPont	183	2.6
Allied Chemical	113	5.1
Union Carbide	73	1.4
Pennwalt Corp.	42	6.7
Kaiser Aluminum	17	1.0
Racon, Inc. (now part of Essex Chemical Corp.)	7	100.0
Total	435	2.6

Union Carbide has since left the CFC-manufacturing business.

The principal competitors of U.S. companies in overseas markets are Imperial Chemical Industries, Farbwerke Hoechst, and Pechiney-Ugine-Kuhlman, of the United Kingdom, Germany, and France, respectively. These three producers share an estimated 60 percent of the European CFC-propellant market (two million tons consumed in aerosols in 1974), with five other producers making up the rest. U.S. producers also compete with essentially all non-Communist producers in one place or another. The industry is highly competitive, for the CFC products manufactured and sold by one firm are chemically identical to, and of essentially equivalent quality as the products of its competitors. Many of the large firms have foreign subsidiaries or affiliates which produce CFCs.

Virtually all U.S. production has been consumed domestically, with estimated exports amounting to only 3.9 percent of production. In contrast, Germany exported 41.6 percent of its production; Italy, 39.5 percent; France, 37.5 percent; and Spain, 24.6 percent. U.S. exports increased considerably in the early 1970s, but stabilized at roughly 35 million pounds, and $23 million, after 1975. Canada, Mexico, and Venezuela were the major U.S. customers in 1977, with shipments to each totaling more than two million pounds of CFCs; however, there were shipments to more than sixty other countries, as well.

On the import side, there have never been significant imports of F-11 or F-12 into the United States.

U.S. producers experienced about a 40 percent drop in CFC sales between 1974 and 1978 as a result of the ozone-depletion controversy and U.S. regulations on aerosols. The Canadian market also dropped at about the same rate, whereas European production showed only a small decline, and Japanese production was unaffected.

The demand reductions led to gross overcapacity among U.S. producers. Allied Chemical shut down its F-11 and F-12 plants at Baton Rouge, Louisiana, in 1975; and, in 1977, Pennwalt wrote off and closed its Thorofare, New Jersey, plant; DuPont mothballed its East Chicago, Indiana, facility; and Union Carbide got out of the CFC business by shutting down its Institute, West Virginia, plant. Altogether, an estimated two hundred jobs and 350 million pounds of CFC capacity were lost. The excess capacity situation in 1975-77 resulted in considerable price attrition. With capacity reductions, CFC prices have begun to rise (for example, DuPont raised various refrigerant prices 8 to 11 percent in March 1978). Prices are likely to rise significantly over the next few years as a result of losses of scale economies, higher input costs caused by the effects of cutbacks on suppliers of precursor chemicals, and substantially increased R & D costs to support research into the ozone depletion theory and to identify and test possible CFC alternatives.

According to industry observers, there are no known plans for new construction of F-11 and F-12 manufacturing facilities over the next few years anywhere in the world. One firm we interviewed canceled a major CFC project it was considering for an undisclosed location. Another producer told us that if the United States bans most uses of CFCs and makes it difficult or impossible to export them, then the firm will simply cease manufacturing them worldwide.

3.2.1.2 Precursor Chemicals. Precursor chemicals are inputs in CFC production; major precursor chemicals are carbon tetrachloride, chlorine, carbon disulfide, hydrofluoric acid, chloroform and fluorspar. Raw material costs represent more than 70 percent of the production cost in a CFC plant operating at full capacity, and, as such, any curtailments in CFC production directly translate into demand reductions for chemical raw material suppliers. Almost all of fluorspar is imported; U.S. production of the other major chemicals and the percents of those chemical outputs employed in CFC production were as follows, in 1974.[15]

Chemical	Production (millions of lb)	Percentage share of CFC production
Chlorine	21,236	8
Carbon disulfide	772	25
Hydrofluoric acid	779	42
Carbon tetrachloride	1,014	100
Chloroform	300	81
Total	24,121	14

These basic chemicals required for CFC manufacture had a production value of 340 million in 1974.[16] Individual roles for the chemicals can be noted. Carbon tetrachloride is a major raw material in F-11 and F-12 production, with carbon disulfide an input into carbon tetrachloride production. Chlorine enters production of all CFCs while chloroform is exclusively used in the production of F-22, which also takes about 10 percent of U.S. hydrofluoric acid production.

Considerable vertical integration occurs in the CFC production process, with much of the precursor chemical production manufactured by firms producing CFCs. The ban on CFC use in nonessential aerosols has already exerted an impact on precursor chemical suppliers, with greatest reduction in demand occurring for carbon tetrachloride (perhaps as much as a 50 percent reduction).

3.2.1.3 Aerosols. The FDA, EPA, and CPSC issued final rules and regulations on the prohibition of nonessential CFC-propelled aerosols on March 17, 1978.[17] An industry source estimated that this would eliminate all but 3 percent of CFC aerosol use by the end of 1979, with production of CFC propellants falling to 13 million pounds from its 1973 level of 510 million pounds.

U.S. industry had ample warning of the oncoming ban, but it was probably consumer pressure that coaxed many firms into early product conversions. Robert Abplanalp, president of Precision Valve Corporation and the inventor of the first workable aerosol valve, unveiled his new valve, called "Aquasol," that used a mixture of water and butane gas as a propellant, just one day after the EPA, FDA, and CPSC proposed their ban in May 1977. Most of the large marketers of personal and domestic aerosols in the United States substituted other propellants for CFMs during 1978.

The situation is not the same in overseas markets where consumers have generally reacted mildly, if at all, to the CFC propellant ozone depletion danger. Aerosol market growth overseas is still very strong. Total U.S. production of aerosols peaked in 1973, with the following distribution of products, in millions of units:[18]

Product category	Use
Hair care	453
Deodorants and antiperspirants	589
Other toiletries	399
Household	690
All other	771
Total	2,902

In 1975 approximately 50 percent of aerosol propellants were CFCs, 45 percent were hydrocarbons, and 5 percent were carbon dioxide. Because of their higher price, however, CFCs represented 90 percent of propellant sales value.

In 1972 the United States accounted for 52 percent of the world aerosol fillings, whereas the share for Europe was 31 percent and that for the rest of the world was 17 percent. The U.S. share has been steadily dropping. In 1976, the respective shares were 40 percent for the United States, 37 percent for Europe, and 23 percent for the rest of the world. Of course, the U.S. share should continue to decline, with an ultimate share of about 5 percent of the world total.

In 1974, employment in the aerosol industry totaled 102,350 persons worldwide, with 28,100 of that work force employed in the United States. When the EPA-FDA-CPSC ban on CFCs for aerosol use was announced in May 1977, the industry's response was remarkably muted. The ban lost its sting for a number of reasons. First, the timetable allowing CFC-propelled aerosols to be sold until April 15, 1979, presented sufficient time for an orderly transition. Second, alternative forms of packaging (for example, pumps, barrier packs, and roll-ons) had already made significant inroads. In 1974, for example, 80 percent of deodorant/antiperspirant products were sold in aerosol form, but, by 1977, the share had fallen to 45 percent. Third, new valve and propellant systems, such as Aquasol, were emerging which some believed may be more versatile, just as effective, and, in the long run, cheaper than existing systems.[19] Fourth, the ability to use less costly propellants such as hydrocarbons provided marketers with a source of profits, so long as the price structure of aerosol products could be maintained. This appears to have been achieved during the past year, and these profits are being used to meet R & D and production line conversion costs. The competitive nature of the consumer good markets, however, will serve to bring prices down after this transitional period of "regulatory windfalls."

Most industry observers believe that the ban on CFCs for use in nonessential aerosols has caused only limited economic hardship, both domestically and internationally. The only really serious impact has been the loss of CFC-propellant sales by CFC producers.

Best guesses are that aerosol sales in the United States in 1977 sagged 10 percent as marketers let their CFC-propelled stocks run down. Although non-CFC aerosols face a range of fears regarding flammability, toxicity, and other safety hazards, the aerosol industry's prospects through 1985, both in the United States and abroad, are felt to be quite bright.[20]

U.S. manufacturers are now slightly ahead of those in Europe in regard to packaging substitutes and production/product conversion from CFCs to hydrocarbons and compressed gases. The inherent cost disadvantage of CFCs as aerosol propellants is a factor that may be quite important in motivating foreign producers to eliminate CFCs in their aerosols.[21] Two of the U.S. firms we interviewed noted that they were already switching from CFCs to hydrocarbons in a range of product lines overseas. (Approximately 25 to 35 percent of the aerosol industry overseas is owned by U.S. companies.) Flows from U.S. parent companies to their foreign subsidiaries of technology, information, and policy on CFC substitutes is likely to be an important mechanism for bringing about reductions in CFC-propellant emissions overseas--both in the activities of the U.S. subsidiaries themselves and by demonstration and competitive effects on foreign producers.

3.2.1.4 Refrigeration and Air Conditioning. The air conditioning and refrigeration sector encompasses a diverse array of products, including dehumidifiers, freezers, ice makers and chillers, as well as refrigerators and air conditioners. Although CFCs as refrigerants represent a relatively small cost for any of these products, most parts of each product are specific to a refrigerant and, as such, refrigerants cannot generally be used interchangeably.

CFCs were initially developed as refrigerants in the 1930s. Upon success, their use was commercialized in the industry. Their wide use in refrigeration and air conditioning can be traced to their desirable low toxicity, low flammability, and excellent heat transfer properties. U.S. production statistics for CFC refrigerants for 1973 reveal an estimated total output of 300 million pounds, of which 17 million were F-11, 169 million were F-12, 90 million were F-22, and 20 million were F-500, F-502, and other CFCs. This quantity represented 28 percent of U.S. production of CFCs for that year.

Refrigeration and air conditioning uses accounted for about 16 percent of the 1975 worldwide nonaerosol emissions of F-11 and 74 percent of the worldwide nonaerosol emissions of F-12. About half of the F-11 and F-12 emissions occurred in the United States. Mobile vehicular air conditioning represented the largest use and emission source among the refrigerant end-use applications.[22]

The reduction of CFC emissions in refrigeration and air conditioning uses will be extremely complex. Reduction options include the redesign of components most prone to leakage, the improvement of operating, maintenance, and repair procedures to reduce leakage and deliberate discharge, improved retrieval and recycling of refrigerants from scrapped equipment, and conversion of F-11 and F-12 systems to less ozone-depleting CFCs such as F-22 or non-CFC refrigerants.

Alternatives to CFC refrigerants currently include ammonia, sulfur dioxide, or methyl chloride, all of which are either highly toxic, flammable, or corrosive. Expensive, time-consuming, and difficult mechanical redesign would be required to convert current systems to other refrigerants. Most systems probably could be redesigned to use F-22 (at a substantial cost), but the viability of this option is currently shrouded in considerable uncertainty regarding the regulatory future of F-22. In our field interviews, various producers stressed the loss of energy efficiency and long lead times entailed by changeovers to F-22.

Regulatory options involving economic incentives are given detailed attention in Peter Bohm's chapter on refrigerant policy.

Air conditioning and refrigeration is, by far, the largest industry that could be directly affected by any regulatory action on CFC usage. It has been estimated that nearly 500,000 jobs in the U.S. economy are directly related to this industry, of which perhaps 20 percent depend on exports to overseas markets. This implies that approximately 83 percent of total CFC-dependent employment lies within this sector. The total value of shipments in the United States for air conditioning and refrigeration equipment in 1976, according to the U.S. Commerce Department was $6 billion (preliminary estimate), equivalent to about 3 percent of the gross national product.[23] If one adds in peripheral suppliers and the service industry, the figure is probably on the order of 5 to 6 percent. It could rise still higher in the future given trends such as increased use of air conditioning in automobiles and growing demand for heat pumps, which operate on a refrigeration cycle.[24]

U.S. air conditioning and refrigeration firms service foreign markets through a variety of arrangements, including wholly owned local subsidiaries, partially owned joint ventures, licensees, distributors, export agents, and exports directly from the United States. The engineering know-how developed over the past one hundred years has allowed U.S. producers to capture more than a 75 percent share of overseas markets. These markets are serviced to a greater extent by the companies' own local production than by exports from the United States. Foreign competition is now emerging, especially from Japanese firms, but this has been limited to refrigeration products. U.S. industry observers believe the foreigners still have much to learn before the become a serious threat to the dominant U.S. firms (Carrier, York, and Trane).

The ability of U.S. firms to bring about reductions in CFC emissions (via product redesign, substitution or refrigerants, etc.) in their overseas operations, should they be so motivated, would generally be easiest to achieve in the case of wholly owned subsidiaries. The dilution of control entailed in joint ventures and the arms-length nature of licensing arrangements would work against the quick and decisive transfer of any CFC control policies or directives from the associated U.S. firms.

U.S. production and trade in refrigeration and air conditioning has been a solid growth sector with a substantial net trade balance, as indicated by these data from the Annual Survey of Manufacturers:[25]

	(Millions of dollars)			
Year	Product shipments	Exports	Imports	Net trade balance
1971	5,494	440	165	275
1976	7,592	1,268	254	1,014

The Middle East (in particular, Saudi Arabia, Kuwait, and the United Arab Emirates) comprises a major growth market for a range of refrigeration and air conditioning products.

The production and trade data highlight our conclusion that the air conditioning and refrigeration industry constitutes the one CFC-related sector where the United States potentially stands to lose a great deal, both at home and in foreign markets, from purely unilateral and relatively extreme regulations on CFC emissions. A near-term unilateral ban on the use of the CFCs F-11, F-12, and F-22 for refrigerant use seems a sure recipe for disaster--from economic, energy consumption, and human health and safety perspectives--and is obviously politically unacceptable. A ban on the use of F-11 and F-12 alone for refrigeration and air conditioning usage, implemented in the absence of similar foreign action and over a five-to-ten-year time scale, would lessen the adverse impact, but could still be quite costly from an international economic standpoint.

Such a regulatory scenario would confront U.S. producers with the prospect of having to produce different air conditioning and refrigeration for home and foreign markets, with relatively shorter production runs entailed for each. The loss of scale economies, energy efficiency, and sales volume, combined with huge expenditures required for R & D and modification of production facilities, in all likelihood, would do extensive damage to the industry and in all probability drive out a number of firms.

Assuming that foreign consumers continue to purchase lower-priced equipment employing F-11 and F-12, then this portends a major reduction in U.S. exports and a virtual bonanza for fledgling foreign producers who might gear up to take advantage of the product price and performance disadvantages of U.S.-based producers. The U.S. manufacturers might also be strongly motivated to transfer their production facilities for F-11 and F-12 related equipment to overseas locations in order to reduce the costs of servicing primary growth markets in the Middle East and developing world. However, a U.S. ban on F-12 use in automotive air conditioners might conceivably have a positive trade effect if the ratio of U.S. market penetration overseas to foreign market penetration in the United States is low and the United States leads in alternate auto air conditioning designs.

Multilateral regulation or selective use restrictions (for example, focusing only on automotive air conditioning) will greatly reduce the scale of international economic impacts as alluded to above. Regulations in the form of requiring equipment component upgrading, or improved maintenance and servicing procedures, or greatly expanded retrieval and recycling of F-11 and F-12 refrigerants would be likely to have only very minor effects on the international competitive position of U.S. producers.

3.2.1.5 Urethane Foams. Rigid and flexible plastic foams are formed from gaseous bubbles and can serve a variety of purposes such as cushioning and insulation. The blowing agent (the gas used to foam the bubbles) is of particular importance, for it plays a major role in determining three properties of the foam: density, flammability, and of particular importance to rigid foams, thermal capacity. CFCs, as blowing agents have the capability of producing low-density, relatively nonflammable, and excellent insulating products. CFC blowing agents are much more expensive than other physical foaming agents such as pentane, benzene, and methylene chloride. Incentives thus exist with regard to employing less costly agents, but the industry has generally found that any cost disadvantages are more than outweighed by the superior performance qualities of CFCs.

The foamed products of potential regulatory concern include rigid polyurethane, flexible polyurethane, and, to a much lesser extent, extruded polyurethane, expanded polystyrene, and polyolefins. Eighty-five percent of rigid foam and 55 percent of flexible foam were estimated as dependent on CFC (primarily F-11) as a blowing agent. Rigid foam was estimated to have consumed about 50 percent of the total volume of CFCs used in foam production, while flexible foams consumed about 40 percent. Emissions (within ten years) from flexible foam exceeded those from rigid by a factor of 3.5.

The disparate emission characteristics of flexible (open-cell) and rigid (closed-cell) foams are explained as follows. The manufacturing of flexible foam uses CFC to blow the cells in the foam, and this CFC is essentially completely emitted during the production process. With rigid foam, a majority (80 to 85 percent of the CFC employed is retained, or "banked," within the cellular structure of the foam and serves to provide thermal insulating properties. These banked CFCs are eventually lost to the atmosphere as a function of product deterioration and method of disposal.

In 1963 U.S. consumption of CFCs as blowing agents in plastic foams was estimated as 21 million pounds, amounting to 5 percent of total CFC-production; by 1975, consumption had risen to 82 million pounds, roughly 9 percent of CFC production. Foam blowing, thus, was second only to the refrigeration/air conditioning industry in the consumption of CFCs for non-aerosol uses.

The share of total production rose to an estimated 12 percent in 1977 as a function of the decrease in CFC-propellant use and increase in foam demand. As CFC propellants for nonessential aerosol use is phased out, of course, the share of remaining CFC production consumed by foams will rise significantly.

Table 3-11 shows the respective U.S. consumption or rigid and flexible foam by major product for 1975 versus 1970. Both types of use show considerable percentage growth in that five-year period, with rigid foam use increasing somewhat more than that for flexible foam.

Very little flexible or rigid foam is exported from, or imported into, the United States because of high transportation costs associated with the low-weight and large-space nature of foam. Distances between production and consumption are generally minimized throughout the world. In 1977 the United States exported only a small fraction of 1 percent of production (mainly to Puerto Rico and scrap sales to Europe) and imported even less. One should note, however, the foams are traded in embodied form (that is, contained in such goods as automobiles, refrigerators, freezers, mattresses, boats, furniture, and so forth). An estimated 10 percent of U.S. exports probably contain some foam.

Factory sales of CFC-blown foams topped $1 billion in 1976, and industries which today use or depend on the foams could well exceed $100 million in sales.[26] A ban on CFC use as a blowing agent for flexible foams would possibly be cushioned by the availability of substitutes such as methylene chloride.

A debate is now raging within the industry with regard to the practicability of methylene chloride. The health and safety of workers exposed

Table 3-11. U.S. Consumption of Rigid and Flexible Foams by Major Product, 1970 and 1975

(millions of pounds)

Type of foam and products using foam	1970	1975	1975/1970
Rigid polyurethane foam			
Construction insulation	71.2	175.0	2.46
Appliance insulation	51.2	73.6	1.44
All other insulation	41.8	63.8	1.53
Furniture	25.3	35.2	1.39
Packaging	7.2	13.2	1.83
All other products	42.1	22.0	0.52
Total rigid polyurethane foam	238.9	382.8	1.60
Flexible foam			
Furniture	207.5	352.0	1.70
Transportation	225.0	341.0	1.52
Bedding	58.8	136.4	2.32
All other products	116.3	94.6	0.81
Total flexible foam	626.8	924.0	1.47

Sources: Bureau of Domestic Commerce, Domestic and International Business Administration, U.S. Department of Commerce, *Economic Significance of Fluorocarbons*: December 1975 (Washington, D.C., 1975) and Midwest Research Institute, *Chemical Technology and Economics in Environmental Perspectives, Task I: Technical Alternatives to Selected Chlorofluorocarbons* (Washington, D.C., Environmental Protection Agency, February, 1976).

to the chemical are a matter of concern. And since gases such as methylene chloride and pentane are flammable, their use might cause cancellation of insurance coverage in processors' plants.

Others stress the necessity of CFCs for low-density foams (seats, chairs, pillows). If it is true that there are no good substitutes, a ban on CFC use could result in the elimination of about one-third of flexible foam production.[27]

If CFC use for flexible foams were allowed to continue, but vapor recovery systems were mandated, then the price of foams would increase significantly. The recovery (about 60 percent) or recycling of CFC blowing agents as a means of emission control is believed to be technologically possible, but currently it is economically prohibitive.

Compared with flexible foams, the essentiality of CFCs as blowing agents for rigid foams can probably not be challenged.[28] Hence, a ban on CFC use would essentially terminate the rigid polyurethane foam industry. One interviewee stressed that his firm would simply get out of the business.

A CFC ban would directly conflict with the energy improvement goals under the Energy Policy and Conservation Act (PL-94-163). The Department of Energy is encouraging the use of rigid foam insulation to save energy in buildings and in manufactured products such as refrigerators. The loss of thermal insulation and weight/space savings would lead to increased energy consumption, and, undoubtedly, to increased U.S. imports of gas and oil. If CFC use for rigid foams were allowed to continue, but strict disposal regulations were mandated (for example, "national rigid foam underground disposal centers"), it is not yet clear what the economic implications might be.

Any CFC regulation-induced cutbacks in rigid and flexible foam production would feed back onto raw materials producers in terms of lower sales volumes and, most probably, higher prices in the United States. These higher prices could serve to make U.S. exports to Europe noncompetitive (assuming the European producers did not face controls) and could also induce U.S. producers to expand production in overseas markets in order to regain competitiveness.

We should also note that products using foams would rise in price either as a result of higher foam prices or loss of thermal insulation requiring greater weight, space, or fuel consumption. Higher-priced exports from the United States, such as refrigerators and automobiles, would most likely translate into a loss of foreign sales volume.

3.2.2 Foreign Trade Effects

Table 3-12 summarizes the foreign economic involvement of the five major CFC-related industries examined in this report. Estimates acquired from industry observers are employed where solid statistical data are lacking.

U.S. exports of CFCs, precursor chemicals, aerosols, and rigid/flexible forms represent only a very small percentage of domestic production. Exports of refrigeration and air conditioning equipment, on the other hand, are substantial in terms of both value ($1.2 billion in 1977) and as a percentage of domestic output (15 percent). Exports of urethane chemicals (precursors of foams) also represent an important share of domestic production (about 20 percent in 1977).

U.S. imports of CFCs, aerosols, foams, and urethane chemicals in 1977 were small, both in terms of value and as a percentage of domestic consumption. In contrast, more than $100 billion in precursor chemicals (for example, hydrofluoric acid, fluorspar, etc.) and almost $300 million of refrigerators and air conditioners were imported in 1977. Precursor chemical imports represented approximately 20 percent of U.S. domestic production.

Table 3-12 indicates that the U.S. in 1977 had net exports of CFC-related goods of a little over $1 billion. The inclusion of trade data on other CFC-related sectors, such as solvents and fluoroplastic materials, along with data on CFC-related goods (that is, flexible and rigid foams) embodied in a broad range of import and export goods, could move this estimate up or down by about 10 to 20 percent. The United States is a net importer of the precursors to CFCs and a net exporter of the precursors to urethanes. The overall favorable balance of trade is, of course, dominated by the huge export surplus in the refrigeration-air conditioning sector (about $1 billion in recent years).

Table 3-12 also characterizes the relative amount of licensing and manufacturing which U.S. companies engage in overseas, as well as the strength of competition they face from foreign-based firms in those markets. Patents, trademarks, and production technologies are licensed to overseas producers to a considerable extent in the refrigeration/air conditioning sector and to a moderate extent in aerosols (valves, cans, and

Table 3-12. Summary Profiles of Sector Foreign Involvement: 1977

Foreign involvement dimensions	CFC-related industries				
	Chlorofluro-carbons	Precursor chemicals	Aerosols	Air conditioning	Urethane foams[a]
1. Estimated value of U.S. exports (1977)	$23.5 million	$12 million	$20 million	$1.2 billion	$175 million
2. U.S. exports as a percentage of domestic production	4.1%	.5%	.5%	15%	20%
3. Estimated value of U.S. imports (1977)	$6.2 million	$112 million	$10 million	$292 million	$20 million
4. U.S. imports as a percentage of domestic production	.9%	20%	.25%	3.6%	2%
5. Estimated net balance of trade (1977)	+$17.3 million	-$100 million	+$10 million	+$1 billion	+$155 million
6. Extent of foreign licensing by U.S. companies	Low	Low	Moderate	High	Moderate
7. Extent of foreign manufacturing investment by U.S. companies	Moderate	Moderate	High	High	Moderate
8. Extent of competition from foreign firms in overseas markets	High	High	High	Low	High

Source: Author estimates, based on data cited in text, including tables 3-10 and 3-11, and information obtained from industry observers.

[a]Data are for urethane chemicals, a subcategory of which represents urethane foams, while the remainder is accounted for by precursors.

so forth) and foams. With the exception of refrigeration/air conditioning, all of the CFC-related sectors face strong and widespread competition in overseas markets, primarily from European and Japanese-based firms.

As was emphasized earlier, the effects on U.S. exports and imports of CFC use regulations depend considerably on whether the regulations are (1) unilateral or multilateral; (2) immediately or gradually implemented; (3) comprehensive or selective in F-11 and F-12 emission-reduction targets; and (4) total (ban) or partial in the amount of CFC reductions desired. The most severe regulatory scenario confronting U.S. producers would be one which is purely unilateral, rapidly imposed, applied to all CFC uses, and aimed at a 100 percent reduction in emissions via a blanket ban. Assuming regulation is required, the most lenient scenario would be one which is imposed multilaterally, imposed very gradually, targeted only at selected emission sources (such as worst offenders like mobile vehicular air conditioning), and which does not ban CFC use, but merely requires a partial reduction in emissions.

It seems useful to focus on the possible trade effects of the most severe scenario for refrigeration/air conditioning, our most important case.

The variations in CFC-use regulations in different markets (for example, U.S. versus rest of the world) would generally compound supply difficulties and multiply costs for both U.S. and foreign producers. Even if no controls were placed on exports, it is doubtful whether U.S. suppliers would remain competitive in export markets without their domestic sales base. But the extent of disruption would be highly product-specific. For some lines of air conditioning and refrigeration, the incompatibilities in CFC use standards would be of little importance. These lines could be custom-produced efficiently for different markets in different configurations (for example, some using F-11 and F-12 and others not) just as they already are custom-supplied for individual buyers. Other lines might be geared only to the U.S. norms and produced (say, using F-22) according to those standards, with little or no competitive disadvantage in overseas markets which are characterized by more lenient or nonexistent standards having minimal disruption.

For still other lines of equipment, all output could probably only be economically mass-produced in one way or the other. And if the incremental costs or the efficiency losses encountered in supplying some markets are sufficiently high, then they would probably have to be sacrificed altogether, entailing potentially serious economic dislocations. Between these extremes are product lines that could be adapted to different CFC-use standards with varying degrees of difficulty--including inadequate production runs, product redesign involving additional R&D expenditures, etc.--with disruptive effects likely but not prohibitive.

It is generally agreed that the conversion of products currently using F-11 and F-12 to alternative refrigerants would result in higher production costs and product prices, at least in the short run. Given that the U.S. demand for air conditioners and refrigerators is relatively price elastic, then both domestic producers and exporters to the U.S. market would probably experience reduced revenues. Many lines of equipment are produced with increasing returns to scale, so decline in output would tend to raise average cost. However, the overseas demand for many types of U.S. air conditioning and refrigeration equipment may be relatively inelastic. This may be especially true for OPEC nation markets such as Saudi Arabia. In many lines, U.S. producers are the only suppliers and substitutes are not readily available. The goods are highly differentiated in regard to style, quality, performance, and brand name. In some cases, the gap between the absolute level of prices for similar goods between U.S. and foreign producers is sufficiently large that competitive positions would not be dramatically affected by a U.S. price increase.

In cases where no import-competing suppliers exist in the target market, it would take years for such a supplier to gear up production and present a challenge. If a good portion of the $1.2 billion in exported air conditioning and refrigeration equipment is, in fact, not very price or substitute sensitive, then it is possible that a CFC regulation-induced price increase could increase revenues and actually improve the balance of trade in the air conditioning/refrigeration sector.

The trade consequences of severe CFC-use regulations for producers of CFCs, precursor chemicals and urethane chemicals would probably be very different in kind. These are homogeneous goods and, as noted, considerable competition exists in foreign markets. It is most unlikely that, faced with a near total ban on CFC use at home, U.S. producers would become significant suppliers of raw materials to the rest of the world. Unit production costs would soar in the United States, and there is little chance that U.S. producers could compete in overseas markets.

Finally, it seems likely that at least some, if not all, of the major CFC-producing and consuming nations will, in time, follow the U.S. lead in CFC-use regulation. As other nations "fall into line," the U.S. position in world trade should improve as a result of CFC regulation-induced innovational advantages conferred upon U.S. producers, who will have pioneered in responding to the regulations. Hence, significant export markets could emerge in CFC emission control information and hardware, reformulated products, etc.

3.2.3 Foreign Investment Effects

Severe CFC-use regulation would obviously reduce the attractiveness of the United States as a location for CFC-related production. But, contrary to the expectations of some governmental policymakers, it does not appear that such regulation would induce any significant shifts in the location of CFC-related production facilities--or even incremental productive capacity--to nations characterized by less stringent CFC regulation, for these reasons:

1. Extended time horizons and managerial perceptions associated with new CFC-related investments appear to be biased against locational shifting. Many of the executives we interviewed privately admit that they expect substantial convergence among OECD nations in regard to CFC regulation over the next five to ten years. With plant lives extending from fifteen years upward, this guarantees economic disruption in the future.

2. Many CFC-related foreign investments (in nations such as Spain, Mexico, Argentina, and Brazil) have been principally motivated by national development policies involving import substitution. Most plants have been designed with small capacities to service local markets behind high tariff walls.

3. Most executives believe that developed nations in North America and Western Europe would impose countermeasures to block the products of

any observed "locational flight" to "CFC pollution havens." Such measures might include selective import duties, border adjustments, quotas, and embargoes. Such "neo-protectionism" in the name of CFC management would confine the output of CFC-related production facilities to local markets.

4. Gains from investing in a "pollution haven" might be counter-balanced by higher costs for other inputs and degrees of risk--labor availability and quality may be poor, supporting infrastructure may be lacking, market sizes may be inadequate, transport costs may be prohibitive, political and expropriation risk may be too high, and so on.

5. The multinational enterprises in CFC-related industries behave as oligopolists, and foreign investments frequently represent a strategic move to buttress the stability of an oligopoly or to reduce the vulnerability of one firm within the oligopoly. The emphasis on strategic and oligopolistic factors strongly implies that flows of CFC-related investment may be rather insensitive to variations in CFC-use regulations in different locations.

If there is to be any shifting at all--and with lead times in new facility planning of five to eight years, shifting will not be undertaken for some time--the most likely candidates would be found in urethane chemicals and refrigeration/air conditioning. There could be considerable motivation to build large, efficient plants in, or close to, rapidly growing markets where F-11 and F-12 are expected never to be banned for foam-blowing and refrigerant uses.

At the moment, however, we do not see CFC-use regulation exerting great influence on patterns of industrial location. The emergence of regulation differentials among nations will represent a shift at the margin in the locational calculus of CFC-related firms. The significance of this shift will depend on the complete structure of costs, risks, and returns which are considered when investment decisions are made.

3.2.4 Summary

To sum up, it appears that even under the most severe CFC regulatory scenario possible, the aggregate international economic dislocations likely to be experienced by the U.S. would be relatively modest. Such adverse impacts would be miniscule compared to the purely domestic economic dislocations (for example, lost sales, plant closings, unemployment) brought on by such "worst case" regulatory action. They would be most serious in the refrigeration/air conditioning sector.

In value terms, the United States would, at the very extreme, probably experience a simultaneous annual decline of perhaps $200 to $300 million each in CFC-related imports and exports. As such, total U.S. imports and exports would each drop by about 0.001 to 0.002 percent. This is a very preliminary assessment, and further analysis is needed to properly assess the potential trade consequences of alternative regulatory scenarios. The EPA must closely consider the potential trade losses in its "essential use" determination for the air conditioning/refrigeration sector.

The competitive disadvantages implied by CFC regulation for U.S. producers are not likely to inspire a substantial migration of CFC-related industry to other nations showing less concern for CFC emissions management. However, the international impacts on specific firms within various CFC-related industries (for example, those which have specialized in international trade) could, of course, be quite substantial. What is modest in macroeconomic terms may be catastrophic for some firms and workers at the microeconomic level. Adjustment assistance may represent the best way to handle these localized adverse effects.

This analysis leads us to conclude that the real policy problem, from an international perspective, is what to do about overseas CFC emissions which lead to depletion of the ozone layer, to damage to public health and environmental quality and to possibly adverse climate changes in the U.S. The problem rests not so much with spillovers of CFC production and consumption into overseas markets attributable to unilateral restrictions by the U.S., but with existing sources of CFC emissions and expansions in CFC-related production designed to service the expected rapid growth in those overseas markets.

3.3 Policy Implications

3.3.1 The Policy Challenge

The conclusion above leads us to the basic policy questions, how can global common property and public good environmental resources, such as the ozone layer and climate, be constructively managed? Besides the CFC problem, similar challenges occur in the worldwide transport of radionucleides, the build-up of atmospheric carbon dioxide, and the dumping of toxic wastes in the oceans. What kinds of institutions and mechanisms might be effective in averting tragedies of the global commons (that is, overuse, misuse, and quality degradation of resources like the stratosphere)? In cataloging possible tools we draw on the work of a number of authors who have examined trans-frontier pollution problems.[29] We enumerate those tools in the following list, categorizing them as multilateral versus unilateral:

Multilateral

1. Principles of conduct
2. International conferences
3. International treaties and conventions
4. International agencies
5. International compensation
6. International litigation

Unilateral

1. Extraterritorial application of laws
2. Import controls
3. Export controls
4. Disclosure requirements
5. Environmental impact statements
6. Withholding of incentives

Let us conclude this chapter by briefly highlighting the nature of the tools available in these two sets.

3.3.2 Multilateral Tools

Most observers agree that the problem of global externalities in theory can be best dealt with multilaterally. This does not mean involving, at least at the start, all nations contributing to, or potentially affected by, the problem. Since a dozen countries account for 80 percent of CFC production, an agreement on limitations among this group would go a long way to solving the problem. Agreement is most likely when the participating nations share the same technical perceptions of the problem, utilize the same technologies in production, have roughly similar desires for such goods as environmental quality, and are extensively linked by a range of transnational networks. Principles which might help to foster such agreement would include Principle 21 of the Declaration of the United Nations Conference on the Human Environment, which reads as follows: "States have . . . the responsibility to ensure that activities within their jurisdiction or control do not cause damage to the environment of other states or of areas beyond the limits of national jurisdiction,"[30] and the OECD's principles of conduct in transfrontier pollution which call for "good neighborship."[31]

Consensus building might also be facilitated by meetings and conferences such as those convened by the United Nations Environment Programme (UNEP) and the EPA regarding the ozone layer.[32] In due course, a treaty or agreement among the major CFC-producing and consuming nations which prohibits or restricts CFC emissions could emerge. Legal precedents and bilateral or multilateral agreements do exist in regard to the problem area of environmental damages which have effects across national boundaries (for example, Great Lakes Water Quality Agreement, Convention on Third Party Liability in the Field of Nuclear Energy, Convention on Civil Liability for Oil Pollution Damage, and Scandinavian Convention on the Protection of the Environment). The history of diplomacy indicates, however, how difficult it is to achieve agreement and effective multinational programs. Others believe that the need for a common jurisdiction over the oceans, stratosphere, and other global resources can be met effectively only by transnational structures--agencies or groups of agencies that, although created by international agreement, operate thereafter with national consensus.[33] Ralph C. d'Arge has suggested that "an international regulatory body for the stratosphere is urgently needed."[34] But precedents for effective management of the stratosphere by an international agency are not encouraging.

Given the pessimism noted above and the fact that the United States appears to be more risk adverse in CFC emissions management than other nations, it may find it useful or necessary to "bribe" or compensate "holdout" nations in order to achieve CFC emission reductions overseas. The victim-must-pay principle naturally strikes most of us as unjust, but unilateral CFC regulation by the United States may change the structure of incentives for other nations. Hence, side-payments in exchange for emissions reductions may leave both the United States and the other nations better off. One other option of a very different kind would be that of private recourse against environmentally harmful CFC-related activities abroad.[35] The way to a private law verdict in a case of transfrontier pollution, however, is not an easy one.[36] The transaction costs are extremely high--the well-known Trail Smelter Case, involving transnational pollution between the United States and Canada, stretched over twenty years. (The case established the principle that "...no State has the right to use or permit the use of its territory in such a manner as to cause [significant] injury by fumes in or to the territory of another or the properties of persons therein...".[37]) d'Arge has concluded that "international court settlements of transnational externalities are not likely to yield satisfactory results" given almost insurmountable problems related to measurement of damages and assessment of damage payments.[38]

3.3.3 Unilateral Tools

If the United States feels strongly that procrastination cannot be permitted in ozone layer (and climate) protection, but finds multilateral efforts to be unsuccessful or bogged down in interminable delay, it may turn to certain unilateral measures aimed at reducing CFC emissions overseas. Such tools are likely to be highly controversial and unpopular, both at home and abroad. Some may even boomerang, straining relations with other nations and inducing retaliatory actions against the United States.

The justification for unilateral measures which attempt to influence CFC-emissions behavior overseas stems from the transfrontier nature of the CFC problem--emissions in foreign nations can damage public health and the environment in the United States. As William J. Baumol has emphasized: "international externalities weaken...longstanding arguments for freedom of trade."[39] Departure from free trade in the name of CFC emissions management, however, may be importantly constrained by certain obligations of the United States under the terms of the General Agreement on Tariffs and Trade (GATT). Additional justification is provided in the landmark 1945 pronouncement of Judge Learned Hand in the Alcoa antitrust case: "It is settled law that any state may impose obligations, even upon persons not within its allegiance, for conduct outside its borders that has consequences within its borders which the state reprehends...."[40] This doctrine, the so-called "effects" doctrine of jurisdiction, has consistently since been applied by the United States in the field of antitrust, and in other areas where it is relevant. It is now enshrined in the semiauthoritative restatement of the Foreign Relations Law of the United States. The "effects" doctrine has not yet been employed by the U.S. government to extend its environmental policy jurisdiction into the territory of foreign states. A case could probably be made, however, that CFC emissions induced by the activities of U.S. companies overseas might have a "substantial and material" effect on U.S. environmental quality and public health. As such, the EPA might wish to seek statutory authority to apply U.S. CFC-use regulations to the foreign operations of U.S. multinationals, a move which would surely be resented both by the U.S. firms and the governments of the nations in which their CFC-related production facilities are located.

Import and export controls may represent more feasible tools. In terms of product safety in general, consumer products intended for import into the U.S. must meet the same standards and related requirements as domestically produced goods. Under the Toxic Substances Control Act (TSCA), entry into the U.S. customs territory may be denied to any chemical substance, mixture, or article containing a chemical substance or mixture that does not comply with U.S. rules or that is being offered for entry in violation of the provisions of manufacturing and processing notices or the regulation of hazardous chemical substances and mixtures, or of an order issued in a civil action brought on the basis of the act.[41] On the other side, the transfrontier dimensions of the ozone depletion problem raise serious questions about the desirability of continued exportion of CFCs or CFC-using products for use abroad that become impermissible at home. The general rule applicable to exports subject to federal product safety legislation is that they need not conform to the safety requirements of the statutes involved but must be labeled to show that they are intended solely for export. In most cases, there has been a clear expression of the view that sovereign foreign states are responsible for setting their own standards for health and safety properties of internationally traded products, and that the imposition of U.S. regulations on other countries cannot be supported. But with CFCs the potential hazards for the U.S. arising from foreign use change the entire picture. Just as the Trading-with-the-Enemy Act serves to prevent the flow of strategic goods, patents and know-how into the hands of potential military adversaries, so too may there be grounds for restricting the flow of CFC-related techniques, products, and plants into the hands of nations which might threaten the environmental "security" of the United States. At the very least, disclosure requirements associated with CFC-related exports could potentially provide some useful "prophylactic effects".[42] Section 12(b) of TSCA requires exporters to notify the EPA Administrator of any export of a toxic chemical. The EPA, in turn, must notify the importing country that data on the chemical is available, or that regulatory action has been taken.

Along with disclosure requirements, some benefits could also flow from applying the National Environmental Policy Act (NEPA) to certain CFC activities overseas. Executive Order 12114, signed January 4, 1979 by President Carter, requires that federal agencies prepare environmental impact statements for major federal actions that have significant environmental effects on the "global commons".[43] Any proposed CFC-related activity abroad which involved prior approval or facilitation by the U.S. government could thus be scrutinized, under the Executive Order, for its ozone-depletion and climate impacts before plans were allowed to proceed. An impact statement could be a prerequisite for obtaining an export license if an export control system was instituted (export licenses and permits, however, were exempted in the 1979 order). Finally, we need to note that a variety of U.S. government programs may currently promote foreign trade, investment, and consumption of CFCs. Examples include (1) special tax benefits on CFC-related export sales, (2) reduced taxes for CFC-related investments under Less-Developed-Country tax provisions, (3) deferal of U.S. taxes on U.S. companies' unrepatriated foreign income from CFC-related sales overseas, (4) direct loans, guaranteed loans and export credit insurance from the Export Import Bank, (5) preinvestment assistance, political risk insurance, and loans from the Overseas Private Investment Corporation, and (6) development assistance, directly or indirectly encouraging CFC consumption, provided by international development financing agencies such as the World Bank, Inter-American Development Bank, and Agency for International Development.

In the evolution of U.S. policy on CFC-induced environmental deterioration, the U.S. may decide that the encouragement of CFC-related activity overseas is inappropriate. Firms and foreign countries have no inherent or vested right to received American agency assistance, and thus the agency could predicate the granting of assistance on the aid being consistent with statutory U.S. policy.[44] Such screening on the part of U.S. agencies would be viewed by some as an unwarranted intrusion into the realm of hard foreign diplomatic and economic realities. But on the issue of managing CFC emissions, however, perhaps the time has come for a truly "environmentally oriented" U.S. foreign policy.

References

1. Environment Committee, Chemicals Group, Organization for Economic Co-operation and Development, The Economic Impact of Restrictions on the Use of Fluorocarbons: Final Report (Paris, OECD, January 24, 1978) and R.C. d'Arge, L. Eubanks and J. Barrington, Benefit-Cost Analyses for Regulating Emissions of Fluorocarbons 11 and 12: Final Report (Washington, D.C., Policy Planning Division of the U.S. Environmental Protection Agency, December 1976).

2. Arthur D. Little, Inc., Preliminary Impact Assessment of Possible Regulatory Actions to Control Atmospheric Emissions of Selected Halocarbons (PB-247115, EPA-450/3-75-073) (Springfield, Va., National Technical Information Service, September 1975); Bureau of Domestic Commerce, Domestic and International Business Administration, U.S. Department of Commerce, Economic Significance of Fluorocarbons: December 1975 (Washington, D.C., 1975); Environmental Resources Limited, Protection of the Ozone Layer: Some Economic on Social Implications of a Possible Ban on the Use of Fluorocarbons (London, December 1976); International Research and Technology Corporation (IRT), Short-Range Marginal Costs for Production of Fluorocarbons 11 and 12 (McLean, Va., July 1976); IRT, The Economic Impact of Potential Regulation of Chlorofluorocarbon-Propelled Aerosols: Final Report (EPA Contract No. 68-01-1918) (McLean, Va., April 1977); Midwest Research Institute, Chemical Technology and Economics in Environmental Perspectives, Task I: Technical Alternatives to Selected Chlorofluorocarbons (EPA-560/1-76-009) (Washington, D.C., February 1976); U.S. Environmental Protection Agency, Chlorofluorocarbon Problem Assessment: Support Document (Toxic Substances Control--Fully Halogenated Chlorofluoroalkanes), 40 CFR Part 762, OTS-060002, May 10, 1977; U.S. Environmental Protection Agency, Office of Program Evaluation, The Economic Impact of Potential Regulation of Chlorofluorocarbon Propelled Aerosols (Washington, D.C., April 1977); and U.S. Food and Drug Administration, Economic Impact Statement of Proposed Rule Making: Certain Fluorocarbons (Chlorofluorocarbons) in Food, Food Additive, Drug, Animal Food, Animal Drug, Cosmetic and Medical Device Products as Prohibition on Use, 1977 (Washington, D.C., 1977).

3. Manufacturing Chemists Association, Environmental Analysis of Fluorocarbons FC-11, FC-12, and FC-22 (Washington, D.C., Alexander Grant and Company, July 8, 1977).

4. U.S. Environmental Protection Agency, Chlorofluorocarbon Problem Assessment: Support Document.

5. Ibid., p. 15.

6. Manufacturing Chemists Association, Environmental Analysis of Fluorocarbons.

7. "Certain Fluorocarbons (Chlorofluorocarbons) in Food, Food Additive, Drug, Survival Food, Animal Drug, Cosmetic, and Medical Device Products as Propellants in Self-Pressurized Containers--Prohibition on Use," Federal Register, vol. 43, no. 43 (Friday, March 17, 1978) II, p. 11319.

8. T.N. Gladwin, Environment Planning and the Multinational Corporation (Greenwich, Conn., JAI Press, 1977; United Nations Conference on Trade and Development, "Implications of Environmental Policies for the Trade Prospects of Developing Countries: Analysis Based on an UNCTAD Questionnaire" (Geneva, 1976) mimeo; I. Walter, _International Economics of Pollution_ (London, Macmillan, 1975).

9. T.N. Gladwin, "Environmental Policy Trends Facing Multinationals," _California Management Review_, vol. 20 (Winter 1977), p. 81-93.

10. For surveys and commentary regarding CFC regulation overseas, see "Nations Fail Global Move on CFC Problem," _Chemical and Engineering News_ (Jan. 25, 1982), pp. 44-46; "CFCs and Ozone: Deadlocked," _The Economist_ (Nov. 28, 1981), p. 92; "Aerosol Makers Brace for Ozone Ordeal," _Chemical Week_ (June 11, 1975), pp. 14-15; "Aerosols: Sky Won't be the Limit," _Chemical Week_ (May 12, 1976), pp. 46-48; "Aerosols: U.S. To Ban Use of Fluorocarbon Propellants?", _Europe Environment_ (June 1975), p. 4; "Britain Sees No Need to Rush in with Fluorocarbon Restrictions," _Chemical Week_ (February 11, 1976), p. 43; "British Aerosol Manufacturers Association, "Fluorocarbons and Ozone--A Review of the Regulatory Position," _Aerosol Communique_ (March 1978), pp. 1-7; "Canada Studies Aerosols," _Chemical Week_ (Nov. 5, 1975), pp. 46-48; "Commentary," _Aerosol Age_ (Jan. 1978), p. 8; "EC Acts on Fluorocarbons," _European Community_ (November-December 1977), p. 46; Environment Committee, Chemicals Group, OECD, _The Economic Impact of Restrictions on the Use of Fluorocarbons_; "Europe Not Convinced by EPA Fluorocarbon Stance," _European Chemical News_ (May 6, 1977), p. 80; "Fluorocarbons as Blowing Agents: The Future Still is in Doubt," _Modern Plastics_ (January 1978), pp. 7-18; "Fluorocarbons Fade in Europe, U.S. Being More Severe," _European Chemical News_ (March 3, 1978), p. 20; R. Gour-Tanguay, "Protection of the Ozone Layers," _Environmental Policy and Law_, vol. 3 (July 1977), pp. 61-62; "Holland Plans Compulsory Fluorocarbon Labelling," _European Chemical News_ (June 10, 1977), p. 6; A.J. Large, "The Spread of International Controls," _The Wall Street Journal_ (Nov. 22, 1976); "Most Aerosols Face a Swedish Ban," _The New York Times_ (Jan. 30, 1978), p. 14; "No Rush Seen To Follow U.S. Aerosol Ban," _Chemical Week_ (June 1, 1977), p. 25; "Ozone Controversy Casts a Global Shadow," _Chemical Week_ (Sept. 15, 1975), pp. 47-48; H.M. Schemeck, "Ban Proposed by '79 in Spray-Can Cases," _The New York Times_ (May 12, 1977), pp. 1, 17; T.B. Stoel, R.I. Compton, and S.M. Gibbons, "International Regulation of Chlorofluoromethanes," _Environmental Policy and Law_, vol 3 (December 1977), pp. 127-132; W. Sullivan, "Little Support Given to U.S. on Action to Protect Ozone," _The New York Times_ (March 15, 1977), p. 16; J.A. Tannenbaum, "The Ozone Issue: Fluorocarbon Battle Expected to Heat Up as the Regulations Move Beyond Aerosols," _The Wall Street Journal_ (Jan. 19, 1978), pp. 16 and 38; and "U.K. Rejects Curbs on Fluorocarbons," _European Chemical News_ (April 23, 1976), p. 10.

11. Large, _The Wall Street Journal_.

12. Bureau of Domestic Commerce, _Economic Significance of Fluorocarbons_.

13. Arthur D. Little, Inc., _Preliminary Impact Assessment of Possible Regulatory Actions_.

14. Ibid., and Manufacturing Chemists Association, Environmental Analysis of Fluorocarbons.

15. Arthur D. Little, Inc., Preliminary Impact Assessment of Possible Regulatory Actions; and Bureau of Domestic Commerce, Economic Significance of Fluorocarbons, and industry contacts.

16. Sullivan, The New York Times (March 15, 1977), p. 13.

17. Federal Register, vol. 43, no. 43, p. 11319.

18. International Research and Technology Corporation, The Economic Impact of Potential Regulation of Chlorofluorocarbon-Propelled Aerosols.

19. "The Aerosol Ban Has Lost Its Sting," Business Week (May 30, 1977).

20. E. Pell, "The Bomb in Your Bathroom," New York Times (June 25, 1978).

21. "Abplanalp Displays New Valve in Europe," Yonkers Herald-Statesman (Sept. 27, 1977).

22. Arthur D. Little, Inc., Preliminary Impact Assessment; and Committee on Impacts of Stratospheric Change, Assembly of Mathematical and Physical Sciences, National Research Council, Halocarbons: Environmental Effect of Chlorofluoromethane Release (Washington, D.C., National Academy of Sciences, 1976).

23. U.S. Bureau of the Census, Trade Statistics: United States of America (1976).

24. "Whirlpool Says Sales Rose 17 percent in Five Months, See Slower Rate for 1978," The Wall Street Journal (June 12, 1978).

25. U.S. Bureau of the Census, Trade Statistics: United States of America (1977) and U.S. Department of Commerce, Exports: Commodity by Country, Report FT-410, Schedule B (December 1976).

26. U.S. Environmental Protection Agency, Transcripts of Proceedings of the Public Meetings in Washington, D.C., on Chlorofluorocarbons sponsored by the EPA, FDA, and CPSC, as recorded by the Acme Reporting Company of Washington, D.C., 1978, p. 312.

27. Modern Plastics (January 1977).

28. Ibid.

29. S.C. McCaffrey, Private Remedies for Transfrontier Environmental Disturbances (Morges, Switzerland); International Union for the Conservation of Nature, 1975; Organization for Economic Cooperation and Development, Problems in Transfrontier Pollution (Paris, OECD, 1974); "OECD Transfrontier Pollution," Environmental Policy and Law, vol. 3 (December 1977); A.D. Scott, "Transfrontier Pollution: Are New Institutions Necessary," Document AEU/ENV/73.10 (Paris, OECD, August 1973); and I. Seidl-Hohenveldern, "Alternative Approaches to Transfrontier Environmental Injuries," Environmental Policy and Law, vol. 2 (April 1976).

30. United Nations Conference on the Human Environment, Declaration and Action Plan (Stockholm, 1972).

31. Environmental Policy and Law, vol. 3 (December 1977); and "Ten Recommendations of OECD's Environment Ministers," *The OECD Observer* (October-November 1974).

32. United Nations Environment Programme, *UNEP Ozone Layer Bulletin* No. 1 (January 1978).

33. L.K. Caldwell, *In Defense of Earth: International Protection of the Biosphere* (Bloomington, Indiana University Press, 1972).

34. R.C. d'Arge, "Transfrontier Pollution--Some Issues on Regulation," in Ingo Walter, ed., *Studies in International Environmental Economics* (New York, John Wiley, 1976), p. 277.

35. *Environmental Policy and Law*, vol. 3 (December 1977).

36. McCaffrey, *Private Remedies*; and A. Rest, "Transfrontier Environmental Damages: Judicial Competence and the Forum Delicti Commissi," *Environmental Policy and Law* vol. 1 (December 1975), pp. 127-131.

37. Trail Smelter Case (United States vs. Canada) 3 UNRIAA 1905 (1935 [Special Agreement], 1938 [Preliminary Decision], and 1941 [Final Decision]).

38. d'Arge, "Transfrontier Pollution," p. 261.

39. W.J. Baumol, "Environmental Protection, International Spillovers, and Trade," in *Wicksell Lectures, 1971* (Uppsala, Sweden, Almquist and Wicksells, 1971).

40. United States vs. Aluminum Company of America, 148 F. 2d 416, 443 (2d Cir. 1945).

41. "U.S. and EEC to Negotiate on Toxic Substances," *Pollution Control Guide* (April 17, 1978).

42. L.D. Brandeis, *Other People's Money* (New York, Frederick A. Stokes, 1914).

43. U.S. Council on Environmental Quality, *Environmental Quality-1979* (Washington, D.C.: U.S.G.P.O., Dec. 1979), p. 582.

44. For a related discussion, see "A Growing Worry: The Consequences of Development," *Conservation Foundation Letter* (January 1978); E.V. Coan, J.N. Hills, and M. McCloskey, "Strategies for International Environmental Action: The Case for an Environmentally Oriented Foreign Policy," *Natural Resources Journal*, vol. 14 (January 1974); and E. Sullivan, "When Environment and Diplomacy Clash," *The Interdependent*, vol. 5 (March 1978).

Chapter 4

THE METHODOLOGY OF BENEFIT-COST ANALYSIS WITH PARTICULAR REFERENCE TO THE CFC PROGRAM

Ezra J. Mishan and Talbot Page

4.1 Introduction

This chapter is a preliminary investigation of some critical concepts of allocative economics to determine the extent to which a benefit-cost analysis of the CFC problem can make an economic contribution to the decision-making process of society. If our conclusions are, in the main, correct, they will act inevitably to weaken the faith that can be reposed in a benefit-cost analysis or, at least, to restrict its range of application.

The facets of a benefit-cost analysis to which we here address ourselves are these:

1. The implications of the concept of economic efficiency
2. Conceptual problems of valuation
3. The legitimacy of using the discounted present value (DPV) method in ranking public projects
4. The question of intergenerational equity
5. The treatment of risk and uncertainty.

These five headings are not arranged in order of importance but follow in logical sequence, as we develop our argument.

4.2 Implications of the Concept of Economic Efficiency

Since benefit-cost analysis is to be regarded as no more than an extension of the conventional allocative analysis to a proposed economic change, often in the form of a proposed project, it is as well to address ourselves at the start directly to allocative economics--to the criterion by which the economist compares one economic arrangement with alternatives, either in the small or the large.

The singularity of the economic method consists of the adoption of the basic economic maxim that the objective data for the economist are the orderings, or the subjective valuations, of the individual members of society, and nothing more.

These valuations are accepted by the economist as relevant data, irrespective of the tastes of the individual or the current state of his information. There is no abstraction such as "the general good" and no entity such as "the state" to be considered by the economist in addition to the welfare of the individuals who comprise society--a view that accords with the philosophic position sometimes referred to as methodological individualism.

Since, however, a large number of individuals are generally affected by any economic change, a further criterion is necessary for ranking the alternative economic situations. A criterion with well-nigh universal support is an actual Pareto improvement--one for which each member of society is made no worse off by the change and one or more members are actually made better off than before. But changes that meet such a criterion are quite uncommon. Hence, the criterion chosen by economists is a _potential_ Pareto improvement: a situation II is ranked above a situation I if, in a costless movement from I to II, the aggregate value of individual gains exceeds the aggregate value of individual losses. Indeed, the excess value of aggregate gains over aggregate losses arising from the specific change is commonly referred to as the social net benefit of that change, and is treated as the economic measure of the resulting change in society's welfare. Costless redistributions of the net gains can be envisaged which, were they implemented, could make each individual better off than he was in situation I.

Clearly, a potential Pareto improvement--which we shall also refer to as the _Pareto criterion_--is consistent with a change that can make the rich richer and the poor poorer yet. For this reason, among others, there have been objections to its adoption, and proposals have been made to elaborate the criterion so as to guard against this contingency. Nonetheless, the standard allocative criterion employed by economists today is no more than this Pareto criterion. A change that is said to increase "economic efficiency" is nothing more than a change which meets the criterion.

The term economic efficiency--when it is not being used by economists simply as a shorthand for $\Sigma \Delta v > 0$ (where the v's are the individual valuations)--entails a norm by which alternative situations may be ranked. The outcome of a political decision about a set of economic alternatives is not necessarily regarded by the economist as efficient. It follows that the norm of economic efficiency is distinct from, and independent of, an expression of the political will: indeed, that political decisions may properly be criticized by reference to the norm of economic efficiency. Rather, the norm of economic efficiency has to rest on an ethical consensus--or what one of the authors has called elsewhere, "a virtual constitution" that is deemed to be impervious to political fashions and the vicissitudes of political office.[1]

The belief that society as a whole would abide by the Pareto criterion can draw upon a number of arguments about the actual operation of the economy: (1) changes which are, in fact, potential Pareto improvements do not generally have regressive distributional effects; (2) a progressive tax system, in any case, provides a safeguard against pronounced distributional consequences resulting from any economic change; (3) over time, a succession of economic changes countenanced by this Pareto criterion will not have markedly regressive distributional effects and will, therefore, tend to bring about an actual Pareto improvement; and (4) a succession of economic changes that meets the Pareto criterion has a better chance of raising the general level of welfare than a succession of changes that meets any other criterion.

An acceptance of the distinction between the political and the economic has the merit of assigning a role to the economist that is independent of the political process. Yet, if the economist gives primacy to "economic efficiency" over considerations of distribution, it is because his craft, when guided by the Pareto criterion, enables him, from time to time, to come up with unambiguous results or with specific numbers. Concern with distributional changes, on the other hand, enables him to come up only with general statements and abstract theorems.

4.3 Conceptual Problems of Valuation

4.3.1 The Use of Compensating and Equivalent Variation

Benefit-cost studies may base their valuations on the compensating variation (CV) concept (that is, with respect to specific changes, the sums which individuals need to pay or to receive in order to restore their welfare to their original levels) or on the equivalent variation (EV) concept (that is, the sums which the individuals have to pay or receive if, spared the specific economic changes, their welfares have to assume the level they would reach if they were actually exposed to those changes). It is now accepted that, in a general analysis in which all prices change, apparently contradictory results can arise according to whether CV or EV is used. It is possible for situation II, represented by a collection of goods, to yield a potential Pareto improvement compared with situation I when based on the set of prices determined by the actual distribution of situation I. At the same time, it is also possible for the original situation I to yield a potential Pareto improvement when compared with situation II, when the comparison is based on the set of prices emerging from the actual distribution of the situation II collection of goods. (For further discussion, see Mishan and also Meade.[2]) In such a situation, the economist has either to decide in favor of CV or EV, or else reject any public project which does not meet the Pareto criterion when measured in terms both of CV and of EV. The latter policy appears the more prudent course, although it clearly favors the status quo inasmuch as the existing situation I is the one effectively adopted in ambiguous cases.

Generally, exercises in benefit-cost analysis are conducted within a partial equilibrium framework. Thus, the public project being contemplated is assumed to require so small a proportion of the economy's total resources that the prices only of the goods immediately under scrutiny are perceptibly affected by the alternative projects. But even within a partial equilibrium framework, contradictory outcomes can still arise when the calculations are done in terms of both CVs and EVs, depending on the magnitude of the individual's response to the welfare effect of the change in question. Thus, wherever given changes in nonmarket goods or bads are valued differently by the individual according to whether the amount he is willing to pay for a good (or to pay for avoiding a bad) differs significantly from the amount he will be willing to accept to forego it, the EV calculation can differ from the CV calculation.

The larger the environmental effects of a project, and the more substantial are the welfare effects on the people involved, the greater the likelihood that a project accepted on an EV test will be rejected on a CV test. Again, the prudent course to adopt may be one of requiring that both tests be made, so effectively favoring the status quo. On the other hand, if there appears to be a consensus bearing on other factors, such as equity or conservation, economists may be able to justify their adoption of the one measure or the other according to the project in question.

4.3.2 Distributional and Other Weights in a Benefit-Cost Analysis

There have been a number of proposals to incorporate distributional weights or merit weights in a benefit-cost analysis. Proposed distributional weights typically vary inversely with the income levels of the various groups affected by the introduction of the public projects under the assumption of diminishing marginal utility of income. Such a procedure effectively transforms money estimates of compensating variations into *utils*. Thus a benefit-cost criterion that is not met in money terms might well be met when the calculation is translated into utility terms, and vice versa. Proposed merit weights typically reflect political decisions. Weisbrod proposes deriving such weights from past political decisions,[3] assuming that all public projects which were adopted, despite their failure to meet benefit-cost criteria over a period, were adopted because of an implicit set of utility weights attached by the political process to the earnings of different income or regional groups. Yet another method of deriving these political weights is by a more direct approach to policymakers.

Whether bureaucratically or democratically chosen, such parameters, purporting to represent "ultimate national objectives," will vary from year to year according to the particular regime in power, or according to the composition of the legislature, or, again, according to political fashions and the exigencies of state. Moreover, since it will soon become recognized, in any representative democracy, that some projects which would be accepted on one set of weights, or national parameters, would be rejected on another set, one may anticipate continued lobbying and political infighting, both by regional and other group interests, over the weights to be adopted. The resulting vicissitudes and conflict would go far to discredit benefit-cost techniques and, possibly, economists also. Choice of the "appropriate" discount rate has been subject to political pressures in the evaluation of public works projects, for example.

The proposal to employ politically determined parameters in project evaluation appears, on the surface, to be one arising from the modesty of the economist who recognizes his inability to place a socially acceptable valuation on a variety of social phenomena that are influenced by an investment project and that alter people's welfare. But it is a modest proposal which issues in more ambitious claims for the _resulting technique_, which is then held to "integrate project planning and national policy" (see Dasgupta),[4] reducing to a single critical magnitude a variety of considerations.

In contrast, in the conventional (unweighted) valuations used in allocative techniques, the calculation of gains and losses is made on a purely economic principle; that is, by placing a value on them by reference only to the subjective valuation of the persons affected by the project. Thus, if the government calls upon the economist to undertake a benefit-cost study, it presumably expects him to employ economic principles and only economic principles. If the economist encounters difficulty in evaluating a particular item, he can leave its calculation out of the analysis and make it clear that he has done so. If, instead, he attempts to derive a value for this social benefit or social loss by reference to values that are implicit in recent political decisions (assuming they are consistent) he is, in effect, presenting the government with a result that depends, _inter alia_, on the government's own preferences or valuations and not on those of the individual citizens whose welfare will be affected by the project. The government having referred the problem to the economist for a solution, the economist, by these means, surreptitiously hands it back to the government.

The government, if democratically elected, may, of course, claim to represent the nation. But it is hardly necessary to remind the reader that the ballot box can produce results very different from those of the market or those reached by an application of the Pareto principle. A majority may well vote in favor of the use of weights or parameters that would justify the introduction of uneconomical projects to be financed by the minority.

By "doctoring" the method of evaluation so as to accommodate current political predilections, economic facts can be concealed from the public, which is then misled into the belief that the proposal has the sanction of pure economic calculation, a belief that is likely to influence the course and outcome of any debate on the subject.

We should add that while arguing for the exclusion of politically determined prices or parameters in project evaluation, no inconsistency is committed in simultaneously acknowledging the existence of political constraints. These do not offer to the economist arbitrary or noneconomic valuations of goods or bads. They act only to circumscribe the range of choices open to the economist. They can best be regarded as information on how the government is expected to act or react to a change in relevant economic circumstances. In accepting these constraints, the economist does not have to endorse the government's policy. Indeed, he may go on record as opposing it. In taking into account the expected actions and reactions of the government, the economist is seeking only to discover whether, in these circumstances, the introduction of the mooted project will yet realize a potential Pareto improvement.

If politically determined prices are admitted, there is no obvious case for limiting the extent of political intervention. If decision makers can attach weights to merit or demerit goods, why not also to the more ordinary goods on the argument that some will have smaller social merit than others? If political decision makers may attach a valuation to accidents or loss of life, why may they not also attach their own valuations to a wide range of other spillover effects? And, if so much can be justified, there seems to be no logical reasons against having political decisions override all market prices and individual valuations. There would then seem to be no reason why each and every investment project should not be approved or rejected directly by the political process, democratic or otherwise.

4.3.3 Intangible Externalities and Merit and Demerit Goods

Assuming the economist intends the term economic efficiency to have reference to an economic criterion that is independent of the political expression of society, and one therefore that can be sanctioned only by an ethical consensus, a question of consistency arises. Although society, in its ethical capacity, accepts the Pareto criterion in ordinary circumstances, there can be circumstances in which society rejects it on ethical grounds.

The economist who ignores all exceptional circumstances of this sort, and continues to base his allocative recommendations entirely on the criterion $\Sigma \Delta v > 0$, is building his allocative propositions upon a "utilitarian base": he restricts himself to the utilities, or welfares, of the individuals affected as expressed in their own valuations, without exception. If he does so, however, his recommendations may no longer claim to be grounded in the ethics of society and, therefore, they may no longer be applicable or relevant to that society. For example, a person B may be willing to sell himself into servitude for the rest of his life to person A for a sum that is smaller than the most person A is willing to pay him. The bargain that could be struck would meet the Pareto criterion, yet the economist who would recommend that the transaction take place would be prescribing a course of action that runs counter to the prevailing ethical consensus in the West.

Clearly, a normative allocation economics is a valid instrument only if it accords with the prevailing ethics of society, and cannot be raised in all circumstances on a utilitarian base. Thus, in addition to the difficult problems of measurement, the economist also has to view his results in the light of his understanding of society's ethics. Adherence at all times to the Pareto criterion is therefore not enough.

But this is not all. This criterion subsumes the ethical validity also of the basic maxim. Yet there can also be occasions on which society would not regard adherence to the basic maxim as ethical either; it would refuse, that is, to be bound by the individual valuations that comprise the data of $\Sigma \Delta v$. The calculation of externalities in a benefit-cost analysis provides a useful example. Thus, a distinction can be made between "tangible" external diseconomics, on the one hand, which cover the range of familiar pollutants that are commonly quoted for illustrative purposes in the economic literature, and on the other hand, "intangible" external diseconomics which comprehend people's responses to a change where no physical discomforts are anticipated therefrom. A well-known example of the latter is that of the "interdependent utilities" hypothesis, in which each person's welfare is also a function, positive or negative, of the level of welfare or, by extension, of the income or possessions of others.

If a person B is expected to suffer as a direct result of the noise or fumes emitted by the automobiles of group A, the cost of the damage he sustains should indeed be entered into the $\Sigma \Delta v$ calculation of net social benefit. In contrast, if the automobiles of the A group have no effect whatever on person B's health, and cause him no inconvenience, his welfare may yet decline in consequence only of his envy of the A group. If this be the case, it is reasonable to suppose that the considered opinion of society is wholly unsympathetic to his claim for compensation; we can expect envy claims to be repudiated. Thus, if our allocation economics is erected upon an ethical base, as it should be, this distinction between "tangible" and "intangible" externalities--ignored in an allocation economics erected upon a utilitarian base--can be crucial in the economic calculation of net social benefit.

Most generally, there may be strong ethical reservation about the motives which impel people to buy certain goods--motives such as resentment, spite, hatred, and exhibitionism. Other motives may seem too petty or trivial to imply any reallocation of resources.

4.3.4 Doubts About the Existence of the Required Ethical Consensus

Allowing that a normative allocation economics is faced with the problem not simply of calculating the money equivalence of the effects on the welfare of individuals arising from different economic changes but also with the problem of prescribing economic changes for a particular society, we reach the following conclusion. The economist, having to base his normative allocative propositions on an ethical consensus is also saddled with the task of determining the ethical judgments of society with respect to a wide range of possible transactions. Unless he is successful in his endeavors, society will (or ought to) ignore his economic recommendations or calculations. As a corollary, then, the economist will have no criterion of economic efficiency to juxtapose against a politically determined allocation.

But there seems considerable evidence that the ethical consensus is dissolving, implying that the task of the welfare economist becomes more difficult, and perhaps, impossible. If the permissive society is becoming "pluralistic" in the sense that a traditional or dominant set of beliefs no longer exists; if there is a trend toward each person judging his own activities and those of others in the light of his own privately constituted conscience, then ultimately the economist will no longer be able to vindicate his prescriptive statements.

Further, inasmuch as the untoward consequences of consumer innovations--including food additives, chemical drugs and pesticides, synthetic materials and a variety of new gadgets--tend to unfold slowly over time, their valuations by market prices may bear no relation whatever to the net utilities conferred over time. Indeed, the very pace of change today with respect to new models and new goods is such that it is no longer possible for the buying public to learn from its own experience to assess the relative merits of a large proportion of the goods coming onto the market. Hence, for many of the public projects in an affluent society the conscientious, normative economist can no longer speak with authority. For he is amply aware that the values to be placed on such basic inputs are part of a highly controversial topic and, moreover, that such inputs are used in a wide range of goods about whose social justification the community may be deeply divided. (For further discussion, see Georgescu-Roegen, Daly, Solow, Page, Sen, Price, Neher, and Burness and Lewis.[5])

It follows that if the circumstances described above prevail, the more restricted conception of the role of the economist, as one whose task it is simply to describe the economic consequences expected to follow from the introduction of alternative projects of policies, may become the dominant one. And the calculations of Σv or of $\Sigma\, \Delta v$, currently used in allocation and benefit-cost analysis, then become no more than a convenient and popular method of presenting the economic effects expected from a proposed policy or project. Such net benefit aggregates, of course, no longer carry independent economic recommendation. They are of value only insofar as they are made use of by the political authority itself as an input into the decision-making process, an input to which any weight (including a zero weight) can be attached.

If the economist wishes, at the end of his analysis, to be able to conclude that one project is better than another, he needs to seek out a broader role by explicitly considering the normative base which ultimately leads to normative policy prescription; moreover, he must be explicit as well about his assumptions concerning the underlying ethical consensus, or lack of it. These observations apply particularly to the discussion and practice of discounting, to which we now turn.

4.4 The Legitimacy of Using the Discounted Present Value Method (DPV) in Ranking Intragenerational Projects

We now consider the case for discounting to the present of the stream of net benefits (positive or negative) of a public project, focusing on intragenerational issues in this section, and on intergenerational issues in the following section.

For methods of project evaluation that rest ultimately on a Pareto criterion, an unresolved difficulty arises if the lifetimes of the people in the community do not overlap at some point of time common to all of them during the period of the net benefit stream in question. That consideration of itself is warrant enough for the introduction of a finite time horizon, extending from $t = 0$ to $t = T$ in a calculation designed to rank alternative public projects. (In the latter part of this chapter, we will consider some of the fundamental aspects of intertemporal equations with the perspective of an unlimited number of generations.)

In order to avoid nonessential elaboration, the practice common in the literature, of first setting aside the problem of uncertainty so as to focus on a critical part of the logic of investment criteria, is followed here, as is also the fiction that market values are equal to social values--in particular that the value of an outlay K on the public project is equal, not to the nominal sum transferred for the purpose, but equal to its opportunity cost.

Although the assumption of "full employment" is popular in the literature of investment criteria, it is of no great consequence. "Unemployment" can be dealt with by attributing lower opportunity costs in any project for which a proportion of labor comes from the existing pools of unemployment, while any employment multiplier effects are conceived to generate benefits. Nonetheless, it will be convenient to make the usual assumption--inapplicable to public projects designed to reduce the level of existing unemployment--that voluntary changes in current savings entail equal changes in private investment.

Let r be the rate of time preference common to all the individuals who are affected by the public project and remain alive over the period in question, and let ρ be the yield on private investment.

Although it is not strictly necessary that the rate of time preference be common to all individuals in any evaluation of the benefit stream, it should be evident that if, say, all of the gainers from the project have a higher (weighted) rate of time preference, than all the losers, or vice versa, the benefit-cost ratio will, in general, vary with the point of time chosen for the evaluation. Moreover, it is entirely possible that for the evaluation taken at, say, the terminal date, the benefit-cost ratio would exceed unity at the same time as the evaluation taken at the initial date would show a benefit-cost ratio below unity. However, since our inquiry goes far beyond this possibility, we may suppose that any weighted rate of time preference is the same both for losers as for gainers or, simpler still, that r and ρ remain applicable in this case, as well. For a number of reasons (of which the most obvious is the income tax paid on the return from investment) ρ is taken to be above r. Although there can be many different r's and ρ's, and each pair can also be dated $t = 0, \ldots, T$, an analysis conducted in terms of such generality adds only elegant complexity which may obscure the main lines of the argument. We shall therefore continue to regard r and ρ as single magnitudes, and not as vectors or matrices, except to comment on the proposals of others.

Writing $PV_a(B)$, then, as a shorthand for the present value of the stream of benefits when discounted at rate a, the four type-(a) criteria to be reviewed are as follows:

$$PV_r(B) > K_0 \tag{1}$$

$$PV_\rho(B) > K_0 \tag{2}$$

$$PV_p(B) > K_0 \tag{3}$$

where

$$p = \sum_{i=1}^{n} w_i r_i + \sum_{j=1}^{s} w_j \rho_j$$

and

$$\Sigma w_i + \Sigma w_j = 1$$

where the w_i and w_j are weights.

$$PV_q(B) > K_0 \quad \rho > q > r \tag{4}$$

Criterion (1), the staple of textbook instruction, is superficially plausible enough. If r is the common rate of time preference, then the community is indifferent between receiving the stream of benefits $(B) = (B_0,\ldots,B_T)$, and receiving its present value $PV_r(B)$. It is, then, convenient to rank the community's preference between any set of alternative investment streams, $B^1, B^2,\ldots,B^g$, each of which results from an initial outlay K, according to the relative magnitudes of $PV_r(B^1)$, $PV_r(B^2),\ldots,$ $PV_r(B^g)$. In particular, any project having a benefit stream that meets criterion (1) tells us that the present value of that stream of benefits exceeds the present value of its costs and, therefore, represents a potential Pareto improvement for the community.

The rationale for criterion (2), treated in Eckstein's paper of 1957, and also advocated in Baumol's two papers,[6] is no less plausible. For it suggests that if funds equal to K_0 are to be spent on a public project, the average yield from the project should be no less than the ρ per annum that the sum K_0 could fetch if it were placed instead in the private investment sector. If, over the period, the benefit stream yields on the average more than ρ, then the $PV_\rho(B) > K_0$ criterion is met, and there is a net gain from adopting the investment project.

Clearly, criterion (3) is a generalization of (1) and (2) extended to cover all the different r's and ρ's in the economy. Since the weights, the w's, are the fractions of K contributed by the separable components of reduced consumption and of reduced private investment, the resultant weighted rate of return represents society's actual opportunity yield per dollar of investing a sum K in a public project. In general, then, p will vary according as whether K is raised by tax finance, loan finance, or as a mixture of both. Although (3) was originally proposed by Krutilla and Eckstein, it was advanced again by Harberger in connection with a rise in interest rates in response to government borrowing which is supposed to check both private investment and consumption.[7] With such a weighted discount rate, Harberger claimed (erroneously, as we shall see) that "the so-called reinvestment problem disappears."[8]

The well known Arrow-Lind paper produced criterion (4) as a modification of the popular criterion (2), $PV_\rho(B) > K_0$ when, for their analysis, ρ can be taken as the highest actuarial rate of return corresponding, say, to the riskiest private investment.[9] Accepting without criticism their argument that the risks associated with public projects, when divided among a large population of taxpayers, are felt by each taxpayer to be negligible--in contrast to the sense of risk apprehended by the private investor--a risk premium of $(\rho - q)$ can be attributed to the private investor. Inasmuch, then, as the investor is indifferent between the riskiest private investment at ρ and a virtual certain return of q on his money, a potential Pareto improvement is effected if funds are removed from this private investment, so foregoing ρ, and placed instead in public investment at a yield greater than q. Hence the proposed criterion $PV_q(B) > K_0$.

However, as they acknowledge in their reply to critical comments,[10] the crucial assumption on which their criterion rested--that the set of public investment projects excludes opportunities in the private investment sector--did not receive explicit emphasis. And if the assumption is lifted, and the government, permitted to undertake private-sector investment, can avail itself again of the yield ρ, the $PV_\rho(B) > K_0$ criterion comes into its own again.

Although criterion (4) is an interesting variation on the type-(a) criterion, in other respects it is, as it stands, subject to a fundamental criticism, which follows.

Since the demonstration that follows applies to any of the four criteria, we can use $PV_p(B) > K_0$ to represent the generic type.

Given the stream of benefits $B_0, B_1, \ldots, B_T$, the above criterion is explicated as,

$$\sum_{T=0}^{T} \frac{B_t}{(1+p)^t} > K_0 \tag{5}$$

By multiplying through by a scalar $(1+p)^t$, we obtain the equivalent inequality,

$$\sum_{t=0}^{T} B_t(1+p)^{T-t} > K_0(1+p)^T, \tag{6}$$

which can be summarized as $TV_p(B) > (K)_p$; where $TV_p(B)$ stands for the terminal value of the stream of benefits when compounded forward to T at the rate p; and $(K)_p$ stands for the terminal value of the outlay K_0 when it is also compounded forward to T and rate p.

If, and only if, $PV_p(B) > K_0$ does $TV_p(B) > (K)_p$: one form of the criterion that is, entails the other. But the latter form is more revealing. For it makes clear that in order for the criterion to be met, the sum of each of the benefits, $B_0, B_1, \ldots, B_t, \ldots$, <u>when wholly invested</u> and <u>reinvested to time T at rate p</u> must exceed a sum equal to K when wholly and continually reinvested at p to time T. Such a criterion is clearly applicable when, <u>in fact</u>, both the benefits and the outlays are to be used in exactly this way. If, however, they are <u>not</u> to be used in this way--and it is unlikely that they will be--then a criterion based on such a supposition can seriously mislead. Certainly, this $PV_p(B) > K_0$ criterion is misleading when it is applied to public investment projects without information in each case about the actual disposal of the returns to the project, and without information about the uses to which the sum K_0 would have been put were it not used as initial outlay for the project.

Let us return now to the criterion $PV_r(B) > K_0$, regarded as a limiting case of the generic $PV_p(B) > K_0$ criterion. Its transformation into the $TV_r(B) > (K)_r$ form, however, enables us to appreciate immediately the sufficient conditions required for its valid application; namely, that all the returns from the project be wholly consumed as they occur and that the sum K_0 be raised entirely from current consumption. Similarly, transforming the other limiting case, $PV_\rho(B) > K_0$, into the form $TV_\rho(B) > (k)_p$

enables us also to appreciate at once that its Pareto validity is assured if, in fact, it is applied to a case in which the benefits, as they occur, are wholly invested and reinvested in the private investment sector at prevailing yield ρ until the terminal date T, and if the sum K_0 raised from the private sector would have been wholly invested and reinvested also at yield ρ until T.

Put otherwise, the correct terminal value of a project's benefit stream, and the correct terminal value of the opportunity cost of its outlay, are both functions of three vectors r, ρ, θ, or, in the simplest possible case, of three variables, r, ρ, and θ, where θ is the fraction of any income or investment return that is reinvested in the private sector. In contrast, a criterion $PV_p(B) > K_0$ makes the terminal value both of the benefit stream and the outlay a function only of p, whether p is equal to r, or to ρ, or to a weighted sum of r and ρ.

To anticipate a little, the above stringent conditions for the Pareto validity of the type-(a) criterion are sufficient. They are not strictly necessary, however. For instance, where the consumption-investment ratio is the same for all the benefits and also for the outlay, then a type-(b) criterion can, as we shall see later, be reduced to the $PV_r(B) > K_0$ criterion.[11]

Such simplifications are very agreeable. But one can go further. Under the terminal value approach, there is no need to discount at all. All that matters are the relevant rates at which returns are to be compounded forward to T. Indeed, once this is done, the terminal values can then be discounted at r, or at ρ, or at any conceivable rate, without any alteration occurring in the ranking or in the criterion.

In order to complete this part of the critique, we must also reexamine criteria based on IRR, the internal rate of return. There is a seeming advantage in being able to use the IRR for ranking projects without reference to the prevailing yields or interest rates in the economy. Nonetheless, it is not possible to accept or reject projects on the basis of IRR alone. For this purpose, the IRR has to be compared with whatever is believed to be the relevant opportunity rate.

In fact, letting λ stand for the IRR, the internal-rate-return criteria corresponding to the DPV criteria (1) through (4) are (1') $\lambda > r$, (2') $\lambda > \rho$, (3') $\lambda > p$, and (4') $\lambda > q$.

As a ranking device, the IRR has fallen into disfavor among economists, chiefly because there can, in general, be more than one IRR for a given investment stream.[12] However, this is the less important reason. The more important reason is that, even in the common case in which all benefits are positive, the unique IRR calculated for an investment stream does not accord with the true average rate of return over time of the value of that stream. In fact, as conventionally defined, the IRR when used as a criterion has the same defect as the DPV criterion; namely, that a reinvestment rate is entailed that has no necessary relation to the actual rates involved in the particular case.

A procedure that is free from the above defects is that proposed by Mishan,[13] one that transforms an investment stream, $-K_0, B_0, B_1, \ldots, B_T$ into the stream $-K_0, 0, 0, \ldots, TV(B)$. Of the initial return B_0, the amount consumed cB_0 is compounded forward at the relevant rates of time preference, say r, to the terminal date T. The remaining amount sB_0 being divided among the different investment opportunities that are actually anticipated according to one of, say, two alternative political directives: either (1) each investment component is compounded, at its yield, to the following year when it is treated as a receipt along with any other income, or else (2) the investment component is taken to yield equal returns for all successive periods up to T, with the original investment component being included at T. Whether assumptions (1) or (2) are adopted, the sum resulting from the investment component at $t = 1$ is designated ΔR_1.

Thus, at time $t = 1$, we have returns $B_1 + \Delta R_1$ to dispose of. Again, the c proportion of this total $(B_1 + \Delta R_1)$ that is consumed at $t = 1$ is compounded to T at the rate r, the remainder being allocated among the various investment opportunities anticipated in consequence of the existing political and institutional constraints. Continuing in this way until T, the original benefit stream is transformed into its terminal value.[14]

A valid ranking of two mutually exclusive projects, X and Y, both of which may be rejected, however, requires not only a common terminal date T, but also a common initial outlay of K_0. This latter requirement is not restrictive. If, say, Y's initial outlay is 20 less than that of X, the 20 left over from the Y investment can be treated as generating a stream of returns in the private sector of the economy, having a terminal value that is to be added to that of the Y stream of benefits.

As for the social opportunity cost of K_0 itself, this is allowed for simply by treating the stream of returns it would generate if left in the private sector, on a par with projects X and Y. For identification, we refer to this alternative as the "reference stream" Z. Using the same rules, this stream compounds to terminal value TV(Z).

In this way, we end up with three terminal values, TV(X), TV(Y), and TV(Z) from which to choose, all generated by initial outlay K_0. No further operation is required for ranking purposes. If both TV(X) and TV(Y) are less than TV(Z), neither public project is acceptable on a Pareto criterion. If, instead, say TV(X) > TV(Y) > T(Z), then TV(X) is chosen on the Pareto criterion. Any further operation that is acceptable, say, reducing the terminal values to present social values, to present benefit-cost ratios, or to internal rates of return cannot alter this basic ranking.

Thus, corresponding present values for X, Y, and Z, are obtained simply by multiplying each of their terminal values by a scalar, $(1+r)^{-T}$. Corresponding present value benefit-cost ratios are obtained by multiplying them by a scalar $(1+r)^{-T}/K_0$. As for the corresponding IRRs, when defined in accordance with the basic concept of an average rate of increase over time of the initial investment K_0, and therefore as that unique value of λ for which $\frac{TV(B)}{(1+\lambda)^T} = K_0$, the resulting equations,

$$\frac{TV(X)}{(1+\lambda_X)^T} = \frac{TV(Y)}{(1+\lambda_Y)^T} = \frac{TV(Z)}{(1+\lambda)^T} = K_0 \tag{7}$$

entail the ranking $\lambda_X > \lambda_Y > \lambda_Z$.

4.5 The Legitimacy of Using DPV in Ranking Intergenerational Public Projects

Extending the simplifying assumption that the time rate of preference r is common to all members of n successive generations that are affected by a public project, the condition under which it is Pareto valid to use

DPV or compounded terminal value (CTV) is the existence of a common point of overlap; that is, a point of time at which each person affected is alive.

This can be illustrated in the simplest case of two persons from different generations, each one being capable of making rational decisions for sixty years, whose rational lives overlap by, say, twenty years. Let person A be alive in this sense from year 0 to year 60, and receive a stream of benefits (positive and negative) that on balance raises his welfare. Let person B be alive from year 40 to year 100 and receive a stream of benefits that on balance reduces his welfare. Since r is the rate of time preference common to both, A's benefit stream can be transformed into an aggregate value of, say, 100 at year 0, or into an equivalent value of $100(1+r)^t$ for any year t up to year 60. Inasmuch as he is indifferent as between all such sums $100(1+r)^t$, for t equal to 0, 1, ..., 60, such sums can be represented by a continuous line sloping upward from year 0 to year 60. Such a continuous line may then be interpreted as a time indifference curve, as shown in figure 4-1, with aggregate net benefit--whether on balance gain or loss--measured vertically on a logarithmic scale. (We retain the simplifying assumption that preference relationships do not change with changes in the vantage point of time. For a discussion of intertemporal decision problems where vantage points shift and the present is more highly valued than either the past or the future, see Page; and for related discussion, see Rawls, Solow, and Strotz.[15]

If r is such that $1.00 is worth $2.00 in twenty year's time, person B, whose stream is equivalent, say, to a net loss of 600 in year 60 is indifferent between this loss and a loss of 300 in year 40, and a loss of 2,400 in year 100.

As depicted in figure 4-1, at any point of time between years 40 and 60, the values placed by persons A and B on their respective net benefit streams are such that the benefit-loss ratio is 4:3. This benefit-loss ratio, being greater than unity, meets a Pareto criterion (A's gain is such that, via costless redistribution, both persons could be made better off), without violating the basic maxim. Adopting this benefit-loss (or benefit-cost) ratio as criterion, discounting to year zero, or alternatively, compounding to year 100, simply multiplies numerator and denomi-

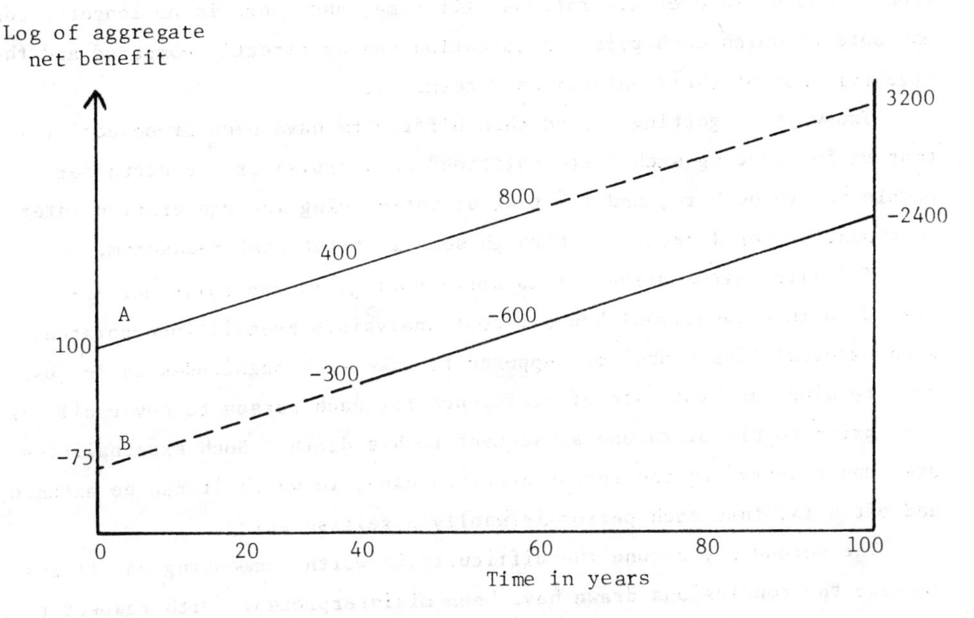

Figure 4-1. Time indifference curves (solid lines) for two persons, and B, from different generations.

nator by a common scalar which, therefore, does not alter the benefit-cost ratio of 4:3.

It follows that if a common point of overlap exists among all individuals of the successive generations, the use of DPV or CTV is, indeed, Pareto valid. Per contra, if there is no common point of overlap then, since the basic maxim is no longer met, neither can the Pareto criterion. There may well be a large number of overlapping generations over, say, a one-thousand-year period. But in such a case, there is no direct way of reaching agreement between all persons of those generations about their respective magnitudes of net benefits at some common point of time. Thus, person A will not be indifferent as between a net receipt of $800 in year 60 and a net receipt of $3,200 in year 100, since he will not be alive in year 100.

Hence, in a time context, the basic maxim requiring economists to accept as their data people's own valuations only of the goods and bads resulting from an economic change poses a problem whenever the time span of the project covers a number of generations. For each person's valuation is now dated over his rational lifetime, and there is no longer a common date at which each person's valuation can be directly compared and the algebraic sum of their valuations determined.

Two ways of getting around this difficulty have been proposed: (1) that of introducing such "externalities" as altruism or a concern for people yet to be born; and (2) that of introducing intergeneration interventions, either directly or through some institutional mechanism.

The first alternative, it is worth noting, is generally not resorted to in conventional benefit-cost analysis. Even if incorporated, such externalities cannot be supposed to take such magnitudes as to justify extending the time rate of preference for each person to cover all the time prior to his birth and subsequent to his death. Such externalities are "unnecessary" in the intrageneration case, in which it can be assumed, and often is, that each person is wholly a selfish being.

The second way around the difficulty is worth commenting on, if only because the conclusions drawn have been misinterpreted. With respect to this second line of reasoning, let us consider, in turn, two possibilities: A, that of government agreements between generations to transform an existing intergeneration stream of costs and benefits by means of transfers so that, in fact, net benefit comparisons can be made at a common point of time, and B, the use of market mechanisms, in particular investment opportunities for transforming an existing intergeneration stream into one that does, in fact, meet a Pareto criterion.

To illustrate the A case, let a situation involving three persons, X, Y, and Z be that depicted in figure 4-2, which clearly has no common point of overlap. Of course, the economist might choose year 60 for the comparison of the three persons. But since he has no warrant for reducing the minus 100 of person Z at year 80 to anything smaller (absolutely) than minus 100 in any year prior to his birth in year 80, he might propose to use minus 100 also in year 60. If he does this, the algebraic total for the three persons in year 60 comes to minus 10, and the project appears inadmissible. If, instead, he uses exactly the same procedure in choosing year 80--and therefore values person X's net gain of 60 in year 60 as equal to a gain of 60 also in year 80 (year 80 being twenty years after X's death), the algebraic total for the three persons is now plus 20, and the project would appear now to be admissible.

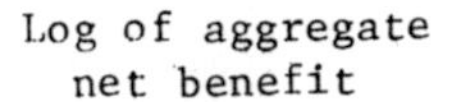

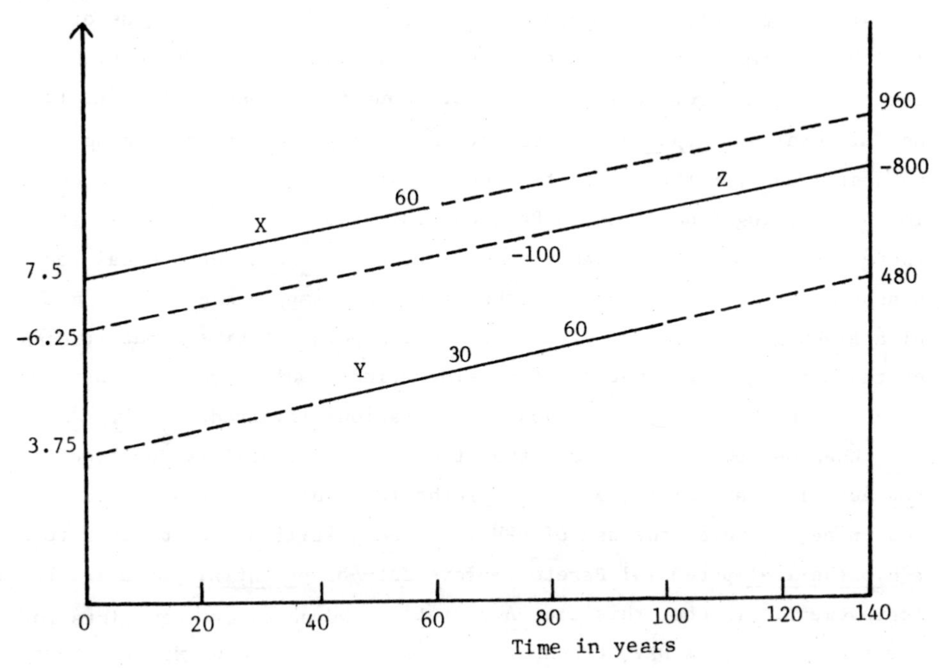

Figure 4-2. Time indifference curves (solid line) for three persons, X, Y and Z.

In this situation, the economist might envisage government intervention taking the following form: instead of X having 60 and Y having 30 in year 60 (Z not yet born), we could transfer 60 from X to Y. And this 90 now received by Y in year 60 is equivalent for Y to 180 in year 80. In year 80, we then transfer 100 from Y to Z. The net result is that X and Z are no better or worse off than before, while Y is better off by 80. A Pareto improvement has thereby been achieved.

Two comments will help us to interpret this proposed way around the problem. First, if costless transfers were indeed possible, then every hypothetical improvement could indeed be converted into an actual Pareto improvement. But nobody takes this easy way out in allocative problems in the context of comparative statics, and there is no warrant for taking such an escape route from the problem when chronological time, particularly generational time, is introduced.

Second, some economists might want to argue, however, that allowing these transfers as between individuals over time to be hypothetical, then a test of hypothetical compensation is met--which is to say, one of potential Pareto improvement. In consequence, if the use of DPV results in a net benefit, a Pareto criterion is met. But this reasoning is facile and misleading. *Hypothetical* transfers between generations can no longer be justified by reference to the four considerations which were adduced for an existing generation in Section 4.3. Those considerations, it was there argued, might reasonably be held to give rise to an ethical consensus in favor of adopting a potential Pareto improvement by the members of a given generation, or of a society at a point of time. But it cannot be taken for granted that those considerations would secure a consensus among the members of all successive generations involved.

When we now face the fact that the possible transfers described do *not* actually take place, but are hypothetical only, this case-A line of reasoning, by which the use of DPV is to be justified, can be seen to meet a *hypothetical* potential Pareto improvement--or *potential* potential Pareto improvement. In sum, this case-A operation, advanced by economists for favoring the use of DPV, transpires to be one that can be split into two hypothetical transfers; one is the familiar hypothetical redistribution among the members of an existing community or generation; the other, a hypothetical redistribution, also, as between the successive generations themselves.

The introduction of investment opportunities, the B case, gives rise to much the same reasoning and reaches the same sort of conclusions. Adopting the rate *r* as reflecting also the rate of return on current investment, the sum 60 received by person X in year 60 could be invested at *r* for twenty years to compound to the sum 120. From the 120 so accumulated, person Z can be paid 100. Thus Z is left as well off as he was without the project (as is the case also with person X who dies twenty years earlier), while person Y is left with 80 in year 80. And this result is believed to meet a Pareto criterion.

The correct interpretation of the above simple example follows that of the preceding example.

Again, there are two sorts of hypothetically costless transfers involved, not just one. The first has reference to the sums assumed to be taken from earlier generations which are invested for the time necessary to produce an algebraic sum of benefits that is positive at some future data, say, the terminal year. The second has reference to the assumed redistribution of this positive algebraic sum among members of the community at that time so as to make "every one" better off. Since the DPV method is not being regarded as contingent upon an agreement, between governments of all generations involved, actually to invest the receipts of earlier generations with the object of presenting later generations with sums calculated to offset the losses they are to suffer, this imaginary transfer between generations is clearly as hypothetical as the subsequent redistribution of net gains among members of the community at any point of time.

A consensus on the acceptability of the ordinary potential Pareto improvement among members of a given generation may be presumed to exist. Moreover, it can be assumed that, at the time of completion of the change, redistribution for actual Pareto improvement is feasible. In contrast, a consensus among members of all generations involved in the long-lived investment project may not be presumed inasmuch as there are no mechanisms which can be counted upon to diffuse the net benefits among this intergenerational community; there is nothing, in effect, to prevent later generations having to shoulder heavy burdens while earlier generations reap benefits. Moreover, the potential redistribution across generations becomes infeasible at the completion of the change if the compensating investment is not undertaken at the beginning of the change.

This argument counters Freeman's assertion,[16] for example, that, in project evaluation, it makes economic sense to discount to the present the value of damages expected to be borne by generations who will be alive many thousands of years from today.

In chapter 6 of this volume, Bailey suggests that the behavioral condition of hyperrationality, along with normal market mechanisms, ensure that a discounting approach leads to actual Pareto improvement across generations. (Hyperrationality involves the assumption that the present acts to preserve its consumption stream intact.) This observation, depending on the plausibility of the behavioral condition, would tend, of course, to strengthen the ethical appeal of the discounting approach. A simple example illustrates the idea and our misgivings about it.

In chapter 5, Smith and d'Arge estimate the benefits associated with CFCs for the single use as a propellant for insect repellent sprays, for personal use in the United States, to be $5 billion. Suppose, for illustration, that these benefits are concentrated in the first year (the "present generation"), and after this first year there will be a perfect substitute at no additional market cost and with no environmental hazard, so that future benefits of CFCs as a propellant for insect repellents are zero. Suppose further that these CFCs released to the atmosphere remain latent for one hundred years, but then, in the hundredth year, there is a 1 percent chance of catastrophic effect in which the entire world population is destroyed. This last supposition is indeed somewhat extreme--it appears that scientists accept nonnegligible probabilities, of 1 percent or so, of enormous catastrophes associated with the continued growth, ten percent or more worldwide, of CFCs--but no one is forecasting ultimate catastrophe from a single and minor use of CFCs over a limited period of production. Nonetheless, it sets out the decision problem more sharply to consider this illustrative case.

Under a discounting or type-(a) criterion, the first step is to calculate the expected value of the potential loss. Assume that the world population a century hence would be about ten billion, and take the value of life as $500,000, a "generous" estimate of what might be found in the risk-benefit literature using a discounting approach. (See Linerooth and Bailey, for example.[17] Some estimates of the value of life are a good deal higher; see Page, Harris, and Bruser.[18]) Then, the expected value of the potential catastrophe is $\$(0.01)(10^{10})(5 \times 10^{5}) = \$5 \times 10^{13}$ or $50 trillion, valued by those living in year 100. Next, this expected value is discounted back to the present at 11 percent, the rate recommended by Bailey, with the resulting present value of $\$1.5 \times 10^{9}$. This present value cost, $1.5 billion, is less than the present benefits of $5 billion associated with the use of CFC spray repellents. In fact, the benefit-cost ratio is 3.4 to one in favor of the environmental gamble.

Some, including ourselves, will find this simple calculation and conclusion unsatisfactory on the grounds that the advantage of a spray repellent over a lotion repellent is too trivial to justify the risk of an ultimate catastrophe. On the contrary, Bailey defends the methodology of this discount approach on the grounds that the future would actually be better off under the gamble than without it. The idea is as follows. Suppose that CFCs were banned for the use of insect repellent spray. Present consumers would be faced with a decline of $5 billion in consumption. If they were hyperrational, they would act to preserve their original consumption pattern, generating $5 billion worth of other consumption expenditures. With the aggregate consumption stream maintained intact, the $5 billion comes out of a national savings.[22] Each year the consumption stream is maintained, so that this $5 billion is compounded forward as an investment foregone. Thus, at the end of a century, there would be $\$(5x10^9)(1.11)^{100}$ worth of less resources, compared with what there would have been without the ban. With the ban, the future avoids the catastrophic risk, valued at $50 trillion, but also is $(5x10^9)(1.11)^{100} = 1.7 \times 10^{14}$ or $170 trillion poorer in resources than it would have been without the ban. Thus, the future would actually be better off without the ban and with the risk caused by CFC emissions. For, in the case of no ban in year zero, the future in year 100 could apply $50 trillion to life-savings programs, reducing the aggregate risk of early death as much as CFCs increase the risk. This risk, being offset, the future would still have $120 trillion left over, and thus would be better off. Moreover, the present would also be better off without the ban. With respect to its own consumption, it would be indifferent, for given hyperrationality, it would act to maintain its consumption stream intact. But because the ban requires compulsion, presumably the present is better off without the ban.[23] Thus the interests of the present and future harmonize. Both are better off without the ban than with it. The decision not to ban is an actual Pareto improvement, compared with the alternative of the ban. If the discounted expected value of the risk had turned out to be more than the present benefits of the spray, and enough more to compensate the present for compulsive regulation, a similar argument could be constructed leading to actual Pareto superiority of the ban.

If hyperrationality described the actual behavior of the economy, we would find the ethical appeal of discounting, at a rate equal to the opportunity cost of capital (criterion type-(a)(2)), greatly strengthened. However, we find the condition of hyperrationality, which harmonizes the interests of present and future, to be implausible.

Consider that the U.S. GNP is growing at about 3 percent a year (without the ban). In a century, we might expect it to increase about 20-fold (1.03^{100}), or to about \$20 trillion. Thus, it is not possible for the ban to reduce GNP by \$170 trillion. The point is that it is possible for a marginal investment to grow at 11 percent for a few years, but it is not possible for it to grow at such a rate for many years if the entire economy is growing at a substantially lower rate. Over a long period, something must give and it appears that this must be the assumption of hyperrationality. Otherwise, it would lead us to believe that if hula hoops were banned in the 1950s, the entire economy would be destroyed a century hence.

We can consider two other behavioral conditions which might be more plausible than hyperrationality for the very long run. For the first, we assume that if CFCs for insect sprays were banned, most of the reduction would come out of consumption and a little out of foregone investment. If foregone opportunities to consume and invest fall into the same pattern as consumer spending of income, we would expect about 90 percent to come out of present consumption and 10 percent out of investment (a "Keynesian savings rule"). Thus, we take 10 percent of \$5 billion and compound that forward at 10 percent of 11 percent for a century. The resulting loss of investment resource a century hence is then $(5 \times 10^9)(.1)(1.011)^{100}$, or \$1.5 billion. If the "Keynesian savings rule" describes the actual behavioral condition of the economy, it is clear that the interests of the future lie with the ban. From the perspective of people in year 100, the ban prevents the catastrophic risk valued by them at \$50 trillion at a modest cost to them of \$1.5 billion. Of course, this is not the whole story. The present is somewhat worse off with the ban, as its consumption is reduced by \$4.5 billion, and each "generation," or year, between 0 and 100 is worse off by a somewhat lesser amount, on the order of about \$1 billion.

The second behavioral condition is the polar opposite case of hyperrationality. For this condition, we assume that the \$5 billion comes entirely out of this year's consumption. Thus, there is no effect of the ban except for reduced consumption the first year and reduced risk in the last year. While this assumption is no doubt unrealistic for major environmental regulations affecting consumer purchases, it has some plausibility for minor changes. For example, it would lead us to believe that if hula hoops were banned in the 1950s, there would be no discernible effect on the economy a century hence. Similarly, if consumers were faced with the prospect of liquid insect sprays instead of aerosols, it seems somewhat plausible that there would be no profound effect on the economy a century hence (except for the change in environmental risk). In fact, it even appears conceivable that if consumers were faced with the slight extra exertion of liquid rather than spray insect repellents, the economy a hundred years hence might actually be modestly stimulated. Thus, this last behavioral condition appears more plausible than hyperrationality applied to the very long run.

The virtue of this last behavioral condition is its simplicity. It shows the conflict of interests between generations most starkly, and it may be the implicit assumption people have in mind when they reject discounting altogether and affirm that the present should not impose a 1 percent risk of ultimate catastrophe upon the future for a mere $5 billion benefit to the present.

However, this last behavioral condition appears to us to be less realistic than the "Keynesian saving" pattern, which is a mixed case lying between the two polar extremes. Here, discounting still has an important role in defining intertemporal opportunity costs of one whole path, or intertemporal distribution, of consumption and risk burden. (For further discussion, see Dasgupta.[19]) In any case, given the behavioral assumptions, it is useful to construct, as far as possible, the intertemporal distribution of costs and benefits with and without regulation. Infeasible implications, such as those derived by straightforward discounting at the short-term marginal opportunity cost of capital, for a century or more, can be avoided. In dealing with long time horizons, discounting exercises depend critically upon the underlying assumptions and empirical conditions and it is easy to be led to nonsensical results.

For the CFC case and many others like it, with immediate benefits and long-delayed costs, we believe that intergenerational conflicts of interest are an inherent part of the decision problem. In these circumstances, the economist must confront the intergenerational distributional implications of such projects.

4.6 The Problem of Intergeneration Equity

Since it is unlikely that a stream of returns from a public project having significant welfare effects on future generations will be transformed via institutional mechanisms into one that can meet a conventional Pareto criterion, the question of intergenerational equity has to be faced squarely. Certainly, an intergeneration consensus for the use of DPV (or CTV) is unlikely, as indicated in the preceding section, since it would entail acceptance by later generations of smaller weights being attached to their valuations than to the valuations of earlier generations.

Thus, although the economist does, from time to time, extend his conventional maximization techniques even so far as to resolve the problem of distribution over generational time, unless the results of his chosen set of assumptions yields a distributional pattern over time that accords with what our present society believes is just, for all the affected generations, his conclusions will, or should, go unheeded. However, one cannot suppose the economist to be wholly uninfluenced by what is held to be just and proper in this respect. For the results of much of the economic literature on the optimal distribution of the product over generational time conforms with the popular belief that a just distribution is one that yields constant per capita income over generational time--at least, whenever the population of successive generations remains unchanged. A somewhat different version of this idea is that the means for future well-being be at least as good as our own, thus focusing our attention on the future condition of the resource base and its ultimate renewability, and the portfolio of catastrophic risks that are passed from one generation to another (see Page[20]).

Whatever our view of the fundamental factors explaining differences in existing incomes, we are likely to agree that an equal per capita real consumption for all generations is eminently fair. Even if we take what seems today to be the less popular view, that one's income is primarily the fruit of one's effort, it is the average income of each generation, not the distribution within it, that is at issue. For, making the minor assumption that the average effort of each generation is about the same, the reasonable supposition that the distribution of relevant characteristics is much the same for one generation as for another impels us to the view that no generation deserves the right to enjoy a higher standard than any other (see Rawls[21]).

Therefore, irrespective of the way each generation chooses to distribute its own outputs among its members, we can agree on each generation's right to a natural resource and capital endowment that, with the same average effort, will produce for it the same per capita real consumption as that of any other generation. In sum, the ethical appeal of equality of per capita consumption over generational time is independent of a belief in the justice of an equal division of the product in any existing society, and is far more compelling. This conclusion is reinforced by an application of Arrow's axioms which imply that a single generation which imposes its will irreversibly, in disregard of the preferences of all generations to follow, is clearly acting as a dictator.

In a trivial sense, the DPV, as a rule of intertemporal choice, is a dictatorship of the present. The present, after all, must choose in the absence of the future. In this trivial sense, every decision rule is a dictatorship of the present. But, suppose the present wishes to be fair to the future's interests, which may differ from the interests of the present because of the difference in vantage points in time, among other things. The present can estimate the future's interests concerning a decision made in the present, such as the control or noncontrol of ozone depletion or climate change. Moreover, the present can try to build its ideas of intertemporal fairness into the aggregation rules which combine the present's preferences with those of the future into a single decision or ranking of alternatives. And, in considering the fairness of intertemporal aggregation rules, some things can be said without any knowledge of the future's actual preferences. Thus, consider the rule illustrated in figure 4-3, which says that infinite majorities should be decisive over finite minorities. In this case, the present generation prefers having project I to not having it, but every other generation prefers not having

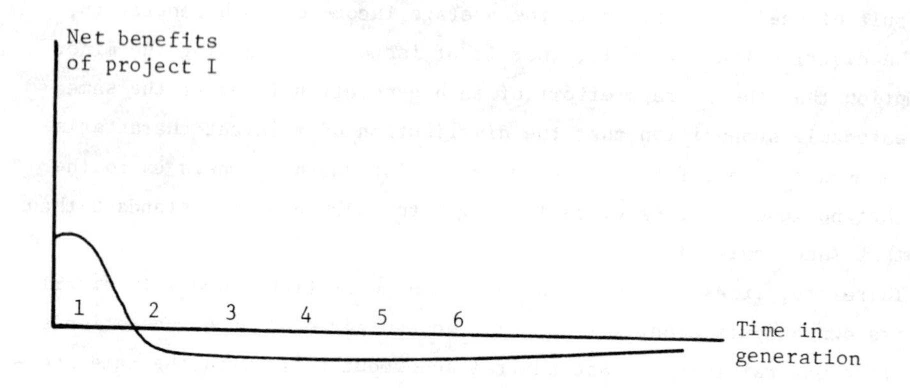

Figure 4-3. An illustrative intertemporal aggregation rule.

it. If the present abides by this rule, on the grounds that it values this version of intertemporal fairness more highly than the particular benefits associated with the project, the decision by the present--say, to forego Freon hairsprays--is clearly not a dictatorship of the present in the Arrow sense.

This aggregation rule may be called the "overtaking rule" because when it applies, eventually there comes a time when every later generation unanimously agrees on some course of action.

How, we may ask, does a discount rule (DPV) fit into this framework? In figure 4-3 it would be easy to draw a stream of net benefits such that, for a discount rate greater than 2 percent, the discounted net benefits were positive. Then the DPV would indicate acceptance of the project, even though only the present generation prefers it to its absence. This, indeed, appears to be a dictatorship of the present. Is this just a happenstance of the way figure 4-3 is drawn and the interpretation of "benefit?"

It appears that the proper procedure is to face the distributional problem directly. Only if the equity issue is to be otherwise ignored does it make sense to consider a lower rate of discount as a kind of _ad hoc_ palliative. One may surmise that some economists are a little uneasy about the possibility of intergenerational inequities resulting from the application of DPV to the time stream of a long-lived public project since, by way of apology to the future, they sometimes propose a low rate of discount. But such a proposal is the sort of concession that springs from doubt. It does not rest on justifiable principle.

It seems to us that a proper and justifiable role for the discount rate is to help define the feasible set of intertemporal paths, from which one must choose an equitable resolution of intertemporal conflicts of interest.[22] Once the intertemporal opportunity set is defined, an equitable intertemporal choice rule need not involve a further discounting procedure. For example, a promising procedure is to make a decision on the basis of net current benefits once a transition period is completed. This approach is virtually the same as following the overtaking principle and has been used by the Council on Wage and Price Stability for its analysis of drinking-water regulations affecting cancer risk.[23] The approach is particularly suited for cases where there are long-term irreversibilities and long-term latencies, and where there appears to be, at most, a single switch in current net benefits once the project start-up costs and latencies are passed in time. The problem of managing CFCs appears to share these characteristics.

4.7 The Treatment of Risk and Uncertainty

The uncertainty of a future event may be split into two phases, that of assessing, where possible, the risk in terms of probability and severity of an event occurring--whether the probability is objective (based on a statistical sample) or subjective (based on personal estimates of likelihood)--and that of determining a method of evaluating the risk of the event in question. The latter phase may be extended to include the choice of an appropriate technique for decision making when there is no known way of assessing the probability of the event occurring.

Wherever the statistics of a chance or risk of an event occurring are known, then, <u>in principle</u>, it is possible to place a value on the consequent increase or decrease in the welfare of each person subjected to that chance or risk, on an ex ante basis. Under these conditions, the Pareto criterion, where its application is justified, can be extended to cover changes in chance and risk. Of course, there will always be the problem not only of assessing the magnitude of the risk, but of ensuring that persons affected by it are aware of it also. (This raises the question of further investment in gathering and disseminating information, but such has never prevented economists engaged in allocative techniques from accepting individual valuations as the relevant data. Another issue involves intertemporal conflicts of interest for the individual, which we touched upon earlier. Society does not allow voluntary contracts made in the present leading to slavery at a future date, thus protecting the future interests of the individual against present interests. The same ethical concern applies to lotteries for future slavery such as the exposure to carcinogens or other irreversible environmental risks. This ethical concern must be balanced against the supposition underlying methodological individualism, that the individual is better able to look out for his future interests than the government.)

Where the risk in question is not known, and no agreement can be reached on the likelihood of its occurrence, techniques designed to deal with the resultant problem go beyond the bounds of allocative economics as described earlier. Such techniques, whatever their virtues, remain without social sanction unless they can be assumed to be understood and approved by society at large.

We may briefly describe some of the popular techniques proposed for dealing with risk and uncertainty, bearing in mind the nature of the CFC problem.

4.7.1 Raising the Discount Rate

The notion of adding some arbitrary percentage points to the rate of discount may be defended as a crude way of coping with uncertainty wherever the uncertainty refers to the magnitudes of the benefits over the future. It would be hard to justify for cases in which the events themselves, or the side effects of the projects in question, are, as yet, unknown.

Moreover, tampering with the discount rate is obviously an awkward form of recourse when employed simultaneously to cope with uncertainty (usually by raising the rate) and with the problem of intergenerational equity (usually by lowering the rate).

4.7.2 Probability Distribution of Net Benefits

Building a probability distribution of net benefits from experts' guesses about future prices is an alternative way of dealing with the same kind of uncertainty. Generally, possible net benefit outcomes can run into many millions. Nevertheless, a sample distribution can be simulated with the aid of a computer set to select, at random, a number of combinations consisting of each of the prices along with one of the price's subjective probabilities, each such combination corresponding with a net benefit figure. A sample of some two hundred or three hundred combinations usually suffice to produce a reliable enough distribution to work with.

Such a technique might usefully be employed in evaluating the opportunity cost involved in controlling CFCs. No comparable calculations, however, can be made for the value of the benefits arising from a reduction of CFCs, inasmuch as existing knowledge of the effects of CFCs is too meager to permit intelligent guesses.

4.7.3 The Uses of Game Theory

Game theory is a technique applicable to cases in which there is complete ignorance about the probability of each of the possible outcomes of an uncertain event or combination of events. As such, it might seem to lend itself to decision making in this instance. However, we reject the decision technique for use here because of the extent of our current ignorance of the CFC problem. Since we are almost wholly ignorant of the phenomenon, there can be no limit to the number of possible outcomes. Even were we able to guess at the nature of some of the results from adopting strategies involving little reduction of current activities, the placing of monetary values at various points of time on a number of the more disastrous of them would be arbitrary and highly controversial.

4.7.4 Risk-Benefit Analysis and Conditional Probability

Other techniques such as risk-benefit analysis (which is, in fact, no more than a variant of benefit-cost analysis where the risk entailed is part of the cost), and the use of strategies based upon conditional probability (where prior events are associated from experience with the likelihood of specific outcomes occurring) have also to be precluded since they, too, depend upon some knowledge of the nature of the risk and upon the social value or cost of the event should it occur.

The CFC problem, in fact, falls into the category of externalities for spillover effects that has grown rapidly since World War II, being the product of recent technical innovation, and having in common certain features that separate them from the more conventional spillover effects--effluent, noise, fumes, congestion, and the like--which feature so large in the economic literature. (See Page for additional detail.[24]) The chief distinguishing features of this new category of spillovers that appear to render them untractable to familiar economic methods are as follows.

First, since the industrial processes and products are novel to this planet, there is very limited experience of the nature or incidence of their side effects. The consequences for humanity of the continuance and spread of these new activities and products are, therefore, as yet under a gigantic question mark. Specific effects are sometimes suspect and give rise to controversy and speculation. For the rest, it is expected and feared that other side effects will emerge over time.

Second, there is, in such cases, an intelligent apprehension that the spillovers associated with these new activities may well take the form of large-scale disasters, possibly having global dimensions. In particular, the damage caused may be irreversible, and possibly fatal, to humanity or to all forms of life on earth.

Third, some part or all of the as yet imperfectly understood damage or hazard of pursuing these new activities may fall on future generations. And there can be general presumption that safe technological methods for dealing with them will be discovered in time.

The question, then, is whether the economist or any kind of scientist can produce meaningful figures purporting to be an economic contribution to the decision-making process when the problem under consideration involves spillovers having the singular features mentioned above.

Thus, in the particular problem under consideration, that of CFC management, the possibility of a number of catastrophic outcomes cannot, at present, be dismissed as being beyond the pale of likelihood.

4.8 Conclusions and Recommendations

In the circumstances surrounding the CFC problem, the conscientious economist has to recognize that the conventional tools may be only of limited service. Nonetheless, some proposals can be made for coping with products or processes, the introduction of which involves the local or global community in some, as yet, unknown degree of hazard. The following are illustrative:

1. Thinking in terms not of the prohibition of a project but of its public regulation, a prudent maxim would have it that the larger the possible catastrophe and the higher the probability of its occurrence, the stricter should be the details of its regulation. Such a maxim, however, is not very useful where the conditions are so novel that we virtually know nothing of the nature of the catastrophes and of the probability of their occurrence.

2. Still thinking in terms of government regulation, it might seem reasonable to suppose that the burden should be placed on the regulatory agency to show--in the words of the Toxic Substances Control Act--that there is a "reasonable basis for concluding that there may be an unreasonable risk." The regulatory agency would then have to demonstrate that something like a large-scale catastrophe is a possibility that cannot be lightly dismissed.

The trouble with this seemingly reasonable proposal, however, is that there may be no way of demonstrating the credibility of one or more possible large-scale catastrophes and, in default of such demonstration, the project in question would be adopted with the possible result that the suspicion of some dreadful calamity would, alas, be vindicated within the lifetime of the existing generation or of some future generation. Thus, while such a rule of procedure might be acceptable enough for a spectrum of limited risks, it is manifestly unacceptable wherever there is a risk of a major and irreversible disaster, even where the degree of risk cannot be calculated and even where there is reason to believe it is small.

3. Arrow and Fisher have discussed the problem of irreversibility in terms of the growth of information over time.[25] On the supposition that information improves continuously with time, they introduce a simple model

designed to indicate the conditions under which there is a balance of advantage in not foreclosing irreversible options. However, the extent of the caution envisaged has to be increased substantially when the problem is placed within an intergeneration context and when the problem is raised in a situation in which the irreversibility contemplated has reference not so much to the loss, say, of some unique wilderness area, but rather to the ecological viability of the planet.

4. Another possible way of proceeding in the face of uncertainty with respect both to the range of outcomes and to their associated probabilities is to compare, for each possible or credible outcome, the consequences, on the one hand, of acting on the basis of what turns out to be unwarranted alarm with the consequences, on the other hand, of acting on the basis of what turns out to be unwarranted complacency. There may be close agreement among scientists that the damaging consequences for humankind of adopting policies based on unwarranted complacency may far exceed in magnitude the consequences, in terms of loss of social gain, of adopting policies based on unwarranted alarm.

Should this agreement exist, it might seem to follow that strict regulation of all suspect activities and products (including a ban on the activities or products) should be enforced until our knowledge of their range of effects on the planet has increased to the point of consensus in detail and a high degree of confidence. Only in the fullness of time, then, should it become evident whether our apprehensions of possible disasters were justified, more than justified, or less than justified.

At this point, a caveat should be entered. However measured, the growth of knowledge, like any other index, is not likely to take the form of a smooth, upward trend. Within short periods of time, say, decades, we can now recognize, with the benefit of hindsight, that what was once believed to be new knowledge, or an advance in our understanding, turned out to be erroneous or misleading. Thus, in the near future, we may come to believe the action of certain items on the biosphere to be less dangerous than we originally thought it was, only to discover later that it was more dangerous. What is more, persistent research may eventually bring to light hitherto unsuspected consequences of these same items that may be potentially more dangerous than those currently suspected.

Issues touched upon in the above proposals combine to raise a crucial question. For the problems of the sort the economist has recently had to face are distinguished by three features: (1) Although there are grounds for suspicion of a possible major disaster, there is an absence of dependable knowledge with respect both to the nature of, and the probability of, the worst outcomes; (2) credible worst outcomes are marked by global irreversibility; and (3) worst outcomes are as likely, or are more likely, to fall on some future generation as on the present.

The crucial question referred to, then, is that of the policy to be pursued during the period necessary for knowledge to accumulate to the extent needed for a decision to be taken with confidence. This question is clearly related to that which faces the pure scientist whenever he is presented with a new hypothesis. His traditional response in these circumstances can be interpreted as one of methodological conservatism--of resisting novelty until it has survived a long gauntlet of attack and opposition.

In all events, for the economist faced with problems having the aforementioned features, it is appropriate to consider, first, two alternative and diametrically opposed social responses, or rules of action, wherever an existing or proposed economic activity may legitimately be suspected of generating dangerous and incalculable spillover effects.

Rule A would countenance the initiation or continuance of an economic activity until the evidence that it is harmful or risky has been established beyond reasonable doubt. Rule B, in contrast, would debar the economic activity in question until evidence that it is safe has been established beyond reasonable doubt. The phrase "beyond reasonable doubt" can excite much controversy, but whatever the interpretation agreed upon, the distinction between the two rules is of the essence.

Concerning the existing tendency, the A rule has generally prevailed in respect of commercial enterprise in the West, at least since the Industrial Revolution, in the belief that the progress of industry, although it inevitably occasions inconvenience, eventually promotes the welfare of society as a whole. Whether or not this presumption could be justified by a sophisticated examination of the evidence is a matter of conjecture. However, it must be conceded that the spillovers which most concerned the public in earlier days were of the more conventional kind and, therefore, in a crude way, at least, subject to economic calculation.

Since World War II, many authoritative voices have challenged this general presumption of economic progress though without making much impression on the public until the last few years. For the seeming success, scientific and industrial, of the past two-hundred years has given rise to an establishment of technocrats, bureaucrats, and enterprises, steeped in the belief that science and technology, given the freedom and the funds, will eventually solve all the problems that have been and are being created by science and technology. Yet that immaculate faith in the omnicompetence of the scientific method to overcome all obstacles has begun to waver. The subsequent history of acclaimed scientific discoveries or technological feats over the last thirty years do not read like a success story. In recognition of the new type of spillovers referred to, the wisdom of being guided by the A rule is no longer self-evident.

If spillovers were such as to be restricted to a single country, and that country, perilously poor, could not depend on outside aid to mitigate the poverty and malnutrition of the bulk of its population, a case could be made for the adoption of the A rule wherever the benefits of introducing an innovation were expected to be substantial.

The position is quite different, however, for a country such as the United States which is a goods-saturated economy. The continued use of freon and other gases is far from urgent in terms of the saving of life. The present value of the net benefits to be sacrificed from dispensing altogether with such luxuries can hardly be an impressive magnitude. And even if it were reckoned at some outlandish figure, say $100 billion or more, application of the A rule to the case in issue would be difficult to support in view of the possible danger and the possible irreversibility of the ecological disaster envisaged. We might well ask just how large the value of the net benefits to be foregone has to be in order to warrant the incurring of a risk of that order. If there were some finite figure for these foregone benefits that would indeed warrant exposing the country's population to such a risk, it is virtually certain to be many times any plausible estimate of the present value of such benefits.

The above methodological maxim, if adopted, is sure to offend some commercial and consumer interests. In the circumstances surrounding the CFC problem, however, the commercial and consumer interests can hardly be very strong. Regulation or prohibition of the items in question would surely be more acceptable even to the less environmentally concerned or less informed segment of the public if they were to be replaced by substitutes that could not seriously be held to occasion much loss of welfare. A policy of replacing push-button aerosols by hand sprays, for instance, can hardly be described as one causing hardship or discomfort. Neither, for that matter, would increased regulation of refrigerant units or control of their disposal be regarded as imposing much of a sacrifice on the consuming public.

The B rule in effect has been applied in the United States to the potential risks of supersonic transport and recombinant DNA, and in those cases, the costs of the B rule, in terms of delay, have been small in comparison with the benefits in terms of precautionary management of potentially irreversible and catastrophic risk. On the other side, where the B rule has not been applied and where later estimates of risk have proved higher than earlier estimates, there is likely to be an enormous amount of unnecessary suffering, as in the cases of Tris and PCBs.

4.9 Final Reflections

We are impelled to conclude that a valid cost-benefit calculation of actions to protect the earth's ozone shield and its climate cannot be undertaken in the present state of our ignorance concerning the relevant physical relationships and therefore in the nature and magnitude of the risks posed by existing economic activities. Nor can the decision techniques devised by economists and others for problems involving future uncertainty shed much light on the issue. It is, of course, proper that continued research into all aspects of the CFC problem should continue. But until such time as there is basic agreement on the range of consequences flowing from the use of all suspect goods and activities, or until such time as processes have been perfected for recycling all substances suspected of directly or indirectly affecting ozone shield and climate, any society having a sense of obligation toward its citizens, and a sense of responsibility for generations yet to come, should adopt the prudent course entailed by the B rule.

The question of the instruments by which to implement action in constraint of such suspect activities is a secondary matter and one open to debate. Although economists, by training, tend to favor taxes rather than blanket prohibitions, there are political advantages in having recourse, in circumstances of such gravity, to the latter and more dramatic instrument and, indeed, for making the period of adjustment as short as possible. Public support tends to rally to a government that is manifestly in earnest about a declared clear and present danger. In contrast, prolonged debates about taxes and subsidies, about possible exemptions and extensions of the *status quo*, are apt to weaken the resolve both of governments and citizens, and to detract from the gravity of the situation.

Finally, although the problem is clearly an international one and the United States should seek ways of persuading other countries to act in concert in the interests of mankind as a whole, any initial failure to achieve international or multinational agreement ought not to deter her from taking unilateral action in an endeavor to diminish the existing risks being run.

The United States is the major user of CFCs, accounting for half the world's total consumption (chapter 3). Unilateral action can therefore make a substantial difference to the global risk while serving to enhance the nation's moral influence, thereby encouraging other nations to follow that example.

References

1. E.J. Mishan, Welfare Economics: An Assessment, deVries Memorial Lectures, North Holland, 1969.

2. E.J. Mishan, Cost-Benefit Analysis (2nd ed., New York, Praeger, 1976); E.J. Mishan, "The Use of CV and EV in CBA," Economica (May 1976); and J.E. Meade, "Review of Cost-Benefit Analysis by E.J. Mishan," Economic Journal vol. 82, pp. 244-246.

3. B. Weisbrod, "Income Redistributive Effects and Benefit-Cost Analysis," in Problems in Public Expenditure Analysis, S.B. Chase, ed. (Washington, D.C., Brookings Institution, 1968), pp. 177-208.

4. P. Dasgupta, S. Marglin, and A. Sen, Guidelines for Project Evaluation (New York, United Nations, 1972), p. 5.

5. N. Georgescu-Roegen, "Energy and Economic Myths," Southern Economic Journal vol. 41 (1975); H. Daly, Steady-State Economics (W.H. Freeman, San Francisco, 1977); R. Solow, "The Economics of Resources or the Resources of Economics," American Economic Review vol. 64 (May 1974); Talbot Page, Conservation and Economic Efficiency: An Approach to Materials Policy (Baltimore, Johns Hopkins Press for Resources for the Future, 1977); A. Sen, "Approaches to the Choice of Discount Rates for Social Cost-Benefit Analysis," Conference paper, Resources for the Future, Washington, D.C., 1977; C. Price, "To the Future: With Indifference or Concern?--The Social Discount Rate and Its Implications for Land Use," Journal of Agricultural Economics vol. 24, no. 2 (1973); P. Neher, "Democratic Exploitation of a Replenishable Resource," Journal of Public Economics vol. 5 (1976), pp. 361-371; H.S. Burness and T.R. Lewis, "Democratic Exploitation of a Non-Replenishable Resource," Social Science Working Paper No. 161 (Pasadena, California Institute of Technology, 1977).

6. O. Eckstein, "Investment Criteria for Economic Development," Quarterly Journal of Economics (June 1957); W.J. Baumol, "On the Social Discount Rate," American Economic Review (September 1968); and W.J. Baumol, "On the Social Rate for Public Projects," in The Analysis and Evaluation of Public Expenditures: The PBB System vol. I (Washington, D.C., Joint Economic Committee, 1969).

7. John Krutilla and Otto Eckstein, Multiple Purpose River Development (Baltimore, Md., Johns Hopkins University Press for Resources for the Future, 1958); and A.C. Harberger, "The Opportunity Costs of Public Investment Financed by Borrowing," in R. Layard, ed., Cost-Benefit Analysis (London, Penguin Books, 1972).

8. Harberger, "The Opportunity Costs," p. 308.

9. Kenneth J. Arrow and Robert C. Lind, "Uncertainty and the Evaluation of Public Investment Decisions," American Economic Review (June 1970).

10. Kenneth J. Arrow and Robert C. Lind, "Reply," American Economic Review (March 1972).

11. S. Marglin, "The Opportunity Cost of Public Investment," Quarterly Journal of Economics vol. 77 (1963), pp. 274-289.

12. Mishan, Cost-Benefit Analysis.

13. E.J. Mishan, "Criteria for Public Investment: Some Simplifying Suggestions," Journal of Political Economy (September 1967); and E.J. Mishan, "A Proposed Normalization Procedure for Public Investment Criteria," Economic Journal vol. 77 (1967), pp. 777-796.

14. Mishan, Cost-Benefit Analysis.

15. Talbot Page, "Equitable Use of the Resource Base," Environment and Planning vol. 9 (1977), pp. 15-22; J. Rawls, A Theory of Justice (Cambridge, Mass., Harvard University Press, 1972); R.H. Solow, "Intergenerational Equity," Review of Economic Studies vol. 41 (Suppl., 1974), pp. 29-45, and R. Solow, "The Economics of Resources"; and R.H. Strotz, "Myopia and Inconsistency in Dynamic Utility Maximization," Review of Economic Studies vol. 23 (1955), pp. 165-180.

16. M. Freeman, "Why We Should Discount Intergenerational Effects," Futures vol. 9, no. 5 (October 1977), pp. 375-376.

17. J. Linerooth, "The Evaluation of Life-Saving: A Survey," (Laxenburg, Austria, International Institute for Applied Systems Analysis, 1975); and Bailey, chapter 6.

18. Talbot Page, R. Harris, and J. Bruser, "Removal of Carcinogens from Drinking Water: A Cost-Benefit Analysis," Social Science Working Paper No. 230 (Pasadena, California Institute of Technology, 1978), app. D.

19. P. Dasgupta, "Resource Depletion, Research and Development and the Social Rate of Discount," paper presented at conference at Resources for the Future, Washington, D.C., 1977.

20. Page, Conservation.

21. Rawls, A Theory of Justice.

22. J. Ferejohn and Talbot Page, "On the Foundations of Intertemporal Choice," American Journal of Agricultural Economics vol. 60, no. 2 (May 1978).

23. Ivy Broder, "Analysis of Proposed EPA Drinking Water Regulations," Washington, D.C., Council on Wage and Price Stability, September 1978.

24. Talbot Page, "A Generic View of Toxic Chemicals and Similar Risks," Ecology Law Quarterly vol. 7, no. 2 (1978), pp. 207-244.

25. Kenneth Arrow and Anthony C. Fisher, "Environmental Preservation, Uncertainty, and Irreversibility," Quarterly Journal of Economics vol. 88 (May 1974).

Chapter 5

UNCERTAINTY, INFORMATION, AND BENEFIT-COST EVALUATION OF CFC MANAGEMENT

Ralph C. d'Arge and V. Kerry Smith

5.1 Overview

Stratospheric pollution is one of a generic class of problems that require policy decisions. The distinguishing features of this class arise from: (1) the uncertainties associated with both the benefits and the costs of alternative policy options; (2) the time horizon required for information to become available that would resolve these uncertainties (through the appearance of the effects of the pollution); and (3) the potential for irreversible changes in the character of the environmental system as a result of progressive increases in the concentration of stratospheric pollution. Any evaluation of the benefits and costs of particular policies to control stratospheric pollution must recognize these dimensions of the problem.

This chapter reports the results of an attempt to identify how the economic criteria associated with conventional benefit-cost analyses might be modified to reflect some of the most important dimensions of the problems with stratospheric pollution. In order to develop these proposed revisions we must first consider the arguments leading to the conventional methods for treating uncertainty in benefit-cost analyses. Generally, these approaches assume that there is a mechanism that reduces the risk to individuals. This reduction can arise from two sources: (1) the

---

Authors' note: We thank Allen Kneese and several anonymous reviewers for helpful comments on drafts of this chapter.

public sector can act to balance the "net" risk experienced by the individual through a diversified set of projects each with potentially different impacts on individuals; or (2) each project's risks may be shared by all residents within the public sector's jurisdiction. In the latter case each member has a very small share of the aggregate risk associated with the action. Both perspectives lead to the conclusion that the public sector can accept a risk neutral position and rely on the expected benefits and costs in making decisions involving these types of uncertainties.

Unfortunately, stratospheric pollution can potentially generate an externality or public "bad" that is imposed on all individuals equally. Consequently it can imply an increasing aggregate risk premium as the number of affected individuals increases. Furthermore, individuals may not be able to "balance" the potential negative effects of stratospheric pollution with corresponding positive impacts of other public programs. Thus, the features of these uncertainties and the effects of stratospheric pollution are not necessarily consistent with a risk neutral posture for public decision making.

In order to consider what might be an appropriate perspective, we have proposed a simple optimal planning model that views the public sector as attempting to maximize the discounted expected net benefits resulting from its allocation decisions (for example, using resources to control stratospheric pollution). By dealing at the aggregate level with total benefits and costs we avoid those questions associated with the relationship between these aggregate net benefits and each individual's share in them. In principle, each individual's attitude toward the risks inherent in the decision could be considered to affect either his share of the benefits or the costs. Our analysis is intended to identify the additional considerations implied by the uncertainty that is associated with policies directed toward controlling stratospheric pollution. Our model introduces uncertainty by recognizing that the decision maker may not know whether the actions taken will lead to large losses. For example, a failure to regulate the emissions of chlorofluorocarbons may lead to a change in the earth's climate, and, for our example, we will assume this implies a substantial loss to the economy. Equally important, it may be impossible to undo the harm once it has occurred. In technical terms we might describe this case as an uncertain irreversibility.

However, there is an important wrinkle to the problem; society may be able to learn more of the nature of the relationship between chlorofluoromethanes and the earth's climate by permitting emissions. How should the public sector deal with these uncertainties? Risk cannot be diversified away over projects or by sharing. Once the change has occurred it is unlikely to be eliminated, yet there may be benefits associated with policies that permit some learning. Clearly, this would conform to one's intuition without formal models. In fact, the models recognize the benefits associated with learning as an explicit component of the decision criteria for allocation decisions designed to maximize the discounted, aggregate net benefits.

Unlike the irreversible investment problems described by Krutilla and Fisher, where all or nothing decisions are warranted, this problem recognizes the benefits from learning. Thus optimal policies must consider both the prospects for learning and the costs which accompany the activity levels (that is, rates of emission) required for the learning to take place.

This is a potentially important amendment to the conventional practice in treating what have been considered uncertain and irreversible actions. Nonetheless, one might reasonably ask what are the ranges of uncertainty? Would the treatment of uncertainty ever affect the nature of the decisions made within a benefit-cost framework? In order to address these issues in specific terms, we considered the treatment of two types of uncertainty in the information necessary to evaluate the rationale for regulating the emissions of chlorofluoromethanes.

The first might be described as estimation uncertainty. Given the rather crude models which had to be used to estimate the costs of a climate change (that might accompany an increase in chlorofluoromethanes), how would the selection of an estimator other than a simple average of the measured impacts affect our sectoral estimates of these costs? Using estimation rules derived to reduce the uncertainty due to that estimation (the so-called James-Stein rule based on a quadratic loss function), we found that the economic costs of a 1°C change in mean annual temperature would be one-half the size of conventional estimates. While the overall estimates of environmental costs due to current emission levels for F-11 and F-12 were not as sensitive as these environmental costs estimates for a climate change, the disparity in the estimates for individual components of these aggregates was substantial. Indeed, consideration of a subset of the categories reported in the analysis of F-11 and F-12 might well lead to different conclusions purely as a result of estimating procedure used to organize these environmental costs.

A second source of uncertainty considered in our analysis arises from the assumptions required on overall emissions of F-11 and F-12 by all nations. In this case the treatment of uncertainty in scenario design affects the judgment that would be based on the desirability of emission controls.

Thus on both analytical and empirical grounds, the treatment of uncertainty and the recognition of the role of learning from the results of allowing some emissions are likely to be important dimensions of the policy analyses of most forms of stratospheric pollution. In what follows we will develop more completely the technical arguments underlying this summary.

5.2. Stratospheric Pollution, Uncertainty, and Public Decision Criteria

The economic analysis of stratospheric pollution offers few methodological differences from the most general treatments of pollution problems. The stratosphere provides an excellent example of a common property resource. If we seek to manage the use of this resource in conformity with efficiency criteria, there is a need to regulate emissions. In simple economic terms, emissions must be controlled up to the point where the costs of regulation at the margin equal the social damages (at the margin) resulting from the selected level of emissions. Since the social damages may extend over a long time horizon, particularly careful attention should be given to the methods used in calculating social damages and to the equity issues which accompany our efficient policies. These considerations may be especially important with respect to their effects upon members of future generations. (See Page's discussion,[1] and chapter 4 of this volume). Of course, beyond recognizing equity effects in the decision process and attempting to impose limits on the "equity costs" associated with meeting efficiency goals, it is difficult to incorporate those items unambiguously into the formal analysis of emissions control policies.[2]

In order to apply our simple notion for efficient regulation of stratospheric emissions accurately, it is necessary to have three well-defined relationships: (1) a valid representation of the relationship between the pollution emissions and the corresponding atmospheric effects, (2) a relationship between the costs of emissions control and the corresponding levels of discharges, and (3) a complete and concise damage function relating societal damages at each point in time to the various patterns of pollution emissions, both over space and time.

The central difficulty with applying this simple rule for regulation of stratospheric pollution is that none of these relationships are known with a reasonable degree of certainty, either in terms of the sign or the magnitude of the individual impacts. Of course, there are an array of estimates of the potential impacts. The best available estimates of such impacts suggest that, on balance, they will be costly to society. Moreover, they are based on only about 20 percent of the world economy.[3] Similar difficulties can be cited with respect to the evaluation of the direct costs and effectiveness of efforts to control emissions.

These limitations imply that the treatment of uncertainty has special importance for decision making with respect to stratospheric pollution. Not only is there substantial uncertainty as to the relationships required for efficient control decisions, but in addition, where there is an understanding of these relations, precise estimates are not available and thereby compound the types of uncertainties associated with decision making.

The major attributes of the stratospheric pollution problem affecting our development of decision criteria for regulatory policy might then be summarized as follows:

1. There are large uncertainties (both in sign and magnitude) in the effects of various levels of emission in the stratosphere. Some changes may involve irreversibilities in the natural environment, although no substantive evidence of this is now available.

2. There are extremely large uncertainties in the translation of ozone layer and climate changes into quantitative biological effects.

3. There are very high uncertainties as to how social communities and the economic system adjust to large-scale changes in ultraviolet radiation and climate or even to small shifts in the biosphere.

4. None of these substantial uncertainties are likely to be reduced to accurate estimates of effects in less than one or two decades.

The decision problem is thus one of analyzing a class of actions in which the transfer function between cause and effect is subject to a high degree of uncertainty at the time of the decision, but where a process of learning over time may be anticipated. While the decision process is sequential, and decisions made during the next decade can be continued or revoked thereafter, it is entirely possible that their effects will not be reversible. Moreover, the biologic and social consequences of these actions may not be observed for one or more human generations after they are undertaken. Thus, most of the economic effects of a gradual buildup of chlorofluorocarbons (CFCs) in the stratosphere would not occur for sixty to one-hundred years (see d'Arge for more details[4]). The monitoring and emissions control costs range from low-cost current techniques (such as fuel desulfurization) to high-cost methods (that is, processing CO_2) and involve detecting subtle changes.

Tables 5-1 and 5-2 provide a partial listing of potential man-induced stratospheric pollutants along with their possible effects and methods of control. These tables illustrate the uncertainties inherent in our problem, given recent downward revisions of expected effects of NO_x, and upward revisions of expected effects of CFCs (see chapter 2). The global economic effects would certainly be different depending on whether the temperature change was positive or negative. Equally important, if not more so, the _interactive_ effects of multiple pollutants are not well understood. The effect of any one pollutant may be augmented or reduced by others, and the damages associated with it thereby altered. Thus, benefit-cost analyses completed for any single pollutant may be seriously biased. Finally, many of the physical and biological effects of the multiple of potential stratospheric pollutants are the subject of vague conjecture.

With this background, it is important to identify the types of projects involved in social intervention (that is, public investment decisions) and their individual implications for the treatment of risk and uncertainty. It is useful to distinguish three broad categories of goods or services subject to public provision: (1) _private goods_, such as electric power, irrigation, highways, and so forth; (2) _intratemporal public_

Table 5-1. Potential stratospheric Pollutants

Source	Identified pollutant	Quantity	Potential effect		
			Physical (surface effect)	Biological (surface effect)	Economic (global)
Industrial activity	CO_2 Waste heat	Current production levels extended over 100 years	+2°C (WHO) +2-4°C (urban local)	(?)	(?)
Aircraft	NO_x	2020 second generation SST's	-.007°C (CIAP)	+UV-B	$4 x 10^9 [a]
Chlorofluorocarbons	CFCs	1973 production effects in 50 years	-.3 to +.1°C (NAS)	+UV-B	$.6-6 x 10^9 [b]
Nitrogen fertilizers		1973 production levels effect over 60 years		+UV-B	$.6 x 10^9 [c]
Weapons tests	Radioactive particles	(?)	(?)	+C°. + 1	(?)
Industrial energy activity production	(Radon-222)	1.7 x 10^4 Cl/yr. in 2000 from coal combustion ash pile release (but one source of Radon-222)		rem est. lung exposure dose	
Electromagnetic spectrum	X-rays; gamma rays	(?)	(?)	+C°	(?)
Industrial or agricultural activity	Particulates		+/-C°(?)	+UV-B	(?)

Table 5-1 (continued)

Source	Identified pollutant	Quantity	Potential effect: Physical (surface effect)	Potential effect: Biological (surface effect)	Potential effect: Economic (global)
Industrial activity	Trace gases CLO_x HO_x	1 to 10 parts in 10^9	+/-C°(?)	+UV-B	(?)
Aircraft (contrails)	Water vapor	Emissions from 10"kg of fuel 1.3 x 10"kg.	+/-C°(?)	+UV-B (?)	(?)
		2×10^8kg.	+/-C°(?)	+UV-B (?)	(?)
Space shuttle	UCI Al_2O_1	CI produced only 1% of F-11 F-12 prod. rates in 1973 with 60 flights	negl. (NAS)	negl. (NAS)	

Note: (?) indicates uncertain quantity or effect.

[a] (CIAP) Climatic Impact Assessment Program, Economic and Social Measures of Prologic and Climatic Change vol. 6 (Washington, D.C., U.S. Department of Transportation, September 1975).

[b] R.C. d'Arge, J. Harrington and L.S. Eubanks, "Benefit-Cost Analyses for Regulating Emissions of Fluorocarbons 11 and 12," Final report to the U.S. Environmental Protection Agency (Washington, D.C., (EPA, February 1976).

[c] National Academy of Sciences, Protection Against Depletion of Stratospheric Ozone by Chlorofluorocarbons (Washington, D.C., NAS, 1979).

Table 5-2. Preliminary Estimates of Control Costs for Suspected Stratospheric Pollutants

Source	Pollutants	Control type	Cost of control
Industrial activity	CO_2	Reforestation & revegetation Processing	(?) Very high
Industrial activity	Waste heat	Conversion efficiencies	(?)
Aircraft	NO_x	Engine redesign Lower flying aircraft	$\$.8 \times 10^9$ moderate[a]
Chlorofluorocarbons	CFCs	Product removal Substitutes	$\$3 - 107 \times 10^9$[b] (?)
Nitrogen fertilizers	NO_x	Substitutes	(?)
Weapons tests	Radioactive	International agreements Removal processes	Very high

[a]Global

[b]United States only

Sources: Cost of control for aircraft is based on data from Climatic Impact Assessment Program, Economic and Social Measures of Prologic and Climatic Change vol. 6 (Washington, D.C., U.S. Department of Transportation, September 1975); cost of control for CFC emissions is based on data from Ralph C. d'Arge, J. Barrington, and L.S. Eubanks, "Benefit-Cost Analyses for Regulating Emissions of Fluorocarbons 11 and 12," Final Report to U.S. Environmental Protection Agency, #68-01-1918 (Washington, D.C., EPA, February 1976); and cost of control for industrial activity is based on data from World Meteorological Organization, World Climate Conference, Proceedings (Geneva, Switzerland, 1979).

goods; and (3) intertemporal public goods. The manner in which uncertainty enters those problems associated with decisions of each type of good can be quite different. Thus, it will be necessary to distinguish the implications of uncertainty for public investment decisions according to these differences.

Pioneering studies of public versus private sector investment considered the implications of different levels of uncertainty for interest and the expected yield rates within the private sector.[5] Given the private opportunity-cost rationale for public investment decisions, there nonetheless remains a practical question which arises in the selection of the appropriate discount rate for public investments. For the market offers not one rate of interest, but many, each with a different degree of risk. Accordingly, it becomes difficult to establish a single rate as the one which is appropriate for the particular investment decision. Krutilla and Eckstein resolved this issue by considering the likely mix of opportunity uses for the project funds,[6] and constructed a weighted average opportunity cost of capital to be used as the discount rate in the project evaluation. The concern with establishing comparability between public investment projects and their private alternatives yielded the conclusion that there may be a rationale for the differential treatment of risk between the public and private sectors. The literature which directly addresses the issue of the appropriate treatment of risk for decisions involving public investments spans a considerable range, often with quite divergent prescriptions. On a heuristic level, it would seem that the selection of the private opportunity-cost approach to evaluating public investments would imply that one also accept a risk-averse posture for the public sector in evaluating its potential investment decisions. That is, assuming that the individual members of the society undertaking particular public investment decision are each risk averse, then it is tempting to conclude that the appropriate attitude for the public sector in making such decisions is the same as its constituency. However, this seemingly plausible argument can be easily contradicted by examination of a variety of special cases. The first of these counter-examples relates to arguments developed by Samuelson,[7] which suggest that a risk-neutral attitude on the part of the public sector is appropriate. The rationale for his suggestion is based on the assumption that the public sector engages in a large number of projects with a potentially diverse array of outcomes. Since there are likely to be losses and gains associated with each of these projects, a loss in one public project may well be canceled by gains in others. So long as all individuals experience an approximately equal share of the impacts of all public projects, the net outcome for each individual may well be small. His argument can be considered as a type of insurance which arises with a large number of projects and no bias in the direction of influence of the risks. Under these circumstances, the discount rate would not be raised to include a premium for risk.

An alternative approach leading to a comparable conclusion can be found in the work of Arrow and Lind.[8] Here, the authors argue that risk-spreading can occur because a single public project may influence a large number of taxpaying beneficiaries, each of whom has a small share in the net returns associated with the project. By assumption, these returns include a random element. The implications of risk can be determined by the payment to avoid risk under alternative assumptions on number of affected individuals and character of good or service yielding the returns. As the number of individuals affected by the project grows, given a constant aggregate risk premium, the individual share of the risk premium will approach zero. Consequently, the Arrow-Lind analysis suggests that public investment decisions should be based on a risk neutral posture.

Both of these arguments are based on an ability to spread risks--either in the form of a diversity of projects, as under Samuelson's approach,[9] or a diversity of individuals, as under the Arrow-Lind model. We shall argue that both frameworks have important limitations for the problems of managing stratospheric pollution. Since these decisions use information that necessarily involves substantial uncertainty, the appropriate public sector attitudes will be especially important to the policies adopted.

In general cases, there are a number of aspects of actual public investment decisions that serve to limit the applicability of each framework. For example, it may well be that the same individuals do not bear the gains and losses for all projects. The decision mechanism may be such as to violate the "safety in numbers of projects" which is so important to

Samuelson's argument. Equally important in the case of the Arrow-Lind theorem, the potential for risk-spreading through constant aggregate risk premium with increasing numbers of individuals may be limited. In the special case of stratospheric pollution, with potential effects on global temperature and ultraviolet radiation, societal risk may be the summation of individual risks. Therefore, each individual's share would not decline with increases in the number of affected parties.

The remaining formal decision rules for the treatment of uncertainty each relate to the special case of actions which involve irreversibility. The first of these was developed by Arrow and Fisher,[10] and is based on a simple two-period model. Their arguments suggest that in the presence of uncertainty and irreversibility, it is desirable to reduce the benefits associated with irreversible actions because they reduce the options available to society. That is, given the presence of uncertainty, their argument suggests that deferring a decision which is irreversible provides a quasi-option value in the form of the additional information gained about the consequences of the action with any time delay. Thus, they suggest that "something of the 'feel' of risk aversion is produced by a restriction on reversibility."[11] Their argument is quite similar to a Bayesian approach to decision making since they consider the information value lost from an early decision.

The second approach to the treatment of irreversibility and uncertainty stems from the early work of Ciriacy-Wantrup.[12] Here, it was deemed essential to maintain a safe, minimum standard of preserved natural and environmental resources. This prescription called for the use of a min-max strategy in decision making. If projects are defined in terms of the losses associated with each action, then his criterion would call for the selection of that action which picks the minimum of the maximum losses that would be realized under any particular action. To illustrate, suppose that society must select one of two actions--to preserve or to pollute the stratosphere. Under each action, there may well be an array of outcomes possible with an associated probability distribution. Under the Ciriacy-Wantrup criterion, the decision maker would search all of the possible outcomes evaluating the losses to society associated with each. Further, he would consider the maximum loss associated with a particular action. This maximum loss becomes the deciding factor irrespective of the probability distributions over the outcomes. We select that action which has the smallest of the maximum losses.

It should be noted that both the Arrow-Fisher argument and the Ciriacy-Wantrup approach are directed toward environmental decisions which involve irreversibility. They are not necessarily intended for those cases in which there is some possibility of reversing the action. Moreover, their application has been to those cases involving all-or-nothing decisions, such as the preservation or development of natural environments.

In the next section of this paper, we develop a simple amendment to the models which have been used to consider irreversible investments,[13] that explicitly introduces uncertainty as to the reversibility of the action under consideration. This modification illustrates the importance of their discrete specification of the decision process.

Each of the three approaches to evaluating the treatment of uncertainty in the process of public decision making has been based on a fairly specific objective function and specific definition of the nature of the activities involved. Within the practices of benefit-cost analysis, a variety of somewhat _ad hoc_ rules have been suggested with an apparently less direct association to economic theory. We shall cite three of these rules to complete this general discussion, but we will not develop any of them in detail. In some cases, the rules may be considered as practical representations of the concepts that emerge from one of the three approaches cited earlier, while, in others, they seem to be rather arbitrary amendments to the practices of benefit-cost analysis which are argued to "move in the right direction" in the treatment of uncertainty. The first of these suggests that the threshold decision ratio of benefits to costs be adjusted up in the presence of uncertainty. Thus, if the benefit-cost ratio required for an acceptable project were 1.25, then this procedure would call for the adjustment of this ratio up by some fixed amount, for example, 0.25. The new ratio would then be applied to those project decisions involving uncertainty.

It is clear from its definition that this rule raises a number of questions in its applications. For example, there is no acceptable mechanism for developing the baseline ratio of benefits to costs. Moreover, it is difficult to conceive of a project that does not involve some degree of uncertainty in the assessment of benefits and costs. Thus, some degree of judgment must be made as to what is appreciable uncertainty and what is not. Of course, one might consider a schedule of uncertainties and associated arbitrary constants to be added to the benefit-cost ratio. This practice identifies an additional shortcoming in the suggestion, which relates to the definition of the constant to be added to the benefit-cost ratio. There appears to be no firm basis upon which to resolve any of these issues.

The second ad hoc approach calls for an arbitrary truncation of the planning period for the project in evaluating benefits. This procedure serves to reduce the contributions made to the discounted benefits or costs from their respective values arising in future time periods. For public decisions involving intertemporal public goods, it may have especially significant impacts. Presumably, it is based on the argument that benefits accruing at future dates are more uncertain than those in the present (as are the costs). There also would seem to be an implicit judgment that most of the costs are incurred initially while the benefits accrue largely over longer time periods. Thus, this practice would serve to reduce the benefit-cost ratio in ways which would differentially penalize those projects where the uncertainty arose from their benefits. In some respects, the outcome is similar to the method which increased the threshold benefit-cost ratio required for acceptability when these ratios are derived from present value criteria. There is also a similarity in the lack of a clear rationale for the truncation of the planning period.

The last approach to treating uncertainty with practical decision rules can be derived directly from a formal treatment of risk and uncertainty (see, for example, Dasgupta and Heal[14]). It calls for the addition of a premium to the existing discount rates to reflect the presence of uncertainty. This premium would also serve to reduce the benefit-cost ratio (assuming the use of present value criteria and that benefits accrue largely over time, whereas costs are largely associated with the early periods). Once again, the magnitude of the premium in the final analysis becomes a matter of judgment. Thus, in each case, we find that these simple amendments to the existing practice of benefit-cost analysis offer *ad hoc* approaches to incorporating uncertainty, and may be subject to pronounced biases when dealing with projects with long time spans.

Given the limited rationale for these *ad hoc* adjustments to the current practices of benefit-cost analysis, we will focus the balance of our discussion on the strengths and weaknesses of the Arrow-Lind and Arrow-Fisher models for treating the issues introduced by risk. If we consider the first of these frameworks, the point noted briefly earlier and discussed in some detail by Fisher is of central importance to the applicability of their results generally to those public decisions involving intertemporal public goods (or externalities), and specifically, to stratospheric pollution.[15] When each individual's share in the net benefits of a particular project does not decline as the number of individuals involved increases, then the project's associated risk cost also does not vanish. Thus, the use of a riskless discount rate may *not* be warranted.

There are additional aspects in which the theorem is limited. Some of the more interesting of these have been discussed by Estelle James in a comment on the Arrow-Lind paper.[16] She indicated that it was useful to separate the characteristics of the risk into a *pure scale* effect and a *relative dispersion* effect. The former involves increases in the spread of values for the random component in net benefits, and is the source of the Arrow-Lind argument for a low social risk premium; the latter involves the shape of the probability distribution associated with the random elements, and corresponds to the arguments we attributed to Samuelson. James argued:

> The effect on dispersion of risk pooling may lead the government to *accept a group* of projects which were *individually rejected*. On the other hand, the scale effect may lead the government to *reject an entire group* of projects which were *accepted individually* due to risk spreading. This inconsistency is avoided and piecemeal and global decisions will coincide when the scale effect (risk spreading) dominates in the evaluation of individual projects and the dispersion (risk pooling) dominates in the group evaluation.[17]

Thus, the James argument offers an opportunity to integrate the two positions. It also suggests the mechanism by which apparently conflicting conclusions of the two approaches can be resolved through the character of the risks associated with the individual projects. Finally, it provides an important commentary on the Arrow-Lind result by suggesting that, not only is the decision rule limited in its applicability to goods with public attributes (as Fisher has argued), but it may also lead to inconsistent decisions when decisions on environmental control are viewed as individual cases versus as an investment portfolio. What is more important for our purposes is the recognition implicit in James's work of the implications of the character of the probability distribution for optimal decision making.

In the final analysis, the treatment of uncertainty and risk in public decision making will involve some measure of judgment. The issue which ought to concern the construction of analytical models to treat these difficulties should be the identification of attributes of the particular process which have direct implications for the properties of alternative decision rules. Too frequently, the literature in this area has tended to adopt either a prescription which completely ignores the character of the probability distribution for the random elements associated with a particular action, as in the case of the Ciriacy-Wantrup safe minimum standard approach, or focuses on a measure of the location parameter for the associated distribution of utilities or net benefits resulting from the action. The James note brings to our attention the sensitivity of these expected value functions to the shape of the underlying probability distributions. It is this issue which we address in the next section.

5.3 The Treatment of Uncertainty in Benefit-Cost Analysis of Stratospheric Pollution Control

The discussion in the previous section suggests that the uncertainty associated with decisions involving the regulation of stratospheric pollution arises from two sources. The first of these is from our limited knowledge of the nature of the transfer function between the cause and effect of emissions leading to increased stratospheric pollution and its attendant effects on the climate. The second arises from difficulty in estimating the social damages associated with the unregulated use of the stratosphere that is implied by the failure to control emissions of pollutants capable of drifting to it. We have argued, after reviewing the literature on the treatment of uncertainty in public investment decisions, that most arguments in the literature on benefit-cost analysis have only limited applicability to decisions involving regulating the use of the stratosphere. These arguments were based largely on the first source of uncertainty. The second, while also important to the methods used in developing decision rules for benefit-cost analysis, seems to be a problem that is present whenever estimates of benefits and costs must be constructed with very limited data.

Our overall evaluations suggested that greater attention must be devoted to underlying probability distributions and to the potential for irreversibility in designing decision criteria for those public actions involved with the stratosphere. In order to organize these suggestions more formally, we have outlined below a simple analytical model of the decision process involved in selecting the optimal level of stratospheric pollution, given that these actions <u>may be</u> irreversible. Our model draws on the work of Kemp and Cropper for exhaustible resources, and, more recently, Smith and Smith and Krutilla for irreversible investment decisions.[18] This framework is not intended to completely describe the problems involved. Rather, we use it to guide the nature of the amendments to conventional benefit-cost practices in the presence of uncertainty in order to reflect some of the special features of the class of problems addressed here.

Consider the problem when society can select the level of the stock of accumulated stratospheric pollution, $S(t)$, by its decisions on goods and services produced and consumed (x). The benefit function to society, $B(x)$, is in terms of the goods and services consumed. These goods, in turn, are related to additions to the stock of stratospheric pollution, S, as in Equation (1):

$$\delta = S(t) = a\,x(t), \tag{1}$$

where $S(t) = \frac{dS}{dt}$

and $a = \text{constant}$

The costs of producing $x(t)$ are given by $C(x(t))$ and, further, we assume that there is the prospect with accumulation of stratospheric pollution that society will incur substantial social costs, L, that will be a function of S, thus $L = L(S(t))$. Unfortunately, these will depend upon whether the stock exceeds an unknown threshold, $\overline{S}$, which we will assume is distributed over the interval (o, ∞) with $g(\overline{S})$ the probability density function. We will also assume that the parameters of g( ), that is, scale and location are related to $\delta(t)$ to reflect the potential effects of learning. Finally, the stock will be assumed to be incapable of reduction, so that decisions are physically irreversible and may lead to large irreversible losses, L(t).

It is possible to write the probability of avoiding social costs as in Equation (2).

$$P(\overline{S} \leq S(t)) = \int_0^{S(t)} g(\overline{S};\ \delta(t))d\overline{S} = \Psi(S(t);\ \delta(t)) \tag{2}$$

We will assume that society makes its choices so as to maximize the discounted expected net benefits from consumption of x(t) recognizing the possible effects. Equation (3) defines the relevant objective function. To simplify notation, we have incorporated the constant, a, into the notation for B( ) and C( ). Thus, $\overline{B}(\delta(t)) = B(\delta(t)/a)$; and $\overline{C}(\delta(t)) = C(\delta(t)/a)$.

$$G = \int_0^{\infty} e^{-\rho t}(1 - \Psi(S(t)\ ;\ \delta(t))[\overline{B}(\delta(t)) - \overline{C}(\delta(t))]dt$$

$$+ \int_0^{\infty} e^{-\rho t}\Psi(S(t)\ ;\ \delta(t)\ ;\ [\overline{B}(\delta(t)) - \overline{C}(\delta(t)) - L(S(t))]dt \tag{3}$$

where:

ρ = discount rate

The first term defines the net benefits when the stratosphere is not irreversibly modified by consumption patterns and the second defines net benefits when it is. This function is maximized subject to the constraint that $\delta(t) \geq 0$. The current value Hamiltonian is given in Equation (4).

$$H_o = (1 - \Psi(S(t)\ ;\ \delta(t))\ [\overline{B}(\delta(t)) - \overline{C}(\delta(t))]$$

$$+ \Psi(S(t)\ ;\ \delta(t))[\overline{B}(\delta(t)) - \overline{C}(\delta(t)) - L(S(t))] + \alpha(t) \cdot \delta(t) \tag{4}$$

Given the benefit and cost functions are well behaved, the necessary conditions for an interior solution are given by:

$$\dot{\alpha}(t) = \alpha(t)\rho - \frac{\partial H_o}{\partial S} \tag{5}$$

or

$$\dot{\alpha}(t) = \alpha(t)\rho - [-\Psi_S L(t) - \Psi L_S(S(t)]$$

$$\frac{\partial H}{\partial \delta} = \overline{B}_\delta - \overline{C}_\delta - \Psi_\delta L(t) + \alpha(t) = 0 \qquad (6)$$

These conditions must be interpreted cautiously, because society cannot reverse its actions (that is, $\delta(t) \geq 0$). Thus, following Arrow's original arguments,[19] and the extension of Arrow and Kurz,[20] activity can only take place in open intervals, where $\alpha(t) \geq B_\delta - C_\delta - \Psi_\delta L(S(t))$. (This model implicitly assumes that investment can take place instantaneously. Capital can be accumulated in blocks virtually "overnight." The open interval is simply the period in which expansion is warranted based on the net shadow value of investment.) The discrete, all-or-nothing (in the first period) character of the investment problem with certain irreversibility and linear cost functions as developed in Krutilla and Fisher is not present in this problem.[21] Rather, during open intervals, we can identify benefits from learning. Combining Equations (5) and (6) for such cases, we have:

$$\overline{B}_\delta = \overline{C}_\delta + \Psi_\delta L(S(t)) + \frac{(\Psi_S L(S(t)) + \Psi L_S(S(t))}{\rho} \qquad (7)$$

Equation (7) provides the amendments to our simple rule for regulating emissions to the stratosphere. We identify their net benefits $(\overline{B}_\delta - \overline{C}_\delta)$ and their costs in terms of the monetary value of the increased likelihood of social costs (that is, $\Psi_S L(S(t))/\rho$ and the increased social costs at each probability level (that is, $\Psi L_S(S(t))/\rho$). We have not directly incorporated the regulatory costs. Only reduced activity (that is, $x(t)$) can control the growth in the stock. However, we have also identified a new element in the analysis with the term $\Psi_\delta L(S(t))$. If we interpret the effect of $\delta(t)$ on Ψ as "learning" and assume a reduced probability of social costs accompanying learning, we have an important addition to existing practice, namely, it may be optimal to consider permitting some emissions so that society can learn more of the underlying mechanisms. Accordingly, in evaluating the treatment of uncertainty in the regulation of emissions to the stratosphere, some recognition should be given to the resolution of some of the inherent uncertainty regarding the physical environment that accompanies permitting the emissions. Of course, this must be balanced against the costs associated with increasing the stock of stratospheric pollutants.

The second aspect of the uncertainty associated with these decisions and costs involved (including those from learning) must be estimated. Once it is recognized that these estimates also provide a source of uncertainty to the ultimate decision process, it is important to consider the risks associated with the various estimation strategies one might wish to use. In particular, in a fashion analogous to the case of uncertainty in the inherent physical mechanisms and the need to consider more fully the probability distributions involved rather than focusing on their central tendencies, estimators may also be selected with similar objectives. This is particularly important to problems where several estimates for constituents of more aggregate sectors must be constructed to evaluate the social costs of stratospheric pollution.

In the mid-fifties, Charles Stein was able to prove that there were estimation rules to evaluate unknown means in those cases with more than two means to be estimated which yield smaller total risk than the individual sample average for each group.[22] If we define risk in terms of a quadratic loss function, James and Stein were later able to suggest that a procedure which makes a preliminary guess that all unobservable means are near the grand mean of these averages and which then adjusts the estimate for each individual average away from the grand mean,[23] based on what is usually designated as the shrinking factor, entails a smaller risk than use of the grand mean. That is, the risk function for these estimates is smaller than the risk function associated with using the observed averages. Since use of these methods offers a more explicit recognition of the role of uncertainty for estimation strategy, in the next section we review the materials used to evaluate the effects of -1°C change in global temperature and the effects of a ban on the production of fluorocarbons F-11 and F-12 using both conventional average and James-Stein estimates of the environmental costs associated with their accumulation in the stratosphere.

5.4 Illustrative Estimates of the Economic Impacts of Climate Change and Chlorofluoromethanes

It is obvious that climate is one of the underlying determinants in economic systems. Human settlements, location and density, value of agriculture, physical health, and even recreational pursuits depend on climate. Increased temperatures may mean decreased heating costs, counteracted by increased air conditioning costs. Lower temperatures may mean a decrease in productivity of agriculture. Ozone depletion may increase skin cancer incidence, pink eye in cattle, and alter insect infestation timing. In what follows, we review: (1) the Climatic Impact Assessment Program (CIAP) estimates of 1°C change in the mean annual global temperature in comparison with James-Stein estimates in each category, and (2) the implications of uncertainty in both the physical environment and the estimates for a benefit-cost evaluation of the regulation of chlorofluoromethanes.

5.4.1 Temperature Change

Consider the case of a 1°C change in the mean annual global temperature which is assumed to occur over a thirty-year interval from 1990 to 2020. (This corresponds fairly well to a CFC climate-warming effect of 0.6°, noted by Forziati in chapter 2, plus some additional warming from the CO_2 greenhouse effect.) The relevant time interval for an actual case would, of course, be a function of the type (or types) of pollutants introduced into the stratosphere and their respective injection rates. Table 5-3 summarizes the estimates prepared under the CIAP study by sector and investigator. The third column identifies the original estimator and the fourth our James-Stein estimator using the sub-sector estimates jointly to derive the sectoral aggregate. The James-Stein estimator, $\hat{y}_{js}$, was estimated using:

$$\hat{y}_{js} = \overline{y} + c(y-\overline{y})$$

where $\overline{y}$ = grand mean of means

y = individual mean for a given sample

c = shrinking factor defined as $c = 1 - \dfrac{(k-3)\alpha^2}{\sum_i (y_i-\overline{y})^2}$

Table 5-3. Alternative Estimates of the Economic Costs of Climatic and Biological Changes (1°C Change in Mean Annual Temperature, No Change in Precipitation, 5 Percent Interest Rate Assumed)

Impact studied	Original Investigator(s)	Original estimate	James-Stein estimate[a]
Corn production	Schulze, Ben-David	-21	211
Cotton production	Schulze, Ben-David	11	85
Wheat production	Mayo, McMillan	92	120
Forest production	Schrueder	2,312	1,075
Marine resources	Bell	1,431	696
Health impacts (U.S. only, excluding skin cancer)	Anderson, Lave, Pauly	2,368	1,106
Urban impacts (wages, U.S. only)	Hoch	3,667	1,657
Total		9,860	4,950

Source: Ralph d'Arge, ed., Economic and Social Measures of Biologic and Climatic Change, CIAP Monograph 6 (Washington, D.C., U.S. Department of Transportation, September 1975).

Note: Cost or benefit measures were developed utilizing the concept of consumer surplus, where applicable, and direct changes in costs or savings where consumer surplus appeared to be not applicable.

[a]Annualized cost--1974 (in millions of U.S. dollars, negative sign denotes benefit).

k = number of samples

α^2 = population variance

A simplified exposition of the estimator with examples is in Efron and Morris.[24] Each estimate was developed independently so there are no corrections for sectoral or categorical interdependence such as the effect of increasing scarcity of rice on the demand for wheat; and there is no consideration of possible future technological changes.

For the categories of economic activity reported in table 5-3, there is a substantial economic cost associated with a climatic change. However, from the perspective of this paper, what seems of greater interest is the impact of estimation strategy on these costs. The total using conventional methods amounts to about $9.9 billion on an annual basis, whereas using the James-Stein estimates, it is half the figure, at about $4.95 billion.

5.4.2 Analysis of Chlorofluorocarbons

CFCs, for the most part, are not purchased directly by households, but are utilized as inputs to produce consumer products or services. As a result, the observed demand relationships for CFCs relate indirectly to consumer valuation through demand for products utilizing CFCs. Given rather specific assumptions, the consumer surplus losses resulting from banning CFCs can be equivalently measured using either final goods demand relationships or the derived demand curves for the CFCs themselves. More specifically, Anderson has recently offered a simple derivation of the relationship between the consumer surplus associated with factor-price changes using factor and product-demand functions.[25] His model assumes output is produced by combining two factors, and that the technology can be characterized by a linear, homogeneous function. For the present purposes, the issue is not one of measuring the effects of a price change where a factor price change can be equated to an equivalent change in the product price and evaluation can then be conducted using either factor or product demands. Rather, it concerns the evaluation of the effects of eliminating (or sharply reducing) use of one factor input. Clearly, in the case of complete elimination, if the factor is essential to the production of the final product (that is, given a two-factor model, x_1 and x_2, and a production function $F(x_1, x_2)$, then x_1 is essential if $F(0, x_2) = 0$ with $x_2 > 0$), then the losses will be equivalent, even though there is substitution between factors for interior points in the production space. Unfortunately, observed data on prices and quantities sold historically may not adequately conform to these required assumptions. In order to derive bounds for the estimated "consumer surplus," both derived surplus and consumer surplus losses were estimated.

Measures of "derived surplus" loss for restrictions in CFC production were developed for F-11 and F-12, along with measures or consumer surplus loss for the major final products using these CFCs in their production.

Other CFCs were not examined, given their relatively minor role (see chapters 2 and 3). Included in the list of final products were refrigerators, aerosol deodorants, auto air conditioners, polyurethane foam mattresses, and mobile vehicle refrigeration systems. According to the IMOS report, these products accounted for about 90 percent of U.S. utilization of major CFCs and more than 98 percent of F-11 and F-12 use in 1972.[26]

Table 5-4 reports the estimates of the present value of derived surplus for F-11 and F-12, and the consumer surplus for major consumer products using F-11 and F-12 in their production. The "derived surplus" estimates amount to about $3 billion. By contrast, the consumer surplus for the major products using them amounts to more than $84 billion. Of course, the latter serves to provide an upper bound for the actual value of consumer surplus. If no substitutes existed for producing the final product without F-11 and F-12, or if CFCs are an essential factor input, then the appropriate measure of economic loss would be the sum of consumer surplus losses in the final markets impacted. However, when there are CFCs that are not essential, and if there do exist substitution possibilities, one must use either the equivalent price change in the final goods market and its demand function or the surplus estimated using the derived demand function.

It has been hypothesized that F-11 and F-12 emissions will induce two global effects:

1. Reduction in stratospheric ozone and increases in UV-B radiation at the earth's surface.

2. A slight rise in surface temperature due to the greenhouse effect.

Both of these global effects, if they occurred at a significant level, would have large-scale ramifications on biological life and, thereby, on the U.S. and other nation's economies. It would seem to be impossible, empirically, to estimate the thousands of interrelated impacts of changes in surface microclimates. In the partial analysis which was undertaken, costs and benefits are estimated for some major sectors of the U.S. economy from ozone depletion or enhancement, and from slight, long-run increases in surface temperature. Given the uncertainty associated with these estimates in table 5-5, we report both the original estimates (largely average fig-

Table 5-4. Estimates of Consumer Surplus and Derived Surplus for Selected Products, United States, 1971 Dollars

(present value 1974, 5 percent discount rate in million dollars)

Commodity	Estimated 1973 expenditures by commodity[a]	Consumer surplus or derived surplus (present value)
F-11	40	2,201
F-12	96	740
Refrigerators	1,386	39,727
Mobile vehicle refrigerator systems	97	3,349
Auto air conditioners	489	36,473
Polyurethane foam mattresses	39	1,007
Aerosol hair care products	134	2,679
Aerosol deodorants and antiperspirants	458	9,165
Aerosol perfumes, cosmetics, and toiletries	346	6,919
Aerosol room deodorizers	60	1,200
Aerosol pharmaceuticals	38	752
Aerosol shaving products	40	795
Aerosol insecticides	258	5,168
Total minus F-11 and F-12	3,345	107,234
Total for aerosol products	1,334	26,678
Total for nonaerosol products	2,011	80,556

Note: Area under the derived curve less equilibrium purchases in 1973.

[a]Expenditures are estimated from the estimated demand relationships rather than actual data since actual price may deviate from predicted prices as given by the estimated demand relationship.

Table 5-5. A Comparison of the Estimates of Environmental Costs by Category Due to Current Levels of F-11 and F-12 Emissions, in the United States, 1974 into Perpetuity
(millions of 1971 dollars)

Category of impact	Cost of benefit[a]	
	Conventional estimate	James-Stein estimate
1. Ozone depletion		
Nonmelanoma skin cancer[b]	206	228
Materials weathering (polymeric materials)	569	500
Fish impact	100	148
Biomass productivity	---	---
2. Temperature change		
Marine resources (13 economic species)	-661[c]	-1,213
Forest products	-11,060	-9,011
Agricultural crops		
Corn	269	-515
Cotton	-16	-728
Urban resources		
Fossil fuel use	-5,719	-2,550
Electricity use	45,617	35,952
Housing and clothing	-11,377	-6,794
Expenditures		
Public expenditures	-696	122
3. Total	17,232	16,139

[a]Costs are expressed as present value of all future costs and benefits resulting from the emission of F-11 and F-12 produced in the year 1973 and maintained at the level into perpetuity. A 5 percent rate of discount was utilized to convert to present values.

[b]Nonmelanoma skin cancer costs are estimated at $325 per case and $1,292 per case. See Ralph d'Arge and others, Economic and Social Measures of Biologic and Climatic Change, CIAP Monograph 6 (Washington, D.C., U.S. Department of Transportation, September 1975).

[c]Negative sign denotes benefit.

ures) and the corresponding James-Stein estimate in the same category. In the present case, the totals for the two sets of estimates are quite close; however, there are considerable differences in the individual categories, particularly for those associated with the environmental costs of ozone-induced changes in temperature. (And some divergent results for fossil fuel and electricity use are reported in chapter 8 of this volume.)

The major question for policy purposes, given the evidence, is whether F-11 and F-12 should be regulated to limit emissions, and at what level these constraints should be imposed. It is clear from tables 5-4 and 5-5 that the present value of net benefits of a complete ban on CFC production is positive if "derived surplus" is used as a measure of social cost, and negative if the sum of "consumer surpluses" is used as the relevant measure, regardless of which estimator is used for the environmental costs (that is, conventional versus James-Stein).

Table 5-6 summarizes these findings and illustrates the effects of surplus measure and estimator of environmental costs (i.e. these are the benefits from a control program for F-11 and F-12). These comparisons lead to several general observations including:

1. A complete ban on F-11 and F-12 may or may not be economically feasible, depending on the availability of substitutes. The benefit-cost ratio for a complete ban may range from approximately 0.2 to more than 5.0.

2. A partial ban on F-11 and F-12 use in products other than as a refrigerant appears to be economically feasible, although a major end-use, hair sprays, has not been included in the benefit-cost comparison.

3. If the hypothesis that CFC emissions affect temperature is untrue, then the economic feasibility of a total ban is questionable.

In figure 5-1, total derived surpluses are contrasted with total environmental costs over various discount rates. While total undiscounted environmental costs are highly nonlinear, the discounting process made them "flatten" out when computed on an annual basis. Of more importance, the degree of "optional" regulation is highly sensitive to both the discount rate selected (none at 8 percent or 75 to 100 percent at 3 percent) and whether CFCs induce a positive temperature change; however, it was not

Table 5-6. Simple Benefit-Cost Comparisons for a Ban on Production of Fluorocarbons 11 and 12 in the United States at a 5 Percent Discount Rate, Present Value 1974

(millions of U.S. 1971 dollars)

Measure of costs and benefits	Benefit estimates[a]	Cost estimates[b]	Benefit-cost ratio
Derived surplus	16,139-17,232	3,345	5.15
Final product consumer surplus[c]	16,139-17,232	107,234	.16
Derived surplus (with omission of temperature impacts)	875-876	3,345	.26

[a]Measured by savings in environmental costs of 1973 level production of F-11 and F-12 in present value terms.

[b]Loss in consumer or derived surplus at 1973 use rates in present value terms.

[c]Includes consumer surplus loss for refrigerators, auto air conditioners, mobile refrigeration systems, polyurethane foam mattresses, aerosol deodorants and antiperspirants, hairsprays and care products, insecticides, pharmaceuticals, cosmetics, perfumes, deodorizers, and shaving products.

particularly sensitive to the estimator used for environmental costs. Hence, that latter source of uncertainty may not be a strategic element in the decision.

For comparative purposes, a set of alternative strategies examined reductions in U.S. production of F-11 and F-12 commencing in either 1978 or 1979, along with various actions by the OECD and other nations. These strategies were examined to provide insight into whether U.S. benefits and costs from reducing CFC emissions might be altered substantially by other nations' decisions on regulation, since environmental costs to the U.S. are directly related to global emissions (and, thereby, only partially to U.S. emissions). The seven strategies examined are:

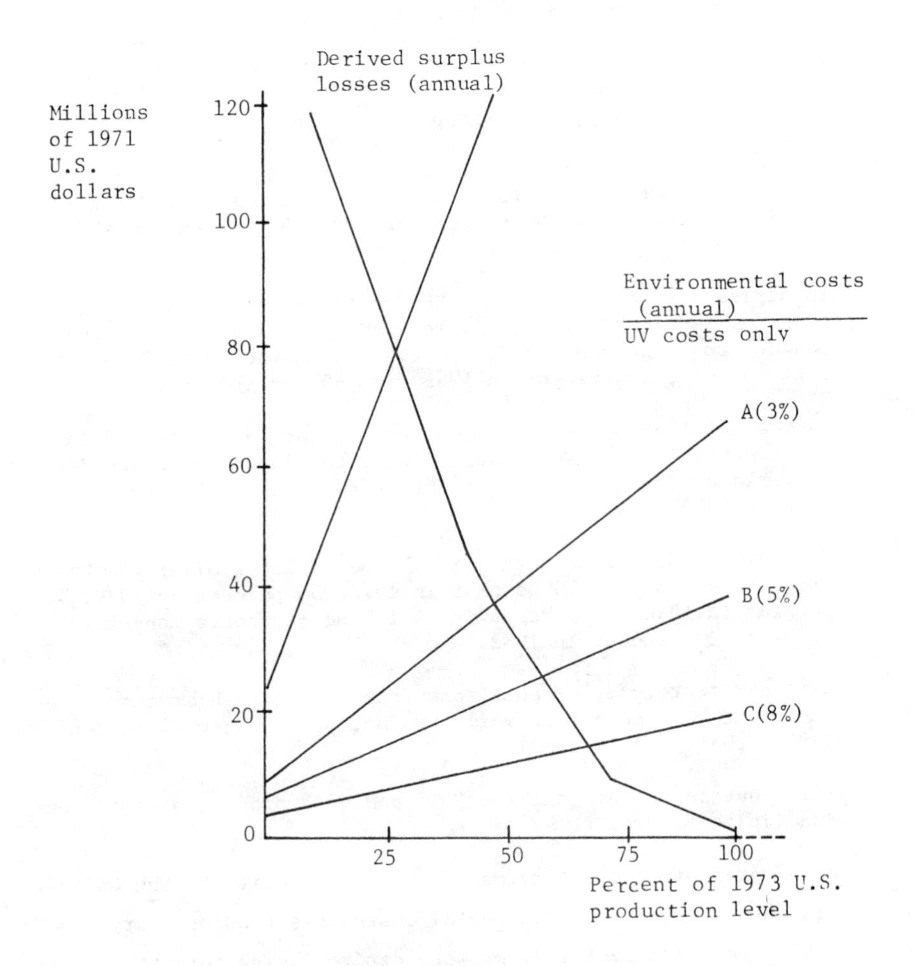

Figure 5-1. U.S. Annual derived surplus losses and environmental costs in 1973. Derived surplus losses are shown as estimated directly from demand curves for F-11 and F-12. Marginal production costs are assumed constant. Environmental costs are shown as measured by skin cancer and materials weathering. Cases A, B, and C do not contain temperature-related costs and coincide with application of 3, 5, and 8 percent discount rates, respectively. Case D includes estimate of temperature-related costs.

A. In strategy A, the United States takes unilateral action. No other nations take action to reduce F-11 and F-12 production. Production levels in the United States are reduced to 50 percent in 1978, 30 percent in 1980, and 10 percent in 1982. A steady state of 10 percent is presumed thereafter.

B. Strategy B is the same as Strategy A, except for a one-year delay in reductions in production levels such that U.S. production is reduced to 50 percent in 1979, 30 percent in 1981, and 10 percent in 1983.

C. In strategy C, U.S. and other OECD nations take identical actions commencing in 1977. World production levels then are reduced to 75 percent in 1978, 62.5 percent in 1980, 57.5 percent in 1981, 55 percent in 1982, and 40 percent in 1983.

D. Same as strategy C, except a one-year delay is introduced into implementation of regulations. Thus, the long-term reduction in world production to 40 percent is reached in 1984 rather than in 1983.

E. In strategy E, all countries in the world act similarly. World production drops to 75 percent in 1978, 50 percent in 1979, 45 percent in 1980, 35 percent in 1981, and reaches a long-term level of 20 percent in 1982.

F. Same as strategy E, except implementation of regulations is delayed one year such that world production drops to 20 percent in 1983 rather than in 1982.

G. All countries maintain 1973- F-11 and F-12 production into the indefinite future.

Table 5-7 presents the estimates of the net benefits to the United States (i.e. the savings in environmental costs less the lost surplus associated with each strategy). They were derived using only the conventional estimation methods. The first apparent conclusion is that the larger the number of countries involved in reducing production of F-11 and F-12, the larger the benefits to the United States. Second, the feasibility of the alternative strategies depends crucially on the accepted rate of discount, <u>and</u> on whether other countries participate in reducing production levels. Third, the feasibility of strategies depends on the method of assessing consumer surplus losses.

Utilizing a 3-percent discount rate, all strategies are economically feasible for the United States acting either unilaterally or multilaterally. At 5 percent, all strategies would be economically feasible for the

Table 5-7. Environmental Costs Savings Less Surplus Losses, Selected Strategies, United States, Present Value in 1974

(millions of 1971 U.S. dollars)

Strategy (see text for description)	Discount rate		
	3%	5%	8%
A	17,201[a] (7,880)[b]	3,827 (-4,720)	-589 (-9,876)
B	16,138 (7,487)	3,127 (-3,790)	-590 (-6,772)
C	30,317 (16,823)	7,984 (-5,270)	254 (-11,430)
D	25,915 (15,238)	6,547 (-3,378)	-5[c] (-8,795)
E	39,949 (33,025)	11,272 (4,297)	1,112 (-5,252)
F	32,583 (28,317)	9,032 (4,640)	855 (-3,274)
G	-57,792	-17,133	-2,447

[a]Derived surplus measure of loss is applied here.

[b]Consumer surplus measure of loss with ten-year lag for refrigeration systems is used in estimates within parentheses.

[c]Belongs in category where both estimates are negative.

United States using the derived surplus measure of loss, but only strategies E and F would be feasible if the consumer surplus measure (with substitution after ten years for refrigerants) was the correct measure of loss. Fourth, strategy G denotes environmental costs to the United States when 1973 production levels of F-11 and F-12 continue into the indefinite future. Note that, because of the extended time horizon for impacts, the present value, using a 3-percent discount, is twenty-five times higher than that measured at 8 percent. Thus, it is not surprising that the economic feasibility of strategies depends crucially on the magnitude of the discount rate; that is, even very substantial future environmental costs have small present values. Fifth, if a relatively low discount rate is given, it will be economically feasible for the United States to reduce F-11 and F-12 production, even though other countries do not do so. The benefits of unilateral action by the United States outweigh the costs.

References

1. Talbot Page, Conservation and Economic Efficiency (Baltimore, Md., Johns Hopkins University Press for Resources for the Future, 1977).

2. John V. Krutilla, and Anthony C. Fisher, The Economics of Natural Environments (Baltimore, Md., Johns Hopkins University Press for Resources for the Future, 1975); T. Sandler and V. Kerry Smith, "Intertemporal and Intergenerational Pareto Efficiency," Journal of Environmental Economics and Management vol. 2 (February 1976); and T. Sandler and V. Kerry Smith, "Intertemporal and Intergenerational Pareto Efficiency Revisited," Journal of Environmental Economics and Management vol. 4 (September 1977).

3. Ralph C. d'Arge, Project Director, Economic and Social Measures of Biologic and Climatic Changes, CIAP Monograph 6 (Washington, D.C., Climatic Impact Assessment Program, U.S. Department of Transportation, 1975).

4. Ibid.

5. John V. Krutilla and Otto Eckstein, Multiple Purpose River Development (Baltimore, Md., Johns Hopkins University Press for Resources for the Future, 1958; and J. Hirshleifer, "Comments," in Public Finances: Needs Sources and Utilization (Princeton, N.J., Princeton University Press, 1961).

6. Krutilla and Eckstein, Multiple Purpose River Development.

7. P.A. Samuelson, "Discussion," American Economic Review vol. 54 (May 1964).

8. Kenneth J. Arrow and Robert C. Lind, "Uncertainty and the Evaluation of Public Investment Decisions," American Economic Review vol. 60 (June 1970).

9. Samuelson, "Discussion."

10. Kenneth J. Arrow, and Anthony C. Fisher, "Environmental Preservation, Uncertainty and Irreversibility," Quarterly Journal of Economics vol. 88 (May 1974).

11. Ibid., p. 318.

12. S.V. Ciriacy-Wantrup, Resource Conservation: Economics and Policies (3 ed., Berkeley, University of California Press, 1968).

13. Krutilla and Fisher, The Economics of Natural Environments.

14. P. Dasgupta and Geoffrey M. Heal, "The Optimal Depletion of Exhaustible Resources," Symposium on the Economics of Exhaustible Resources, special issue of The Review of Economic Studies vol. 41 (1974).

15. Anthony C. Fisher, "A Paradox in the Theory of Public Investment," Journal of Public Economics vol. 2 (November 1973).

16. Estelle James, "A Note on Uncertainty and the Evaluation of Public Investment Decisions," American Economic Review vol. 65 (March 1975).

17. Ibid., p. 200.

18. M.C. Kemp, "How to Eat a Cake of Unknown Size," in M.C. Kemp, ed., Three Topics in the Theory of International Trade (Amsterdam, North Holland, 1976), ch. 23; M.L. Cropper, "Regulating Activities with Catastrophic Environmental Effects," Journal of Environmental Economics and Management vol. 3 (June 1976); V. Kerry Smith, "Uncertainty and Allocation Decisions Involving Unique Natural Environments," American Journal of Agricultural Economics vol. 6 (May 1979); V. Kerry Smith and John V. Krutilla, "Endangered Species, Irreversibilities, and Uncertainty: A Comment," Journal of Environmental Economics and Management vol. 6 (September 1979), pp. 175-186.

19. Kenneth J. Arrow, "Optimal Capital Policy with Irreversible Investment," in J.N. Wolfe, ed., Value, Capital, and Growth: Papers in Honour of Sir John Hicks (Chicago, Ill., Aldine Publishing Co., 1968).

20. Kenneth J. Arrow, and M. Kurz, "Optimal Growth and Irreversible Investment in a Ramsey Model," Econometrica vol. 38 (March 1970).

21. Krutilla and Fisher, The Economics of Natural Environment.

22. Charles Stein, "Inadmissibility of the Usual Estimator for the Mean of a Multivariate Normal Distribution," in Jerry Neyman, ed., Proceedings of the Third Berkeley Symposium on Mathematical Statistics and Probability (Berkeley, University of California Press, 1955).

23. W. James and C. Stein, "Estimation with Quadratic Loss," Proceedings of the Fourth Berkeley Symposium on Mathematical Statistics and Probability (Berkeley, University of California Press, 1961).

24. B. Efron and C. Morris, "Stein's Paradox in Statistics," Scientific American (February 1977).

25. J.E. Anderson, "The Social Cost of Input Distortions: A Comment and a Generalization," American Economic Review vol. 66 (March 1976).

26. Council on Environmental Quality and Federal Council for Science and Technology, Fluorocarbons and the Environment, Report of Federal Task Force on Inadvertent Modification of the Stratosphere (IMOS), June 1975.

Chapter 6

COSTS AND BENEFITS OF CHLOROFLUOROCARBON CONTROL

Martin J. Bailey

6.1 On Estimating Costs and Benefits

This chapter presents some estimates of the costs and benefits of restricting the amount of chlorofluorocarbons (CFCs) released into the atmosphere, and of the uncertainties surrounding them. Furthermore, in two technical sections, it addresses issues of more general concern. One of those sections covers the appropriate discount rate applicable to private investment made for environmental reasons, to be used in benefit-cost analysis of environmental regulations and programs. The other covers the minimum value of safety--the benefits of reducing hazards to life and health.

The first section of the chapter, dealing with CFCs, presents estimates of the prospective damage avoided by restricting CFC emissions, for two types of damage. Both types of damage come from increases in ultraviolet radiation at the earth's surface caused by ozone depletion that is expected to result from CFC emissions. The first type is higher skin cancer rates in humans; the second is damage to paints and plastics exposed to the sky. The estimates include discounted present values in 1978 dollars of the benefits of avoiding these two kinds of damage for each of two regulatory strategies, and also of the costs to the private economy of these strategies. The balance between these benefits and costs appears to justify partial cutbacks of nonessential uses. Certain major effects of CFC releases were not covered in the estimates: they include nothing concerning changes in temperature and climate, and they include nothing on damage by ultraviolet radiation to crops, forests, and fisheries.

The second section of the chapter notes the controversy about discount rates for public project evaluation, and points out that the question is different for private sector investment mandated by environmental regulation. Whereas the resources for a public sector project may come both from reductions in private investment and in consumption, the resources for private-sector environmental investment come entirely from reductions in other private investment. Even after allowing for a doubt about the effect of current (noninvestment) costs of environmental regulations, this section concludes that the appropriate discount rate is the pre-tax private-sector rate of return.

The third section presents the case for concluding that the benefits of protecting life and health are greater than the earnings protected and the medical costs avoided. It concentrates on the protection of life, and compares the benefit in terms of willingness to pay for safety with the discounted present value of expected lifetime earnings. For nonsuicidal persons who buy life insurance, it shows that the benefit must exceed this discounted present value by a definite minimum amount.

6.1.1 Assumptions About Central Values

Based on the recent reports of the National Research Council,[1] the paper by Wofsy,[2] and a private communication from Donald Wuebbles, we developed three alternative scenarios for estimated reduction of ozone in the stratosphere. (Results seem generally consistent with the estimates presented by Forziati in chapter 2.) Details on the factors entering the estimates appear as tables 6-1 and 6-2. The three scenarios are:

1. Continued growth of CFC emissions at about 10 percent a year would deplete the ozone by 15 percent around the year 2000, and 47 percent around the year 2020 (with a limiting depletion of 75 percent nearly attained after 2050). As our base case, we assume that by 2020 CFCs will become scarce and expensive, and their release will effectively cease.

2. Continued zero-growth CFC emissions at the 1974 levels would deplete the ozone by 6 percent by the year 2000, 10 percent by the year 2020, with a limiting depletion of around 20 percent nearly attained by 2120.

3. Cessation of all CFC emissions in 1978 would deplete the ozone by 3 percent around the year 1990, which would be the peak depletion, and then decline to 1 percent after 2080, steadily becoming negligible thereafter.

Table 6-1. Data Base Estimates for Benefits of Control, for Selected Years 1980-2060

Year	U.S. population (millions)	Skin cancer case rate per 100,000 (No O_3 depletion)	Materials protection costs, U.S. ($\$10^6$ of 1978)	Ozone depletion in percentage for CFC emissions that		
				Grow at 10% per year cease 2020	Cease 1978	Stay level at 1974 amounts
1980	222	220	84	2.5	2.5	2.5
2000	260	290	320	15	2.9	6.3
2020	290	290	720	47	2.4	10
2040	308	290	1400	48[a]	2.0	12
2060	323	290	2700	41	1.7	15
2080	338	290	5200	34	1.4	16
2100	353	290	10100	29	1.2	17

[a] Peak depletion is 50 percent in 2030.

Table 6-2. Summary of Parametric Estimates

Variable i	Estimate i	Uncertainty f_i	Sources
Percentage change in ozone	Table 6-1	3.2	d
Rate of induced change of ultraviolet: $\frac{d(\ln UVB)}{d(\ln O_3}$	2	1.34	e
Rate of induced change of cancer case rate CCR: $\frac{d(\ln CCR)}{d(\ln UVB)}$	1	2	e
Cancer case rate per 100,000	290	1.42	f
Population	Table 6-1	1.14	g
Cost per cancer case	$4,400	1.38[a]	h
Derived using:			
Cost per nonfatal case, 1978	$1,400	1.42	i
Deaths per 1,000 cancer cases	9	1.3	j
Disbenefit per death, 1978	$330,000	2	k
Ultraviolet-induced materials weathering; cost of control (MWC)	Table 6-1	1.5	1
Rate of induced change of cost of control of materials weathering: $\frac{d(\ln MWD)}{d(\ln UVB)}$	2	1.2	m
Rate of interest for discounting	.11	2.61[b] 2.36[c]	n
Per capita annual growth rate of productivity and real income	.02	1	q

Sources:

[a]Derived using the sum of variances in natural numbers, rather than logarithms.

[b]Uncertainty factor for cumulative discount factor for benefits of skin cancer reduction, averaged over the years of peak benefits. Includes the uncertainty of the extrapolated growth rate of productivity and real income.

Table 6-2 (continued)

[c]Uncertainty factor for cumulative discount factor for benefits of materials weathering reduction, averaged over the years of its peak benefits. This factor is smaller than for the discount factor for benefits of skin cancer reduction, because the materials weathering effects come earlier. Includes the uncertainty of the extrapolated growth rate of productivity and real income.

[d]Adapted from National Research Council, Halocarbons: Response to the Ozone Protection Sections of the Clean Air Act Amendments of 1977: An Interim Report (Washington, D.C., National Academy of Sciences, 1977).

[e]Federal Task Force on Inadvertent Modification of the Stratosphere (IMOS) (Washington, D.C., CEQ and Federal Council for Science and Technology, June 1975).

[f]Extrapolated from National Research Council, Halocarbons: Environmental Effects of Chlorofluoromethane Release (Washington, D.C., National Academy of Sciences, 1976).

[g]U.S. Bureau of the Census, Projections of the Population of the United States, 1977 to 2050, Current Population Reports, Population Estimates and Projections, Series P-25, no. 704 (Washington, D.C., Bureau of the Census, July 1977).

[h]Our estimates.

[i]Our extrapolation from Robert Anderson, Lester G. Lave, and M.V. Pauly, "Health Costs of Changing Macro-Climates," in Economics and Social Measures of Biologic and Climatic Changes, Monograph No. 4 (Washington, D.C., DOT, Climatic Impact Assessment Program, 1975).

[j]Our extrapolation from Elizabeth L. Scott, "Progress on the Analysis of NCHS Data on Non-Melanoma Skin Cancer," paper presented to the EPA Workshop on Biological and Climatic Effects Research, University of Maryland, September 19, 1977.

[k]Martin Bailey, Reducing Risks to Life: Measurement of the Benefits (Washington, D.C., American Enterprise Institute, 1980).

[l]Our uncertainty.

[m]Adapted from A.R. Schultz, D.A. Gordon, and W.L. Hawkins, "Materials Weathering," In Economic and Social Measures of Biologic and Climatic Change, Monograph No. 3 (Washington, D.C., U.S. Department of Transportation, Climatic Impact Assessment Program, 1975), our uncertainty.

[n]See 6.2.6 in text. Our uncertainties based on section 6.2.6.

[q]Our extrapolation. Uncertainty consolidated into that of discount rate.[a]

Based on the Federal Task Force (IMOS) Report,[3] we estimate that for each 1 percent depletion of the ozone, the amount of ultraviolet light reaching the earth's surface increases by 2 percent, for small changes. More precisely, if the original amounts of ozone and ultraviolet flux are Q_o and U_o, the new amount of ozone, after depletion, is Q_1, then the resulting ultraviolet flux will be:

$$U_1 = U_o \exp [2(\ln Q_o - \ln Q_1)] \qquad (1)$$

Based on analysis and figures reported in National Research Council,[4] we estimate a continued rise in the incidence of skin cancer from the 1970 figure of 140 per 100,000, and then a leveling out by the start of twenty-first century at around 290 cases per 100,000 population, in the absence of ozone depletion. The rise reflects both a rising fraction of old people in the population and a continuation of the up-trend in the age-adjusted incidence caused by life-style changes involving more exposure to the sun with less body cover. We use U.S. Bureau of the Census projections of future population,[5] extended through the twenty-first century at their assumed one-half percent growth rate (see table 6-1). We assume that induced increases of cancer cases lag twenty years behind increased ultraviolet flux, and we base the number of induced cases on the population at the time of the increased flux. We use the Federal Task Force (IMOS) Report estimate that the percentage change in incidence equals the percentage change in ultraviolet flux.[6]

Anderson, Lave, and Pauly estimate the cost per skin cancer case in 1971 at from \$125 to \$1,292, depending on various assumptions about incidence, lost workdays, and so on.[7] Their figures were especially sensitive to their assumption about incidence, which ranged from 50 to 200 cases per 100,000 population. For our purposes, it is essential to use a cost figure that is based on the same incidence that we stated above in our assumptions. For intermediate assumptions in other respects, the resulting figure is \$480 per case in 1971 dollars, which becomes \$850 in 1978 dollars, when adjusted both for prices and for higher real incomes. Their estimates included only lost workdays and direct medical costs. A proper measure of the benefits of avoiding increased skin cancer incidence would estimate what we would all pay in advance to avoid the risk of that contingency. Based on the reasoning and data in Bailey,[8] we estimate that this willingness to pay is $1,400 per case, for nonfatal cases.

In the years around 1970, the death rate from all skin cancers was about 1.5 per 100 cases (National Cancer Institute data in Scott[9]). We estimate that this mortality rate will drop to 0.9 percent in the twenty-first century. Based on Bailey,[10] we estimate that, in 1978, the average person in the United States was willing to pay $330 to obtain a reduction of 0.001 in the risk of premature death. (As indicated in the introduction, a fundamental discussion of willingness to pay for reductions in risk appears in section 6.3, which extends our discussion here.)

Based on Shultz, Gordon, and Hawkins,[11] we estimate that the percentage change in damage to outdoor paints and plastics is about double the percentage change in ultraviolet flux. We use the data of Shultz and coauthors on the market for additives to plastics and paints, as reported in table 6-1.

We have no data on prompt deaths caused by excessive exposure to the sun (sun poisoning, fatal sunburn cases, and so on), nor have we a way to estimate the increase in prompt deaths that could be expected with depletion of the ozone layer. Similarly, there are no estimates of the effects of increased ultraviolet flux on crops, pastures, and other elements in ecological food chains on which humans depend.

For the discount factors to be applied to future costs and benefits of control of CFC emissions, to obtain capitalized present values, we use 11 percent per year. To allow for higher willingness to pay for benefits in future years due to higher per capita incomes, we use an annual growth rate of two percent of real income per capita, and assume that the income elasticity of demand for these benefits is unity.

6.1.2 Assumptions About Uncertainties

Because the relationships are multiplicative, along the causal chain from emissions of CFCs through depletion of ozone to induced skin cancers, other effects, and their costs, the uncertainty of an estimate is best considered as a multiplicative factor. If estimate A could be in error by a factor of two, then the range of its possible values is from A/2 to 2A. The possible range of values is then the square of the error factor.

Furthermore, where the error in one estimate is unrelated to the error in another, the best estimate of the error factor for the product of the estimates is a root mean square error using logarithms. If f_i is the error factor for the ith term in the sequence of multiplications, and F the error factor for the product, following National Research Council,[12] we use:

$$\ln F = \sqrt{\sum_i (\ln f_i)^2} \quad (2)$$

(The variance is the sum of variances for variables that are summed.)

The National Research Council estimated F for the effect of CFC release on ozone depletion at 2.1 for a 95 percent confidence interval (2σ),[13] that is, the range from the upper limit to the lower was estimated as a factor of 4.4, which is $(2.1)^2$. However, to obtain this range they examined only thirty-five from among more than one hundred reaction coefficients, in effect treating the remaining ones as known precisely. In comparison with their earlier uncertainty estimates, based on seven reaction coefficients only, their consideration of the added twenty-eight coefficients doubled the range of recognized uncertainty. At the same time, new laboratory data on some of the coefficients implies an appreciable understatement of the present uncertainty factor F. There is also the possibility that new reactions will be discovered, that nonhuman sources of the chemicals catalyzing the destruction of ozone will be found to be more significant to the ozone balance than is now supposed, and so on. Therefore, we use an uncertainty range of 10 in place of the 1977 estimate of 4.4, which implies a factor F (for the effect of CFC release on ozone depletion) of 3.2 in place of 2.1.

Our other uncertainty factors we take from the cited sources wherever possible, and where necessary have introduced our own to reflect the uncertainties of our own estimates. Table 6-2 summarizes the parametric estimates reported above and their uncertainties, drawing on some of the estimates of table 6-1 as underlying data.

As indicated in the notes to table 6-2, some variables are sums of other variables; in this case, the variance of the sum is the sum of the variances, in natural numbers rather than logarithms. In mixed operations like those used here, where the uncertainties themselves are subjective in most cases rather than being based on statistical samples, elaborate efforts to reconcile the different variance formulas are scarcely justified. Our procedure was to use the uncertainty range (that is the "confidence limits") in both additive and multiplicative cases. That is, in Equation (2) for f_i, we used the ratio of the upper confidence limit to the lower confidence limit; in the ordinary variance formula we used the numerical difference between the two limits. Thus the limits obtained for the result of a subset of effects could be combined with those for another subset in a straightforward way, using whichever formula is appropriate for combining them.

The resulting uncertainty factor F, for the benefits of controlling emissions of CFCs, appear in table 6-3. For each type of benefit, these factors are multiplicative, because the predominant operation used in arriving at these benefits is multiplication (although certain of the variables multiplied together are sums of variables with independent uncertainties). For the sum of the benefits, although the factor is obtained using the variance formula in natural numbers, we consider the factor to be approximately multiplicative also. The variance formula in this instance had to include a covariance term, in addition to the sum of squares, because the uncertainties about the ultraviolet flux and the cumulative discount factor are common to both, so that, if these variables are higher than expected, both the benefits together will be higher than expected.

6.1.3 A First Look at the Benefits of Control

With the assumptions stated above, the effects on the numbers of skin cancer cases of adopting certain control policies are those shown in table 6-4. The pattern for materials weathering costs is similar, although the first benefits appear in 1985 rather than 2005.

The discounted present values of these benefits, using the assumptions in table 6-1, along with our estimates of their ranges of uncertainty, appear in table 6-5. The expected total benefit of stopping all emissions of CFCs in 1978 is $3.4 billion, present value, although it could be as high as $15 billion or as low as $770 million, for these two benefits. Stopping the growth of emissions in 1974 has about two-thirds the benefits of stopping emissions entirely. In both cases, about two-

Table 6-3. Uncertainty Factors F for Present Value of Benefits of Emission Control

Reduction in cancer	6.28
Reduction in materials weathering	4.67
Sum of these benefits	4.39

Table 6-4. Reductions in Skin Cancer Cases (in thousands of cases)

Year	Stop emissions in 1978	Stop growth of emissions in 1974
2005	14	11
2025	360	270
2050	1400	1000
2075	1100	690
2100	920	420

Table 6-5. Present Values of Benefits of Control (in millions of 1978 dollars)

Values	Stop emissions in 1978			Stop growth of emissions in 1974		
	Cancer	Materials	Total	Cancer	Materials	Total
Upper limit value	6,900	10,700	15,000	5,200	7,000	10,000
Central estimate	1,100	2,300	3,400	820	1,500	2,320
Lower limit value	180	500	770	130	320	530

thirds of the dollar benefits are caused by reduced materials weathering, and about one-third by reduction of skin cancer.

6.1.4 Costs of Control

Reducing emissions of CFC generally requires that we use less of them as aerosol propellants, refrigerants, and so on. In certain uses, such as refrigerators and air conditioners, there is the possibility also of tightening their closed systems so that no CFCs escape during use, and then making sure that the CFCs in worn-out units are recovered. Discontinuing a use deprives consumers of its benefits, although these benefits may be small if effective substitutes are available at a competitive cost.

The best available measure of the value to ultimate consumers of a component of a final good, taking account of input substitutes, is to use the area under the derived demand curve for this component. This measure is precisely accurate if the final product is sold in a perfectly competitive market.[14]

The net cost to consumers of reducing the availability of CFCs from the equilibrium amount is the difference between the gross cost, given as noted by the area under the pertinent range of the derived demand curve for CFCs, and the resource cost of CFC output over the same range. For our cost function, we will use that estimated by d'Arge, Eubanks, and Barrington,[15] based on their estimates of the derived demand curves for the principal CFCs. Their data gave the function:

$$S(Q) = 147.004 - 2.934Q + .0146Q^2 \tag{3}$$

where the units of S(Q) are millions of 1971 dollars; Q ranges from zero to 100, and represents the percentage of equilibrium production permitted. The equilibrium level of 1974 represents the base level of full output in this equation. Adjustment for the price level to convert it to dollars of 1978 revised this equation to the values:

$$S(Q) = 233.16 - 4.6552Q + .023236Q^2 \tag{4}$$

By this equation, if output drops to zero, the loss will be $233,163,000 per year. With output at 100, the loss drops to zero. Our assumptions on costs appear in table 6-6.

Growth of demand has been about 10 percent a year, prior to 1974; although demand growth may have slowed down in the United States, the use of CFCs in the rest of the world has been growing faster, and may continue at a higher rate than 10 percent a year in the absence of controls. Moreover, some of the slowdown in the United States is caused by controls already in place, and we are considering the costs here of all controls. Some studies have considered it a serious possibility that future growth of demand would be as high as 22 percent a year.[16] Hence, a 10 percent

Table 6-6. Cost Assumptions

Variable i	Estimate i	Uncertainty f_i
Base year cost to consumers of output cutback	Eq. (4)	2.0
Growth rate of demand	0.10	2.0
Discounted present value of cost[a]	Table 6-7	2.7

Sources: For base year cost to consumers of output cutback, see Ralph C. d'Arge, L. Eubanks, and J. Barrington, Benefit-Cost Analysis for Regulating Emissions of Fluorocarbons 11 and 12 (Washington, D.C., EPA, December 1976). Uncertainty based on their standard errors, and on variations in their estimating equations. For the growth rate of demand, see Federal Task Force, Fluorocarbons and the Environment (Washington, D.C., CEQ and Federal Council for Science and Technology, June 1975).

[a]See Equation (2) uncertainty.

rate is a reasonable estimate, although perhaps at the high end of the range for the United States itself, for which we make cost and benefit estimates. In contrast, the estimate of base-year cost to consumers, based on derived demand curves for the principal CFCs, can more easily be much too small than much too large. Economical substitutes for CFCs might be developed for use in refrigeration and air conditioning. However, if not, the cost of control based on econometrically estimated derived demand is an extreme underestimate.[17] Hence, the asymmetries of these uncertainties roughly balance each other, and we use the final uncertainty factor F in the last line of table 6-6 symmetrically.

The discounted present values of costs appear in table 6-7 for the two control policies considered thus far. Stopping growth in 1974 costs a little less than two-thirds as much as does stopping all use in 1978, a similar relationship to that found for the benefits. However, comparison with table 6-5 shows that the costs of the relatively drastic control policies under consideration are more than double the benefits at the central estimates of each. Although stopping the growth of emissions may

Table 6-7. Present Value of Costs of Control
(in millions of 1978 dollars)

Value	Stop emissions in 1978	Stop growth of emissions in 1974
Upper limit value	22,000	13,000
Central estimate	8,000	4,900
Lower limit value	3,000	1,800

not seem drastic at first glance, after a few years it requires cutting into what are now considered "essential" uses of CFCs, those most difficult and costly to curtail. (See chapter 3 for a detailed discussion of CFC uses which indicates those that are most essential.) These uses have grown rapidly along with the less essential uses, and can be expected to grow in the future. Thus, with the passage of time, the no-growth policy becomes almost as drastic as stopping emissions entirely.

6.1.5 Optimal Control Strategy

If no other benefits of control were expected, our estimates would imply that only a modest cutback of the least essential uses of CFCs would be justified. However, we noted in the section on assumptions that depletion of the ozone will increase the number of severe sunburn cases and associated deaths, and that it will damage agriculture and the biological food chains on which humans depend. In addition to what these imply as benefits of control, there is the benefit of reducing the threat of rising temperatures on the earth's surface, with its implied drastic changes in climates. This effect depends primarily on the "greenhouse" effect of CFCs which is unrelated to ozone depletion. According to the National Research Council and chapter 2 of this volume, if CFC release continues to grow, their effect on surface temperatures will eventually outweigh the already serious effect from carbon dioxide, and will raise the earth's surface temperatures by several additional degrees.[18] This change would shift the climate belts, melt part of the polar ice caps and so raise the sea level, and have many related effects that are hard to predict. The economic costs of these changes would be enormous. These possibilities increase the likelihood that drastic cutbacks in CFC use, approaching a total ban worldwide, will prove to justified.

Although present knowledge provides only a shaky basis for policy, it indicates that it is wise to restrict the use of CFCs for those uses, such as aerosol propellants for personal and household sprays, for which substitutes are readily available at no appreciable cost of inconvenience to consumers. It also appears wise for the U.S. government to press for a worldwide ban on such uses.

An interesting alternative would be to press for a worldwide tax on CFC use. The central estimates of the benefits of control above, for skin cancer and materials weathering, imply that, for small cutbacks, the discounted present value of the known benefit would run between 25¢ and 30¢ per pound. A tax of this size would increase the price of F-11 and F-12 by two-thirds to three-quarters; because of the unmeasured benefits, we have high confidence that a tax at least this high is justified. It can be shown that Equation (4) implies that industry and consumers would respond to this tax by cutting back by some 50 to 55 percent on the use of these CFCs relative to unrestricted use.

As a further step, it is important to obtain more information about the effects of CFC release on climate and on the biosphere. It could easily turn out that these effects would justify much more drastic restriction of CFC use worldwide, and that they would be of compelling concern to people and governments in all parts of the world. At present, the opinions of the U.S. government on this issue are viewed with skepticism by many or most other governments, because of the poor data base and uncertain theory about what may be the most serious effects of CFC release. This problem can be overcome mainly by producing reliable, convincing information on these effects.

6.2 The Discount Rate for Environmental Control Programs

In this section, we consider the appropriate discount rate to use in benefit-cost analysis of programs that require or induce private firms to invest in new equipment, or to change materials used in production, for environmental reasons. In the case of restrictions on CFCs, because of the delay of decades between the costs of control and the eventual benefits, the choice of a discount rate will have a large effect on the comparison of costs with discounted benefits.

An extensive literature has investigated the appropriate discount rate for benefit-cost analysis of government investment, and it remains controversial. One school of thought argues that the rate used for this purpose should be the same as the pre-tax rate of return to investment in the private sector; a second school of thought argues that the rate for benefit-cost analysis should be lower, because government investment has little or no risk, when diversified, and because the risk premium in private investment represents a true social cost.[19] However, the issue of the appropriate discount rate for benefit-cost analysis of mandated or induced private investment (and other costs) to serve environmental ends is separate and distinct, and is relatively little studied. Although some of the same issues may arise, they must be addressed separately for this case; the arguments used in the controversy over the discount rate for government investment are sometimes inapplicable. For example, the issue of the source of funds for government investment--whether they reduce private consumption or private investment--is distinct from the corresponding issue concerning mandated or induced private investment. Government investment may be financed either by taxes or by issuing new government bonds; private investment must be financed privately, except for tax incentives and the like that may help pay for it. The source of the resources must therefore be looked at separately. The risk question must also be looked at separately; investment by private firms involves private risk, more directly than for public investment.

The choice of discount rate is sometimes said to involve a question of equity between present and future generations. If the control program affects mortality rates in the distant future, as is the case with control of CFCs, that fact may also be said to involve equity questions. To consider such equity questions properly, however, one must examine the whole "portfolio" of wealth, knowledge, and environment that we transmit to future generations, and consider how this portfolio is affected by a particular measure.

6.2.1 Wealth We Give to the Future

We give several things to the next and later generations. First and foremost, we give them life; to many we deny life inasmuch as the typical number of children in a U.S. family is two, rather than the biologically possible twenty to thirty. We further choose how much to consume currently and how much wealth to accumulate--to save--to give to the future. This wealth includes not only the tangible wealth ordinarily reckoned in saving and investment, but also the stock of knowledge of all kinds, and the quality of the environment. We are the sole temporary trustees of the welfare of future generations who, despite the risk of disastrous setbacks through nuclear wars, are very likely to be generally much wealthier than we are.

Despite the expected large wealth of future generations, most economists writing about cost-benefit analysis and intertemporal choice argue that the present generation should save more. There is a good case for this claim, because the tax system reduces the incentive to save and invest. Many also argue that the present generation has an irrational or unethical "myopia" that reduces saving and investment below what they should be. This claim raises a bona fide ethical question we need to explore further, while noting that it is separate and distinct from the claim that it might be unethical to use the private sector rate of return to discount future benefits of environmental investment.

The average rate of return in the private sector is a measure of opportunity cost for investment in environmental improvements. As such, it has no more ethical content than the opportunity costs for other resources used in this investment; for example, labor, materials, and so forth. One can easily be misled into overlooking this point by the practice of using the rate of return to discount future benefits, reducing them to present values, and so appearing to downgrade them. We discount because we want to measure the costs and benefits of an investment in units of present wealth, to provide a common standard for projects with different future time profiles. If we merely wish to know if the benefits exceed their opportunity costs, however, we need not use this standard; we could instead accumulate the costs forward to the years in which the benefits appear. The relatively large wealth measured by the accumulated value of these costs, obtained in this way, is what society gives up by choosing the environmental investment instead of ordinary productive investment. It can scarcely be unethical to state the sacrifice honestly and accurately. Discounting both the costs and benefits back to the present, compared to accumulating the costs forward, has no effect on the ratio of the benefit to its accumulated opportunity cost, because discounting divides both by the same factor.

To clarify what we mean, for present purposes, by the private sector rate of return, we have to decide whether the resources for environmental investment come from private consumption or private investment; we examine this question in the next section, where we show that they come from private investment. That is, environmental investment changes the mix of assets in the wealth we give the future, without changing its total resource cost. If we measure opportunity costs correctly and choose those investments of all types with the greatest present value per dollar invested, we will manage our wealth well and give the future the most we can for a given level of saving.

The management procedure of using the correct measure of opportunity cost is separate from the question of whether we save as much as we should. Some have argued that it is immoral to have a positive rate of time preference, a factor that enters into our decisions to save, and thereby into the equilibrium private sector rate of return. Although there are good reasons why the United States should save and invest more, we doubt that this is one of them. The idea of a zero rate of time preference, combined with a concern with the welfare of all future generations, leads to absurd results.[20]

6.2.2 Environmental Investment and National Savings

We now examine the precise effects of mandated environmental investments and commodity specifications on total national saving and investment. National saving, the part of national income that is neither consumed privately nor used in the current operations of government, can be increased either by increasing income or by reducing consumption. (For simplicity, we disregard the possibility of reducing the size and cost of the government.) Income could be increased if there were slack resources--unemployment. Consumption can be reduced either by increasing the incentive to save or, possibly, by increasing taxes. There is no reason to believe that mandated environmental investment can reduce unemployment further than it can prudently be reduced by monetary and fiscal policy; that is, that it can employ resources that would otherwise have to remain slack.

Hence it would be a mistake to consider unemployed resources as a source of environmental investment; this investment, and other costs of environmental control, diverts employed resources either from consumption or ordinary private investment.

Resources for environmental purposes can come from consumption if households voluntarily consume less. Consumption is the most thoroughly and successfully researched of the major economic aggregates; it is a function of income and the incentive to save: the net, after-tax rate of return. As we have just noted, environmental investment contributes nothing at the time of the investment, compared to available policy alternatives, to income (employment). Therefore, its effect on total investment depends on its effect on the incentive to save. Unfortunately, this effect goes the wrong way: because environmental investment adds nothing to the profits of the investor, and hence nothing to the rate of return paid to the saver, it reduces the incentive to save. With real income given and fixed, as we have noted that it is, the reduced incentive to save must result in reduced total savings.[21] Hence, total investment, where the total includes environmental investment, is reduced by a policy of requiring firms to invest in environmental control equipment; this follows because saving equals investment, so that a decline in saving implies a decline in investment. That is, other private investment is reduced by more than the amount of environmental investment.

6.2.3 Running Costs of Environmental Controls

The above discussion concentrates entirely on environmental investment which represents only part of the cost of environmental controls. Such controls have capital costs and running costs; the above logic applies only to the capital-cost portion. The issue of running costs is of some interest in connection with the question just considered concerning the volume of national savings--the wealth transmitted to the future. Suppose that control costs for CFCs will exceed benefits for the next forty years, after which there will be a new benefit stream accruing mainly to people who are now minors or not yet born. The running costs of environmental controls--the operating costs of special equipment, the changes of materials used to more expensive or lower quality materials, and so forth--lower the real incomes of all households, regardless of income source. The benefits of the controls add to the environmental wealth passed on to the next generation. Hence, if the current generation responds to the reduction in real incomes by consuming less, as we would normally expect, part of the wealth passed on will be financed by reduced consumption, not by a reduction in other wealth.

Whether consumption would stay at the same fraction of income, in response to the control costs for CFCs, depends on whether current and near-future households are strictly hyperrational. If they are, they would respond to the accumulation of environmental wealth by reducing their private savings, relative to the alternative of keeping real income higher and letting the CFCs be released. Households would view the accumulating environmental wealth as part of their permanent incomes, replacing the direct measureable loss, and so would consume no less. That is, none of the cost of the enhanced environmental wealth would be paid for by reduced consumption.

Although the case of strict hyperrationality seems far-fetched, the evidence gathered so far partially supports it.[22] Hence, it is a reasonable first approximation to argue that all the costs of CFM control would come out of national savings, as previously argued. Further evidence may change this conclusion, but it is the best available starting point for present analysis.

6.2.4 Why We Use the Discount Rate

Consider a project with known benefits extending far into the future--say, a hydroelectric project--whose project resources would be drawn from private investment. Suppose for simplicity that its risk, properly considered in portfolio terms, is the same as that of the private investment it would replace. Suppose, finally, that its future benefits, discounted at the rate of return of the alternative private investment, are less than its cost, whereas at some lower discount rate (such as the government bond rate), they would exceed the cost. Then the choice is not one of equity toward future generations but simply one of good management. Future generations would benefit more from the private investment than from the project under study; completing the project would divert resources from a more productive to a less productive use.

The case is essentially the same, although less obvious, in the more difficult analysis of environmental investments that affect future health and life expectations. They will have ways in the future to increase longevity, as we have now, which they will be better able to afford if we increase their wealth by ordinary investment. An environmental investment now that will save lives fifty years from now (as in the case of control of CFCs) will then yield a benefit, measured by the future willingness to pay for reduced risk. Then, if such benefits, plus all other benefits, discounted to present values using the private investment rate of return, should be found to be less than the cost, it would again be simple mismanagement to insist on the environmental investment. Letting the alternative private investment go forward in its place would give to the people fifty years from now more wealth. They could then use that wealth to save more lives with the programs then available than our environmental investment would save.

6.2.5 Investment Risks and the Discount Rate

The risk factor in an environmental investment depends on whether the risks of the portfolio of all wealth is larger with or without the proposed environmental investment. As a first step in evaluating this portfolio risk, consider first the risk in the investment itself, separate from portfolio considerations. There are the technical uncertainties about such things, in the case of CFC control, as the upward mixing rate of CFCs into the stratosphere, the relative importance of CFCs versus other chemicals (including especially those from "natural," nonhuman sources) in the degradation of the ozone layer, and the morbidity and mortality effects of increased ultraviolet radiation. In addition, we do not know precisely what future medical progress there will be in the detection and cure of skin cancer; and we do not know precisely how much people fifty years from now will be willing to pay to reduce the death rate. These technical risks and valuation risks are typical of all investment, public or private; however, few private investments go forward with so many technical questions unanswered (and with so long an investment period) as is the case for control of CFCs. Hence, in simple _prima facie_ terms, this environmental investment looks extremely risky.

Now what of portfolio considerations? Although the environmental investment is privately financed (except to the extent of government participation through tax incentives, and so forth) to a first approximation it has no effect on the variability of private profits. The principal risk consideration, from a portfolio standpoint is the variability of possible outcomes for the quality of the environment and for its impact on households.

The risk premium that is appropriate for a given investment depends, for the representative household, on its effect on the risk of the entire portfolio of the household's wealth. To illustrate, suppose that portfolio variance V is a sufficient measure of portfolio risk, and suppose there are just three assets, whose amounts held by the household are A, B, and C. Asset A is a perfectly riskless asset, asset B is the given investment, and asset C is a representative bundle of the existing wealth of the community. The household's portfolio W consists of the amount it holds of the three assets $W = A + B + C$, where we choose units so that each asset has a present price of one. Its future value we denote by $W' = Ap_a + Bp_b + Cp_c$; with mean, $\overline{W}' = A\overline{p}_a + B\overline{p}_b + C\overline{p}_c$; and variance, $V = E[W' - W')^2]$. Because asset A is riskless, $p_a = \overline{p}_a$ and the household's portfolio variance become:

$$V = B^2\sigma_b^2 + BC\sigma_{bc} + C^2\sigma_c^2 , \tag{5}$$

which depends on σ_{bc} as well as σ_b--the covariance effect, which is well known in the portfolio theory literature. To obtain the relevant risk measure for asset B, we consider substituting a unit of asset B for a unit of asset A, and obtain:

$$\frac{dV}{dB} = 2B\sigma_b^2 + C\sigma_{bc} \tag{6}$$

For a given investment in future health and longevity, such an investment in the control of CFC emissions, its covariance of outcome with other wealth is zero or positive unless human capital has a negative covariance of future value with tangible wealth, or unless the given investment has a negative covariance of outcome with that of other investment in human capital. The first of these covariances is positive because, during economic setbacks when physical capital loses some of its value, more people are unemployed or have reduced productivity. The second is also positive because willingness to pay for health and safety is a positive function of earnings, which measure the (risky) payoff of investment in human capital. Hence, on covariance grounds, the given investment must have a positive risk similar to that for private investment even if it is a small part of the representative portfolio.

Turning to the term $2B\sigma_b^2$, we have already noted the large technical uncertainties about the ultimate effects of CFC emissions. These imply that the term σ_b^2 is relatively large. The term B reflects the importance of the cost of CFC control as a share of total wealth. If it were large, this factor would magnify the risk.

In conclusion, qualitatively it appears that environmental investment is similar to ordinary private investment in terms of portfolio risk. The covariance effect is broadly the same as for ordinary private investment, and the own-variance effect would, if anything, suggest a higher portfolio risk for environmental investment than for (readily diversified) private investment.

It follows that there is no case for using a low discount rate for discounting the prospective benefits of control of CFCs (or for the other small cases). Risk considerations would suggest instead using about the same discount rate for this investment as the average pretax private-sector rate of return of investment. Christensen estimates this rate of return as averaging 11.3 percent for noncorporate investment and 15.4 percent for corporate investment for 1929-69;[23] the overall average would be about 13.3 percent. For 1949-69 the figure would be slightly higher.[24] See also Harberger,[25] whose slightly lower estimate would apply if some of the resources for environmental control were provided by reduced current consumption. The upshot is that the right discount rate for this problem probably ranges between 12 to 14 percent, and assuredly ranges from 10 to 16 percent, by traditional standards of measurement.

6.2.6 Externalities, Overregulation, and the Discount Rate

Although we find that, by traditional standards, the right discount rate for benefit-cost analysis of private-sector regulation is from 12 to 14 percent, this conclusion is modified by externalities and by new information that justify regulation. Especially, it can be changed as a side effect of gross mismanagement of such regulation. In practice, the agencies charged with protecting the environment and the workplace against hazards to health and safety shun benefit-cost analysis and adopt the most stringent regulations that are feasible. Denison shows that the costs of this practice have already mounted enough to slow measured U.S. economic growth appreciably.[26] Its continuation will slow measured growth drastically. It will also lower the rate of return to private-sector investment.

As a roughly representative example, consider an industry whose mandated investment to improve worker safety and to limit its emissions of pollutants is half of that industry's gross investment. Inasmuch as the mandated investment adds nothing to profits, if the equipment involved has about the same economic life as has this industry's other equipment the rate of return on its total investment is cut in half. It is quite conceivable that regulation will produce an effect of this size on private-sector rates of return. If so, the appropriate discount rate for analysis of particular mandated investments will also be cut in half (although, by assumption, the analysis will have no effect on the regulatory decision).

However, although the regulatory climate of the 1970s forced some industries to tie up half or more of their gross investment in mandated equipment (or features of equipment) to reduce pollution and to improve worker health and safety, this climate is changing. If all of industry had to tie up half its gross investment in such equipment, U.S. economic growth would become negative, and the typical household's tangible standard of living would fall. Before the process went far, workers would join industry in forcing a change in regulatory goals, as they have already done in the case of emission standards for automobiles. One cannot foresee where the resulting political equilibrium will come out, but it seems unlikely that it will entail a drastic slowing down of U.S. economic growth. If growth is not slowed down drastically, the private-sector rate of return to investment will not be much reduced.

The rate of return will assuredly be reduced to some extent, however. An ideal regulatory process would require some private investment in antipollution and safety equipment. Differently stated, the rate of return to private investment is lower than it seems, because private investors have not borne the full costs of their activities; they have not had to pay for the costs of the pollution they cause, nor for unanticipated long-term damage to worker health. The socially correct discount rate is therefore lower than the historic rates of return given by Christensen.

To estimate the correct adjustment, I assume that unrestrained regulation would mandate that half of gross investment be devoted to pollution control, health, and safety. The sketchy evidence available indicates that mismanagement of regulation roughly quadruples its private-sector cost: excessive stringency doubles the cost, and lack of fine-tuning of specific regulations doubles the cost again (for an example, see Bailey[27]). If these assumptions are correct, whereas unrestrained regulation would cut the private-sector rate of return in half, ideal regulation would reduce it by one-eighth. For example, unrestrained regulation would reduce a private-sector rate of return of 0.12 to 0.06, while ideal regulation would reduce it to 0.105. Applying this reasoning to the numbers at the end of the preceding section, we find that the correct discount rate, adjusted for ideal regulation, is probably in the range of 0.105 to 0.122, and assuredly is in the range from 0.09 to 0.14. However, the adjustment itself is shaky and uncertain, so that the "assured" range should be widened, say, to something like 0.08 to 0.15.

6.3 The Minimum Value of Safety

An extensive literature deals with the "willingness-to-pay" concept of the value of safety: that is, of the value of reducing hazards to life and limb, for use in cost-benefit analysis.[28] Although the willingness-to-pay concept is still in some ways controversial, there is no well-defined alternative that commands appreciable support. It is widely understood that this concept is different from the human capital concept (the discounted present value of expected future earnings), and that willingness-to-pay is the logical counterpart to other values used in cost-benefit analysis.[29] A still unsettled question is whether every person's shadow-price value per life, implied by this willingness to pay for lowered risk of premature death, necessarily exceeds his human capital.

We have shown elsewhere that in certain cases a person willing to buy life insurance necessarily has a shadow price value per life greater than his human capital.[30] In this section, we show that this result is general, for all nonsuicidal persons. That is, it is true if the person prefers to be alive: at each level of family consumption, his utility for the prospect of his living is higher than for the prospect of his premature death. Intuitively, this result makes sense because a person values his leisure time as well as his cash income. The proofs do not depend on this point, however.

6.3.1 Some Fundamentals

Extending the above remarks, we concentrate our attention entirely on the willingness-to-pay measure of the value of reduced risk, and on its qualitative relation to human capital; that is, to discounted future lifetime earnings. Also, we use the expected-utility framework without addressing the controversy about its appropriateness. In this section, we set out some concepts and earlier results that we need for our future steps.

The shadow price of a life implicit in the willingness to pay for reduced risk is $-dY/dp$, where Y is income, and p is the probability of survival. If, for example, a person is willing to give up $200 to increase his chance of surviving this year (while keeping an unchanged expectancy thereafter) from 0.994 to 0.955, the shadow price is $- \frac{-\$200}{.001} = \$200{,}000$. For a group of 1,000 persons with this same willingness to pay, the compensating variation measure of the benefit to them of expecting one less death this year in their midst is $200,000; in this sense, the sum is the value of one expected life saved. If, instead of sacrificing income, the person sacrifices consumption to buy a safety device, such as a smoke detector, we should understand $-dC/dp$ to represent the same shadow price concept as does $-dY/dp$.

If a person's utility function is everywhere convex (from below), then he buys no insurance and he is a miser, or, given a fair game, he gambles everything he has on the longest possible long shot. It will be apparent later that we can set no lower bound on such a person's shadow price of life. We can say nothing further about him; our interest turns instead to the risk averter, whose utility function is concave (see also our concluding remarks). For present purposes we can make utility a function simply of (discounted) lifetime consumption, including bequest, denoted by C; Utility = U(C). Further, if he survives the coming year, then he will live out a life of normal length, the prospect of which yields the utility $U_a(C)$. If he dies the coming year, his dependents will have consumption C' during the period he otherwise would have lived; this prospect yields him the utility $U_d(C')$. If his chance of surviving the year is p, his expected utility is:

$$E(U) = pU_a(C) + (1-p)U_d(C') \tag{7}$$

We assume that he attempts, in all his choices, to maximize E(U). Suppose he can sacrifice part of consumption C, without affecting C', to increase p; the largest such sacrifice he would accept would keep E(U) unchanged:

$$dE(U) = U_a(C)dp + pU'_a(C)dC - U_d(C')dp = 0$$

which implies:

$$-\frac{dC}{dp} = \frac{U_a(C)-U_d(C')}{pU'_a(C)}$$

and:

$$\lim_{p\to 1}\left[-\frac{dC}{dp}\right] = \frac{U_a(C)-U_d(C')}{U'_a(C)} \tag{8}$$

The final expression in Equation (8), simplified by the convenient choice of origin that sets $U_d(C') = 0$, becomes a standard one for the shadow price of life, usually with the implicit assumption that C' = 0. For this case, this shadow price is shown in figure 6-1, where we show the utility function as concave at every positive level of consumption, and where we select the origin as noted and assume that $U_d(0) = U_a(0) = 0$. At consumption level C_1 , the ratio of U to its derivative, that is, to its tangent, given in Equation (8), is the distance from -X to C_1, that is, it is $C_1 + X$, where -X is the intercept with the horizontal axis of the line tangent to U_a at C_1. For the assumed concave utility function, it is clear that X must be positive; that is, that the value of the ratio in Equation (8) is greater than C_1.

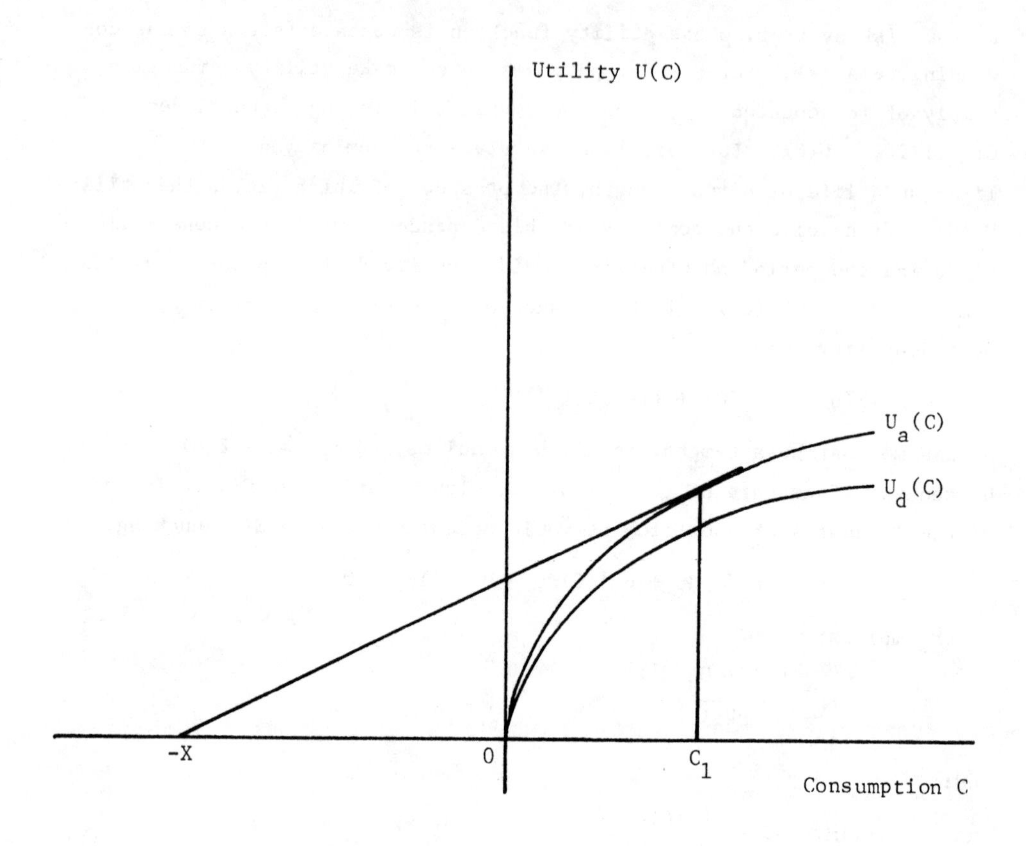

Figure 6-1. Shadow price of life, standard case

If we could be sure that the utility function passes through the origin, as shown in figure 6-1, there would be nothing further to discuss. Every nonsuicidal person with such a utility function is willing to pay for safety enough to imply a shadow price of life greater than C_1 (which includes intended bequest). That is, for p close to one, the shadow price of life is greater than discounted lifetime income for every such person.

However, Conley points out that, with consumption less than the subsistence level, the person would starve, so that every consumption level below that level is equivalent to zero consumption.[31] Denote the minimum subsistence level of consumption by C^o; the utility function for this case

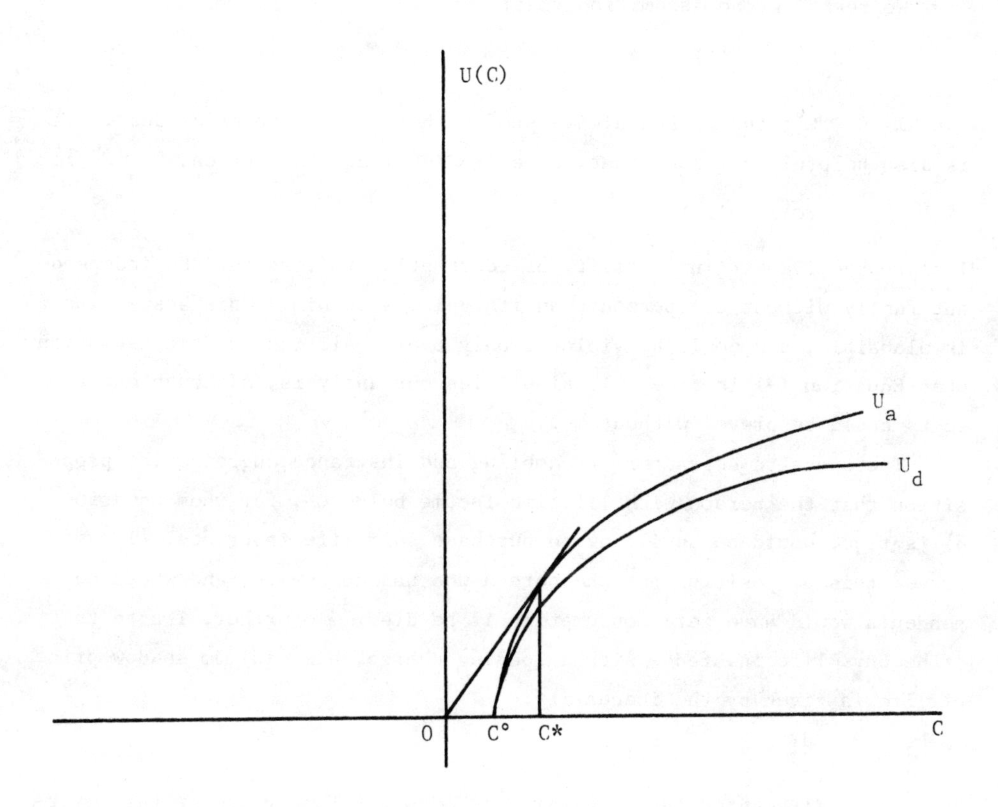

Figure 6-2. Conley's case: definition of C*

appears in figure 6-2. Here, $U_a(C^o) = U(0) = 0$. For consumption levels C below C*, at which a ray through the origin is tangent to U_a, the ratio in Equation (8) is less than C. A line tangent to the utility function to the left of C* intersects the horizontal axis to the right of the origin. For consumption levels C above C*, the case resembles figure 6-1, in that the shadow price of life is greater than C. Conley conjectured that the typical person expects lifetime income and consumption above C*, and so has a shadow price of life greater than this lifetime income. Cook disputes this conjecture,[32] and in response I have shown elsewhere that Conley is correct for nonsuicidal persons who buy life insurance,[33] if without such insurance their deaths would leave their dependents with zero consumption prospects.

We reaffirm the assumption that:

$$U_d(C) < U_a(C), \tag{9}$$

for all $C > C^o$; the person always prefers being alive to being dead. It is also helpful to assume that, at every level of consumption, $C > C^o$:

$$U'_d(C) \leqq U'_a(C) \tag{10}$$

The prospective marginal utility of consumption is less for the independent family without the person than its value with him. This assumption is plausible, and could be violated only over small ranges of consumption when Equation (9) is true. It simplifies our analysis, although our results could be proved without it.

The opposite character of gambling and insurance suggests the proposition that the person with lifetime income below C*, for whom Equation 8) is true, would be unwilling to purchase fair life insurance. I have proved this proposition for the person who has no wealth, and whose dependents would have zero consumption if he dies.[34] Further, I note that, if he buys life insurance with a loading charge, his minimum shadow price of life is given by the inequality:

$$-\frac{dY}{dp} > Y + (L-1)Y' , \tag{11}$$

where Y is discounted future earnings; Y' is the face value of the insurance policy; and L is the ratio of insurance premiums to expected claims.[35] The load factor raises the minimum shadow price above discounted lifetime earnings, by the indicated amount.

6.3.2 Generalization for Wealth and Welfare

My proof of Equation (11) used the assumption that a person's death reduces his dependents' consumption to zero unless he buys life insurance. Two further cases are of obvious interest; first, he has wealth that will survive him, and second, if he does not, a welfare program will save them from starving. Both these changes of assumption affect the willingness to buy life insurance, and so require reexamination of the proof.

The discussion of the shadow price of life given in Equation (8) and of the dividing line C* in no way involved the division of income between personal service income and income from wealth. Hence, if combined income from all sources is above C*, the person's shadow price of life is greater than this <u>entire</u> income. However, the purchase of life insurance no longer implies that this is necessarily the case, unless the person bought life insurance before he had income from wealth.

For a person with wealth, only his personal service income is lost if he dies permaturely, so that in Equation (7) the consumption level C' is nonzero, where C' is the consumption level due to the wealth that survives him. If he buys no life insurance and survives, the consumption level C he can have is:

$$C = Y + C' \quad , \tag{12}$$

where Y as noted above is his discounted future personal service earnings.

The concavity of the utility function for consumption above C^o directly implies that the shadow price of life exceeds C - C' if C' equals or exceeds C^o. It follows immediately from Equation (12) that the shadow price exceeds Y. This result points the way to the general result we seek, for cases in which C' is less than C^o. Recalling our explantion of figure 6-1, we write the shadow price of life as:

$$-\frac{dC}{dp} = C + X \tag{13}$$

where X is the positive or negative quantity that satisfies Equation (13).

Substituting from Equation (12), we have:

$$-\frac{dC}{dp} = Y + C' + X \tag{14}$$

which implies:

$$-\frac{dC}{dp} > Y \leftrightarrow C' + X > 0 \tag{15}$$

Hence, if we can establish that C' + X > 0 for every C' > 0, if the person buys life insurance, we will thereby generalize the earlier results. Our remarks following Equation (12) mean that Equation (15) can fail to hold only if $C' < C^o$. In this case, we have:

$$U_a(C') = U_d(C') = 0 \tag{16}$$

so that if we choose C' as the origin for the representation of income and utility, the logic used earlier, after Equation (8), can still be used.

That is, we can write:

$$U_a(C) = U_a(Y + C') = V_a(Y)$$

$$U_d(Y' + C') = V_d(Y') \tag{17}$$

where Y' is again the face value of his life insurance, if he buys any. Inasmuch as C' is a given constant, V_a and V_d represent U_a and U_d with the origin transposed, and by Equation (16) we have:

$$V_a(0) = V_d(0) = 0 \tag{18}$$

Constancy of C' implies that:

$$-\frac{dC}{dp} = -\frac{dY}{dp} \tag{19}$$

which we interpret as saying that insurance premiums are deductions from ("paid out of") earned income (not from accumulated wealth). The financial risk of premature death involves loss of Y, not C', and the loss of utility is the difference between $V_a(Y)$ and zero. Hence the choice problems the person faces concerning whether to buy life insurance and whether to accept a risky job for a given wage premium have the same logical form exactly as they had in the preceding section. That is, the proofs referred to in that section apply equally to the present problem just as if C' were zero, simply by replacing $U_a(C)$ and $U_d(C)$ by $V_a(Y')$, and $V_d(Y')$, respectively, and using those proofs to show that Equation (11) holds for every person who buys life insurance.

Proof of this claim is straightforward though laborious, and is omitted here. It is also straightforward to show that the result is even stronger if C' comes from a welfare program, not accumulated wealth, and so is available only if earned income drops more or less to the starvation level (or to zero due to the household head's death). Therefore the shadow price of life exceeds the discounted value of lifetime earnings for every nonsuicidal person who buys life insurance (and for those who would buy it if the premiums, while still exceeding claims, were lower). Furthermore, Equation (11) can be used to obtain a higher numerical minimum for this shadow price, above discounted earnings. How much higher than this minimum the shadow price actually is can be learned only by studying these persons' actual choices.

References

1. National Research Council, Halocarbons: Effects on Stratospheric Ozone (Washington, D.C., National Academy of Sciences, 1976); National Research Council, Halocarbons: Environmental Effects of Chlorofluoromethane Release (Washington, D.C., National Academy of Sciences, 1976); and National Research Council, Response to the Ozone Protection Sections of the Clean Air Act Amendments of 1977: An Interim Report (Washington, D.C., National Academy of Sciences, 1977).

2. Steven C. Wofsy, M. F. McElroy, and N. D. Sze, "Freon Consumption Implications for Stmospheric Ozone," Science vol. 187 (1975).

3. Federal Task Force on Inadvertent Modification of the Stratosphere (IMOS), Fluorocarbons and the Environment (Washington, D.C., Council on Environmental Quality and Federal Council for Science and Technology, June 1975).

4. National Research Council, Halocarbons: Environmental Effects, pp. 81-94, and 100.

5. U.S. Bureau of the Census, Projections of the Population of the United States, 1977 to 2050, Current Population Reports, Population Estimates and Projections Series P-25, No. 704,(Washington, D.C., Bureau of the Census, July 1977).

6. Federal Task Force (IMOS), Fluorocarbons and the Environment.

7. Robert Anderson, Lester G. Lave, and M. V. Pauly, "Health Costs of Changing Macro-climates," in Economics and Social Measures of Biologic and Climatic Change, Monograph No. 4, (Washington, D.C., U.S. Department of Transportation, Climatic Impact Assessment Program, 1975) p. 4-3.

8. Martin J. Bailey, Reducing Risks to Life: Measurement of the Benefits (Washington, D.C., American Enterprise Institute, 1980).

9. Elizabeth L. Scott, "Progress on the Analysis of NCHS Data on Non-melanoma Skin Cancer." Paper presented to the Environmental Protection Agency Workshop on Biological and Climatic Effects Research, University of Maryland, September 19, 1977.

10. Bailey, Reducing Risks to Life.

11. A. R. Shultz, D. A. Gordon, and W. L. Hawkins, "Materials Weathering," in Economic and Social Measures of Biologic and Climatic Change, Monograph No. 3 (Washington, D.C., U.S. Department of Transportation, Climatic Impact Assessment Program, 1975).

12. National Research Council, Halocarbons: Effects on Stratospheric Ozone.

13. National Research Council, Response to the Ozone Protection Sections.

14. See D. Wisecarver, "The Social Costs of Input-Market Distortions," American Economic Review vol. 64 (1974) p. 359; and Ralph C. D'Arge, L. Eubanks, and J. Barrington, Benefit-Cost Analysis for Regulating Emissions of Fluorocarbons 11 and 12 (Washington, D.C., U.S. Environmental Protection Agency, December 1976).
15. d'Arge, Eubanks, and Barrington, Benefit-Cost Analysis.
16. See the Federal Task Force (IMOS), Fluorocarbons and the Environment; and Wofsy, McElroy, and Sze, "Freon Consumption," p. 535.
17. d'Arge, Eubanks, and Barrington, Benefit-Cost Analysis.
18. National Research Council, Halocarbons: Environmental Effects.
19. See A. R. Prest and R. Turvey, "Cost-Benefit Analysis: A Survey," Economic Journal vol. 75 (1965) p. 683; and Martin J. Bailey and M. C. Jensen, "Risk and the Discount Rate for Public Investment," in Michael C. Jensen, ed., Studies in the Theory of Capital Markets (New York, Praeger, 1972) p. 269.
20. Mancur Olson and Martin J. Bailey, "Positive Time Preference," Journal of Political Economy, vol. 89 (1981) p. 1.
21. Martin Feldstein, "The Welfare Cost of Capital Income Taxation," Journal of Political Economy suppl. vol. 86, no. 2 (1978) p. S29; and Martin J. Bailey, "Saving and the Rate of Interest," Journal of Political Economy, vol. 65 (1957) p. 279.
22. Paul A. David and J. L. Scadding, "Private Savings: Ultarationality, Aggregation, and 'Denison's Law'," Journal of Political Economy vol. 82 (1974) p. 225.
23. L. R. Christensen, "Entrepreneurial Income: How Does It Measure Up?" American Economic Review, vol. 61 (1971) p. 571.
24. Ibid., table 1.
25. Arnold C. Harberger, "On Measuring the Social Opportunity Costs of Public Funds," in Proceedings of the Committee on Water Resources and Economic Development of the Discount Rate in Public Investment Evaluation (Denver, Colo., Western Agricultural Economics Research Council, 1968).
26. Edward F. Denison, "Effects of Selected Changes in the Institutional and Human Environment Upon Output Per Unit of Input," Survey of Current Business, vol. 58 (May 1978) p. 21.
27. Bailey, Reducing Risks to Life.
28. We are indebted to Talbot Page for his helpful comments on this part of the chapter.
29. E. J. Mishan, "Evaluation of Life and Limb: A Theoretical Approach," Journal of Political Economy, vol. 79 (1971).
30. Bailey, Reducing Risks to Life; and Olson and Bailey, "Time Preference."
31. B. E. Conley, "The Value of Human Life in the Demand for Safety," American Economic Review, vol. 66 (1976).

32. Philip J. Cook, "The Earnings Approach to Life Valuation" Mimeograph (Durham, N.C., Duke University, 1977).

33. Martin J. Bailey, "Earnings, Life Valuation, and Insurance" (College Park, Bureau of Business and Economic Research, University of Maryland, 1977); and Martin J. Bailey, "Safety Decisions and Insurance," American Economic Review Papers and Proceedings (May 1978).

34. Bailey, "Earnings."

35. Bailey, "Safety."

Chapter 7

POLLUTION, CLIMATE CHANGE, AND THE CONSEQUENT ECONOMIC COST CONCERNING AGRICULTURAL PRODUCTION

Harry K. Kelejian and Bruce C. Vavrichek

7.1 Introduction and Overview

In this chapter we develop and apply a model to assess the agricultural impacts of atmospheric pollution, and the social implications. Our model consists of a chain of physical, economic, and social processes which together determine the effects of various pollutants. The first link in this chain relates pollutants' release to their concentrations in the atmosphere. Next, the model relates these pollutant concentrations to changes in world climate, both directly through the greenhouse effect and indirectly through the ozone effect. We first consider the effects on temperature only and then link these temperature changes with associated precipitation changes. Given these changes in climate, we next analyze the resulting effects on world agricultural markets, specifically on agricultural productivity and crop production decisions. Finally, we consider the implications of changes in agricultural markets on measures of producer and consumer welfare.

Our economic model considers the production of several crops in several regions. More than one region is considered so that we may incorporate the trading of agricultural crops and so calculate the distribution of the effects of pollution changes across geographical regions. More

---

Authors Note: We thank J. Kahn for invaluable research assistance, and acknowledge helpful comments received from Martin J. Bailey, John H. Cumberland, James R. Hibbs, Irving Hoch, Ezra J. Mishan, Wallace E. Oates, Mancur J. Olson, Talbot Page, V. Ramanathan, and Ivar E. Strand. Computer support for this research was provided in part by the University of Maryland Computer Science Center.

than one crop is considered so that we can allow for the shifting of crop production intensities as the climate changes in various regions. Consideration of only one crop or region could lead to very biased conclusions about the economic consequences of pollution-related activities.

There are two major advantages of the comprehensive formulation of our model. First, in terms of making regulatory decisions concerning pollution release rates, the model allows for direct calculation of the estimated consequences of various policy options. Second, the need for additional research in specific areas can be suggested and evaluated, by comparing research strategies within our unified model.

The application of our model suggests that various pollution scenarios may affect the United States quite differently than they affect the rest of the world. For example, if the annual rate of chlorofluorocarbon emissions were to grow at 10 percent until the year 2000, wheat and corn production would not be greatly affected in the United States, but they would show considerable decline in the rest of the world. A second implication of our simulations is that the addition of the partial effects of CFCs, carbon dioxide, and nitrous oxide on crop production provides little indication of their combined effects. As an illustration, we estimate that if CFC production were to increase by 10 percent annually, carbon dioxide by 5 percent, and nitrous oxide by 2.5 percent, the sum of the partial effects of these pollutants on wheat production in the United States would be +0.40 percent. However, the total combined effect of these pollutants (where release of all three pollutants occurs simultaneously) is -13.1 percent.

7.2 The Technical Relationship Between Pollution, the Stratosphere, and Climatic Change

We begin with the effect of stratospheric pollution on climate. We first approximate the relationship between quantities of pollutants released into the atmosphere and change in global mean surface temperature, and then approximate the relationship between global temperature change, temperature change in latitude belts, and rainfall.

We have developed these approximations because of considerable disagreement among physical scientists on the quantitative effect of pollutant release on the climate. That disagreement reflects differences in model specification, data limitations, and the inherent randomness and lack of stationarity of the underlying physical processes. Our approximations consist of linear relationships, and though the linear specification is simplistic, it is thought that the errors introduced by this functional form will not do serious injustice to the existing scientific research. Further input on these relationships from physical scientists should lead to more accurate formulations. In our analysis, we primarily rely on Ramanathan,[1] a 1977 report by Kellogg,[2] works by the National Research Council,[3] and a study by the Council on Environmental Quality.[4] Whenever published results seem unresolvedly contradictory, the work of Ramanathan is used.

7.2.1 The Effects of Three Pollutants on Global Mean Surface Temperature

This section considers, in turn, the effects on temperature of carbon dioxide, chlorofluorocarbons (CFCs), and nitrous oxide. For each chemical group, the relationship between release rates and concentrations in the stratosphere is examined, and then changing concentration levels are related to changes in global mean surface temperature. In the process, we approximate linear greenhouse and ozone effects for each chemical group. As noted in the introduction to this volume, the greenhouse effect is the primary direct effect of atmospheric pollutants, as they absorb the earth's heat and remit it to the earth's surface. The ozone effect arises because some of the pollutants, particularly the CFCs, react with ozone to reduce its density, diminishing the ability of the ozone to insulate the earth. As compared to the greenhouse effect, the effect on surface temperatures of a reduction in the ozone shield is likely to be small. A summary overview of these effects, which we draw upon in our detailed analyses, appears as table 7-1.

7.2.1.1 The Effect of Carbon Dioxide on Temperature. The primary source of increased carbon dioxide (CO_2) in the atmosphere is the burning of fossil fuels. According to Woodwell,[5] prior to the Industrial Revolution, the CO_2 concentration was stable, and the burning of fossil fuels

Table 7-1. Summary of Anthropogenic Influences on the Global Mean Surface Temperature

Effect on mankind	Atmospheric concentration	The "greenhouse" effect on surface temperature (°C)	The "ozone" effect on surface temperature (°C)
Raising the CO_2 content of the atmosphere	+25%[a] by 2000 AD	+0.5 to 2	0
	+100%[a] by 2050 AD	+1.5 to 6	0
Adding CFCs to the troposphere	0.9 ppbv[b] by 2000 AD	+0.1 to 0.4[e]	-0.01 to -0.04
	2.5 ppbv[c] by 2050 AD	+0.25 to 1[e]	-0.025 to -0.1
	3.5 ppbv[d] by 2000 AD	+0.4 to 1.5[e]	-0.04 to -0.15
Raising the N_2O content of the atmosphere	+100% to 2050 AD	+0.25 to 1[f]	Effect uncertain

Source: W.W. Kellogg, "Effect of Human Activities on Global Climate," World Meteorological Technical Note No. 156 (Geneva, World Meteorological Organization, February 1977) p. 22, as modified by Ramanathan.

[a]Relative to the 1973 concentration level.

[b]This would result if CFC production continued at the 1973 level until the year 2000; ppbv equals parts per billion by volume.

[c]This would result if CFC production continued at the 1973 level until the year 2050.

[d]This would result if CFC production increases 10 percent per year until the year 2000, where the base year is 1973.

[e]Estimated by Ramanathan, and reviewed and extended by the National Research Council, National Academy of Sciences, Panel on Atmospheric Chemistry, Halocarbons: Effects on Stratospheric Ozone (Washington, D.C., NAS, 1976).

[f]Estimated in 1976. This now appears to be an upper limit on the increase in N_2O in this time period.

was, at most, a minor source of CO_2. The CO_2 concentration in the atmosphere has changed from its pre-Industrial Revolution steady-state value of 290 parts per million by volume (ppmv) to a 1975 value of 330 ppmv.[6] We assume that other sources of CO_2 (vegetation, marine life, and decaying organic matter) are responsible for the previous steady-state concentration and that any increase is attributable to the burning of fossil fuels.

To estimate the relation between change in the atmospheric concentration of CO_2 from 1975 through the year 2000 in units of ppmv (expressed as $\Delta CONC_{CO2}$) and the total release of CO_2 into the atmosphere during that period in units of billions of metric tons (expressed as REL_{CO2}), we use the following information from Mitchell.[7] Approximately 45 percent of all CO_2 produced does not remain in the atmosphere, but is deposited in carbon reserves, primarily in the oceans. Thus, of the total 1975 fossil fuel production of CO_2 of 20 billion metric tons, only 11 billion metric tons became part of the CO_2 in the atmosphere. According to Mitchell, this 11 billion tons represented 0.385 percent of the total CO_2 in the atmosphere. In turn, this is the equivalent of a change in the concentration of CO_2 of 1.27 ppmv (330 ppmv x 0.00385 = 1.27 ppmv). Thus, the release of 1 billion metric tons of CO_2 into the atmosphere would increase the atmospheric concentration by 0.116 ppmv (1.27/11 = 0.116). (The atmospheric residence time for CO_2 is well in excess of one hundred years. Thus, all of the CO_2 released into the atmosphere during the period 1975 through 2000 will remain in the year 2000. In this particular case, release and retention of CO_2 are the same. For the other two chemical groups to be considered later, the atmospheric residence times are shorter, and release and retention are considered separately.)

We formulate the relation between $\Delta CONC_{CO2}$ and REL_{CO2} linearly as

$$\Delta CONC_{CO2} = b(REL_{CO2} - \overline{REL}_{CO2}) \tag{1}$$

where $\Delta CONC_{CO2}$ and REL_{CO2} are as defined before, b is a constant, and $\overline{REL}_{CO2}$ is that level of REL_{CO2} which would leave the atmospheric concentration of CO_2 unchanged ($\Delta CONC_{CO2} = 0$). According to the discussion above, b = 0.116, and $\overline{REL}_{CO2} \doteq 0$, thus, we estimate Equation (1) as

$$\Delta CONC_{CO2} = 0.116\ REL_{CO2} \tag{2}$$

To relate this cumulative change in the atmospheric concentration of CO_2 from 1975 to 2000 to the resultant temperature change measured in degrees Centigrade during this same period due to the greenhouse effect (ΔT^G_{CO2}) we use the Kellogg-Ramanathan results in table 7-1. There, we see that a 25 percent increase in CO_2 concentration during the period 1973-2000 is estimated to result in a 1.25°C increase in temperature (using the midpoint of the temperature interval as the most likely value). The change in concentration from 1973 to 2000 is the 1975-2000 period change ($\Delta CONC_{CO2}$) plus the difference between the 1975 concentration (330 ppmv) and 1973 concentration (325 ppmv). Assuming a zero temperature change if the CO_2 concentration were to remain at its 1973 level, and noting that 25 percent of the 1973 concentration is 81.25 ppmv (325 ppmv x 0.25) we can approximate the relation between $\Delta CONC_{CO2}$ and ΔT^G_{CO2} as

$$\Delta T^G_{CO2} = \frac{1.25^\circ C}{81.25 \text{ ppmv}} (5. + \Delta CONC_{CO2}) \quad (3)$$

or

$$\Delta T^G_{CO2} = 0.0770 + 0.0154 \, \Delta CONC_{CO2} \quad (4)$$

where in Equation (4) both variables are cumulative for the period 1975-2000. (An alternative estimation of the relation in Equation (4) is obtained from the National Research Council Study that predicted a 25-year temperature increase of 0.5°C given a year 2000 CO_2 concentration of 390 ppmv.[8] These data yield the estimated relation, $\Delta T^G_{CO2} = 0.0083 \, \Delta CONC_{CO2}$. We have chosen to use Equation (4) for uniformity of our analysis.) Again following the Kellogg-Ramanathan results, we assume the temperature change caused by the effect of CO_2 on the ozone layer (ΔT^{OZ}_{CO2}) to be nil, and obtain the total temperature effect of a change in CO_2 concentration for the period 1975 through 2000 to be

$$\Delta T_{CO2} = \Delta T^G_{CO2} + \Delta T^{OZ}_{CO2} = 0.0770 + 0.0154 \, \Delta CONC_{CO2} \quad (5)$$

Through Equations (2) and (5) the estimated temperature effect of the 1975-2000 release of CO_2 into the atmosphere can be approximated.

7.2.1.2 The Effect of CFCs on Temperature. An initial set of ozone-depleting halocarbons was considered in this study; this set included the CFCs, particularly F-11 (CCl_3F) and F-12 (CCl_2F_2), as well as carbon tetrachloride (CCl_4) and methyl chloroform (CH_3CCl_3). The total concentration of these chemicals in the atmosphere in 1975 was 0.1 ppbv.[9] Although a large quantity of carbon tetrachloride (CCl_4) was produced, little was released into the atmosphere. Methyl chloroform has a high release rate into the atmosphere, but a short lifetime there. Further, in terms of the greenhouse effect, which is determined by the total amount of pollutants suspended in the atmosphere, methyl chloroform is a relatively minor pollutant. In contrast, not only is a high percentage of the production of CFCs released into the atmosphere, but the quantities that are released reside there for a long period of time. Further, F-11 and F-12 jointly account for the bulk of CFC production (Gladwin, Ugelow, and Walter, in table 3-4, show that F-11 and F-12 account for 80 percent of world production of CFCs). Hence, we concluded that we could obtain a useful approximation to the relationship between the CFC concentration in the atmosphere and the CFC release rate by considering only F-11 and F-12. We assume a twenty-five year lifetime for F-11 and F-12, followed by instantaneous decay.

To estimate a linear relation between the twenty-five year (1975 to 2000) cumulative change in CFC atmospheric concentration, as measured in parts per billion by volume ($\Delta CONC_{CFC}$), and their twenty-five year cumulative release, as measured in 10^6 metric tons (REL_{CFC}), we use data obtained from the National Research Council.[10] These data relate to three cases. In the first case it is calculated that total (world) quantities of CFCs released from 1951 through 1975 were 2.854 million metric tons of F-11 and 4.084 million metric tons of F-12. These add to 6.938 million metric tons and correspond to an atmospheric concentration of CFCs of 0.1 ppbv. The second case projects a CFC concentration of 0.9 ppbv in the year 2000 assuming continued CFC production at the 1973 level. The final case projects a CFC concentration of 3.5 ppbv in the year 2000 assuming CFC production grows annually at 10 percent of its 1973 level. Data relating to these three cases appear in table 7-2. We further assume that a zero production rate of CFCs for twenty-five years would imply a zero concentration in the year 2000.

These data together imply the estimated relation

$$\Delta CONC_{CFC} = -0.100 + 0.0361\ REL_{CFC} \qquad (6)$$

where both variables are cumulative from 1975 to the year 2000.

Table 7-2. Concentration and Quantities of CFCs

Case	Concentration in ppbv	Twenty-five year cumulative release in metric tons of F-11 and F-12
Year 1975	0.1	6.938×10^6
Year 2000, continued 1973 levels of production	0.9	17.415×10^6
Year 2000, production increasing at 10% annually	3.5	82.896×10^6

To estimate the relation between this change in CFC concentration and the twenty-five year cumulative change in temperature via the greenhouse effect (ΔT^G_{CFC}) we again use the data in table 7-1. If the CFC concentration is increased to 0.9 ppbv (3.5 ppbv) by the year 2000, the expected temperature increase will be 0.25^oC (0.95^o), using central interval values as expected temperatures. These data yield the estimated relationship,

$$\Delta T^G_{CFC} = 0.00769 + 0.2692\ (\Delta CONC_{CFC} + 0.1) \quad (7)$$

where $\Delta CONC_{CFC} + 0.1$ is the level of the CFC concentration in 2000. Thus,

$$\Delta T^G_{CFC} = 0.0346 + 0.2692\ \Delta CONC_{CFC} \quad (8)$$

Turning to the CFC-induced change in surface temperature which results from changes in the ozone layer, the National Research Council notes two off-setting effects of reduced stratospheric ozone.[11] First, more solar radiation reaching the earth's surface will tend to produce a warming effect. Second, a reduction in solar absorption in the stratosphere will cause a cooling of the stratosphere, which, in turn, will cause a cooling of the earth's surface. Though these effects essentially are offsetting, there is a small net cooling effect. Ramanathan (see also table 7-1) concurs with these findings and approximates this cooling as approximately 10 percent of the greenhouse (warming) effect of CFCs,[12] suggesting that we can account for the ozone effect by scaling Equation (8) by -0.1. (Further confirmation of these estimates appears in Forziati's discussion in chapter 2.) The total effect of CFCs on surface temperature (ΔT_{CFC}) is the sum of the greenhouse and ozone effects:

$$\Delta T_{CFC} = 0.03114 + 0.2429\ \Delta CONF_{CFC} \qquad (9)$$

<u>7.2.1.3 The Effect of Nitrous Oxide on Temperature</u>. Nitrous oxide (N_2O) is a naturally produced compound which is reasonably prevalent in the atmosphere and potentially important in determining the climate through its greenhouse effect. In the atmosphere, N_2O reacts with oxygen to form two other compounds (NO and NO_2) which are directly responsible for 70 percent of the so-called natural ozone destruction constantly occurring in the stratosphere.[13] Nitrogen fertilizer is a recently discovered source of nitrous oxide in the atmosphere, and it should increase in importance given increased applications of fertilizer, over time. However, recent research suggests the nitrous oxide effect has been overstated, so that the previous "most likely" estimate of the impact of nitrous oxide on climate and ozone now appears to be the "upper limit" of this effect.[14] In particular, the ozone-depleting effect appears small and of uncertain sign, so that it seems reasonable to treat that effect as essentially zero.

The 1975 concentration of nitrous oxide in the atmosphere was 250 ppbv, corresponding to a natural production (and release) rate in the oceans and soil of 1.55×10^8 metric tons per year.[15] The atmospheric lifetime of nitrous oxide is thought to be approximately fifteen years,[16] so that the 1975 concentration (250 ppbv) corresponds to a 23×10^8 ($= 1.55 \times 10^8 \times 15$) metric ton quantity of N_2O in the atmosphere.

To estimate the relation between cumulative change in N_2O concentration by the year 2000, as measured in ppbv (as expressed by $\Delta CONC_{N2O}$), and cumulative fifteen-year release of N_2O, as measured in 10^8 metric tons (as expressed by REL_{N2O}), we use the above information as well as noting that the concentration would drop to zero if the cumulative N_2O release were zero for fifteen consecutive years. These facts imply the estimated relation

$$\Delta CONC_{N2O} = -250. + 10.9\ REL_{N2O} \qquad (10)$$

The relationship of $\Delta CONC_{N2O}$ to the cumulative temperature change by the year 2000 attributable to the greenhouse effect of N_2O (ΔT^G_{N2O}) is estimated using the data in table 7-2. With the knowledge that Kellogg's original estimate of the greenhouse effect of a change in N_2O concentration was too large, we choose as most likely a temperature increase of 0.3°C by the year 2050 owing to a 100 percent increase in N_2O concentration. Further, we linearly interpolate this result to state that a one-third increase in N_2O concentration would by 2000 imply a temperature increase of 0.1°C. The two data points for $(\Delta T^G_{N2O}, \Delta CONC_{N2O})$ of (0.0°c,0) and (0.1°, 83.3 ppbv) imply the relation

$$\Delta T^G_{N2O} = 0.0012 \ REL_{N2O} \qquad (11)$$

Given the uncertainty in the direction of the ozone effect on temperature, we assume the effect is approximately zero, i.e.:

$$\Delta T^{OZ}_{N2O} \doteq 0 \qquad (12)$$

resulting in the overall estimate

$$\Delta T_{N2O} = 0.0012 \ REL_{N2O} \qquad (13)$$

7.2.2 A Multivariate Formulation of Climatic Change

Since agricultural production depends on precipitation and variation in precipitation as well as temperature, we are concerned with how those climate variables relate to mean temperature changes. Climatologists have been more secure with "one-dimensional" models, which predict _global_ mean temperature and precipitation, than with "two-dimensional" or "three-dimensional" models, which respectively deal with climatic variables disaggregated by latitude, or by latitude _and_ longitude. Despite the speculative nature of the case, however, we felt it necessary to construct at least a two-dimensional model, since the distribution of pollutant impacts is crucial in determining economic and social impacts. Here, we drew on a survey of twenty-four eminent climatologists by the National Defense University.[17]

In that survey, five global temperature change predictions were distinguished, and for each a scenario of changes in other climatic variables was constructed. The five global temperature change predictions and the distribution of responses by the twenty-four climatologists are presented in table 7-3. Table 7-3 reports, for example, that 10 percent of the climatologists believe that Northern Hemisphere temperature will decrease

Table 7-3. Distribution of Predicted Global Change in Temperatures in Survey of Climatologists

Temperature change category	Predicted change in mean Northern Hemisphere temperature from present by the year 2000 (in °C)	Midpoint of temperature range (in °C)	Relative frequency
I Large cooling	0.3° to 1.2° colder	-0.75°	0.10
II Moderate cooling	0.05° to 0.3° colder	-0.175°	0.25
III Same as last 30 years	0.05° colder to 0.25° warmer	+0.10°	0.30
IV Moderate warming	0.25° to 0.6° warmer	+0.425°	0.25
V Large warming	0.6° to 1.8° warmer	+1.2°	0.10

Source: National Defense University, Climatic Change to the Year 2000: A Survey of Expert Opinions (Washington, D.C., Defense Advanced Research Projects Agency, Department of Defense, Fort McNair, 1978).

by some amount between 0.3°C and 1.2°C by the year 2000. (For simplicity, we will treat the midpoint of a temperature range as representative of the range.)

The climatologists also predicted a latitudinal temperature pattern for each of the temperature change categories, employing four latitude categories. The average values of their responses are presented in table 7-4, yielding a geographic distribution of temperature change for a given change in global mean temperatures.

Precipitation predictions were obtained from all twenty-four climatologists for each of the temperature scenarios; each climatologist was then given a set of three choices to respond to questions of this sort: "If the global temperature changes by an amount X, by how much will precipitation change in latitude belt Y?" For annual precipitation the choice categories were, "As compared to 1941-1970," annual precipitation will increase by more than 10 percent, change by less than 10 percent, or decrease by more than 10 percent. The climatologists indicated their choice of these three options for each of the latitude belts.

Table 7-4. Latitudinal Temperature Changes for the Five Temperature Scenarios

Latitude belt	Latitude range	Temperature Scenario (°C)				
		I (Mean=-75°)	II (-.175°)	III (+.10°)	IV (+.425°)	V (+1.2°)
Polar	65°-90°	-1.75°	-1.00°	+0.25°	+1.25°	+3.20°
Higher mid-latitude	45°-65°	-1.05	-0.50	+0.25	+0.65	+1.50
Lower mid-latitude	30°-45°	-0.85	-0.35	+0.25	+0.45	+1.43
Subtropic latitude	10°-30°	-0.50	-0.30	+0.20	+0.40	+0.78

Assuming that the change in precipitation, measured in percentage points, is a normal random variable allowed us to calculate a mean and variance for each distribution. For example, one of our cases has this distribution:

Change in precipitation (ΔP)	Relative frequency
$+10\% \leq \Delta P$	0.2
$-10\% \leq \Delta P \leq +10\%$	0.5
$\Delta P \leq -10\%$	0.3

(14)

We treat the frequency as a probability and write:

$$\text{Prob}(\Delta P \geq 10) = .2, \text{ and Prob}(\Delta P \leq -10) = .3 \tag{15}$$

applying our normality assumption, and denoting the mean and variance of ΔP as μ and σ^2, we transform Equation (16) to:

$$\text{Prob}\left(\frac{\Delta P-\mu}{\sigma} \geq \frac{10-\mu}{\sigma}\right) = .2, \text{ and Prob}\left(\frac{\Delta P-\mu}{\sigma} < \frac{-10-\mu}{\sigma}\right) = .3 \tag{16}$$

Since $(\Delta P-\mu)/\sigma$ is a standard normal variable, it follows from the standard normal table that:

$$(10 - \mu)/\sigma = .840, \text{ and } (-10 - \mu)/\sigma = -.525 \tag{17}$$

which together imply that $\mu = -2.3$ percent and $\sigma = 14.7$ percent.

The results calculated above were obtained for latitude belt "higher mid-latitude" and temperature category I, as defined in table 7-4. They state that if mean global temperature were to decrease by 0.75°C, the precipitation in the higher mid-latitude belt would be expected to decrease by 2.3 percent. Similar calculations were carried out for each combination of temperature category and latitude belt, with results presented in table 7-5. Using the means of the precipitation changes in table 7-5 and the temperature figures from table 7-4, the relation between the change in temperature and the resulting mean percentage change in precipitation can be plotted as shown in figure 7-1.

The temperature-precipitation relationships in table 7-5 refer to latitude belts as a whole. Some reverse relationships may be superimposed on those general patterns within the interiors of mid-latitude continents, based on numerical modeling.[18] The large standard deviations in table 7-5 suggest that the climatologists' survey does not preclude such relationships. For uniformity of analysis, however, we use the predictions from the climatologists' survey with the caveat that disaggregated information may reveal different relationships within subregions. Within the constraints of that caveat, the general pattern appearing in figure 7-1 is one of positive correlation between temperature change and precipitation, the two generally increasing together.

The climatologists were also asked how much the variability in precipitation would change. Would precipitation, for example, be more uniform over time, or would there be periods of drought and periods of heavy precipitation? Again, there were three categories of response: that "precipitation variability will" increase by more than 25 percent, change by less than 25 percent, or decrease by more than 25 percent. Means and standard deviations were calculated for the distribution of responses by procedures paralleling those employed for table 7-5; results of this are shown in table 7-6. We can consider the results of those tables in tandem. Thus, if we plot mean values from table 7-5 against those of table 7-6 we can infer that precipitation variability will decrease as temperatures increase in the higher mid-latitude belt. Similar relationships for the other latitude belts can easily be derived. Given this information on climate change, we now consider resulting effects on agricultural production.

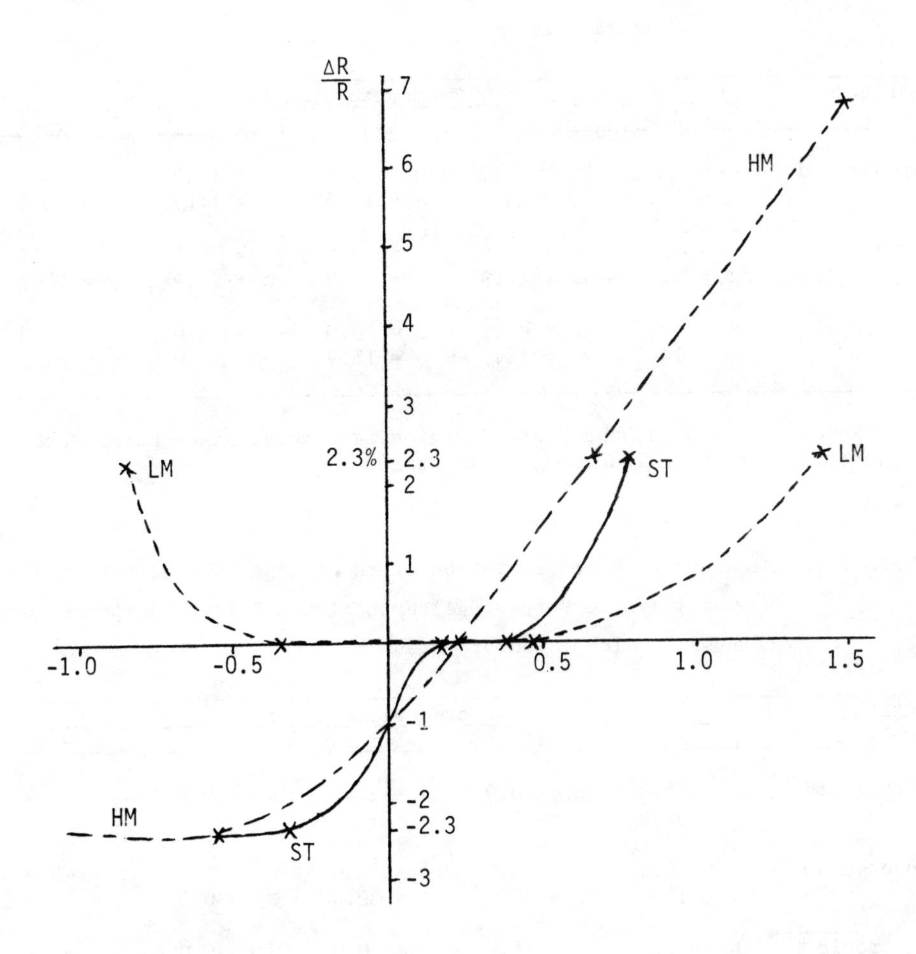

Figure 7-1. Percentage change in precipitation as a function of the change in temperature for the three latitude belts. HM, Higher mid-latitude belt; LM, Lower mid-latitude belt; ST, Subtropic latitude belt.

Table 7-5. Mean and Standard Deviation of Annual Precipitation Change (in Percentage) by Latitude Belt in Response to Changes in Global Temperatures

Latitude belt	Temperature Category I	II	III	IV	V
Higher mid	μ = -2.3 σ = 14.7	μ = -2.3 σ = 14.7	μ = 0.0 σ = 11.9	μ = +2.3 σ = 14.7	μ = +6.8 σ = 13.0
Lower mid	μ = +2.3 σ = 14.7	μ = 0.0 σ = 11.9	μ = 0.0 σ = 11.9	μ = 0.0 σ = 11.9	μ = +2.3 σ = 14.7
Subtropic	μ = -2.3 σ = 14.7	μ = -2.3 σ = 14.7	μ = 0.0 σ = 11.9	μ = 0.0 σ = 11.9	μ = +2.3 σ = 14.7

Note: Henceforth, the "Polar" region is ignored since normal crop production in that region is virtually zero.

Table 7-6. Mean and Standard Deviation of the Change in Precipitation Variability (in Percentage) by Latitude Belt in Response to Changes in Global Temperatures

Latitude belt	Temperature Category I	II	III	IV	V
Higher mid	μ = +5.9 σ = 37.0	μ = 0.0 σ = 30.0	μ = 0.0 σ = 30.0	μ = 0.0 σ = 30.0	μ = -5.9 σ = 37.0
Lower mid	μ = +5.0 σ = 37.0	μ = 0.0 σ = 30.0	μ = 0.0 σ = 30.0	μ = 0.0 σ = 30.0	μ = -5.9 σ = 37.0
Subtropic	μ =+13.5 σ = 46.0	μ = +5.9 σ = 37.0	μ = 0.0 σ = 30.0	μ = 0.0 σ = 30.0	μ = -5.9 σ = 37.0

7.3 The Economic Model

In this section we develop the model used to assess some of the economic and social implications of climatic changes induced by stratospheric pollution. Our presentation is a general one, seen as useful in future applications, as well as in our specific empirical work. That empirical work, as reported in section 7.4 below, involved some necessary delimitations and simplifications.

7.3.1 Supplies of Agricultural Crops: An Overview

In any geographic region, the supplies of agricultural crops depend on several factors, including attributes of the climate, the amount of land used in the production of each crop, and the quantities of other production inputs, such as fertilizers. This dependence can be formalized by noting that the quantity produced of an agricultural crop is equal to the average crop yield per unit of land, Y, multiplied by the quantity of land used in its production, L. Of these two components, the average crop yield depends on climate characteristics of the area (including, we will assume, temperature, T, and rainfall, R) and on nonland factors of production. The effect of climate on crop yield is of particular concern since climate is affected by atmospheric pollution policies. Land use will also depend indirectly upon climatic conditions. The reason for this is that land allocation among crops, and the overall level of land use will be determined by the profitability of the various crops which, in part, will be determined by their yields. Therefore, atmospheric pollution policies, via their climate effects, can be expected to affect both the yield per unit of land of the various crops as well as the allocation and overall level of land use in agricultural production. We will now formalize some of these ideas, beginning with a discussion of the yield equation.

7.3.1.1 The Yield Equation.

The general form of the yield function used in this study is:

$$Y_i^j = \phi_i^j (T^j, R^j) + \gamma_i^j (X_i^j) + B_i^j,$$

$$i = 1, \ldots, N \; ; \; j = 1, \ldots, M \; , \qquad (19)$$

where

Y_i^j is the yield per unit land of the i-th crop in the j-th region;

T^j and R^j are, respectively, measures of temperature and rainfall in j-th region;

X_i^j is a measure of the quantity of fertilizer applied per unit of land to crop i in the j-th region;

and B_i^j is a term representing the net effect of all of the other factors which affect crop yields.

The number of crops is denoted by N, and the number of regions by M.

Finally, ϕ_i^j and γ_i^j are functions describing, respectively, the effects of the climate and fertilizer variables on the corresponding yield.

We assume that an increase in fertilizer application will lead to an increase in the corresponding yield, with the further usual assumption of diminishing marginal productivity. Concerning the temperature and rainfall variables, we assume that $\phi_i^j(T^j, R^j)$ is concave and that the marginal products of both T^j and R^j first increase (at a decreasing rate) and then decrease (at an increasing rate). For illustrative purposes the assumed relationship between the yield and T^j for a given value of R^j, say R_o^j, is as outlined in figure 7-2.

7.3.1.2 The Supply of Land Equation. Although total land in a particular region is fixed, it may have a variety of uses. For crop production, then, available land need not be considered fixed. To make a unit of land available for agricultural production, its services must be bid away from other potential users. Small quantities of land can be bid away at a relatively low price because of the low marginal product of land elsewhere when its supply is great, but larger quantities can only be bid away at a higher price. This "opportunity cost" basis for land pricing suggests viewing the supply of land to agricultural users as an increasing function of its price. Let P_L^j denote the price of land in region j relative to the price of the N-th crop (the numeraire); let L_i^j denote the quantity of land used in the production of crop i in region j. Then, the supply of agricultural land is assumed to be related to its real price as

$$P_L^j = h^j(\sum_i L_i^j) \tag{20}$$

where h^j is an increasing function, and ΣL_i^j is the total land used in agriculture in region j.

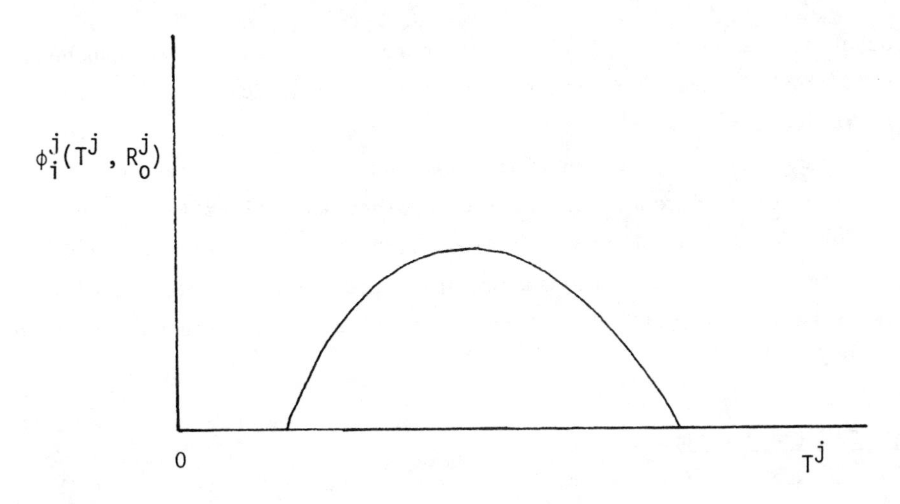

Figure 7-2. Yield and temperature

7.3.1.3 Crop Supply. We can now express the supply of crop i in region j, S_i^j, as the product of its average yield and land used in that crop's production:

$$S_i^j = Y_i^j . L_i^j \tag{21}$$

The direct dependence of S_i^j upon the climate variables is evident from Equations (19) and (21).

7.3.2 The Demand Equations

We specify the demands for agricultural crops as functions of their prices, a measure of the overall price level in the region, and per capita personal income (all measured relative to the numeraire), as well as the size of the regional population. Functionally, this relation is represented as:

$$\frac{D_i^j}{POP^j} = d_i^j(P_1^j, \ldots, P_{N-1}^j, PIND^j, I^j) \tag{22}$$

where relating to region j, D_i^j is the total demand for crop i; P_k^j is the real price of the k-th agricultural crop, k=1, . . . , N-1; $PIND^j$ is the real price level for all goods; I^j is real per capita income; and POP^j is the size of the regional population. The demand for a particular agricultural crop is therefore assumed to depend not only on its own price, but other prices as well. Differences in tastes across regions can be accounted for by parameter differences or functional form differences across the regions. At this time we do not specify the functional forms of the demand functions in Equation (22). We do, however, assume that agricultural crops do substitute for each other as well as for "other goods" in general, and so the demand for each crop is assumed to be positively affected by each of the substitute price terms in Equation (22). We also take the own price effect to be negative and the income effect to be positive.

7.3.3 The Trade Assumptions

We assume that interregional trade occurs so as to ensure equilibrium in world markets. In the absence of transportation costs and other impediments, such as tariffs (transportation costs and other impediments to trade are considered below), this implies that all regional real prices will be the same for each crop so that:

$$P_i^1 = P_i^2 = \ldots = P_i^M = P_i, \text{ for } i=1, \ldots, N \qquad (23)$$

Correspondingly, the market equilibrium condition is that total world demand must equal total world supply for each commodity:

$$\sum_j D_i^j = \sum_j Y_i^j \cdot L_i^j, \text{ for } i=1, \ldots, N \qquad (24)$$

7.3.4 Behavioral Assumptions for Producers

We assume that in each region crop producers behave in such a way that the aggregate demands for the factors of crop production in that region maximize aggregate real profits over all the crops produced in that region. We therefore determine the factor demand equations for each region j in terms of the aggregate profit model,

$$\pi^j = \sum_i (Y_i^j L_i^j P_i) - P_L^j \sum_i (L_i^j) - P_X^j \sum_i (X_i^j L_i^j) - C^j \qquad (25)$$

where P_X^j is the real price of fertilizer, which is assumed to be exogenously given.

The first term in Equation (25), $\sum_i (Y_i^j L_i^j P_i)$, represents the sum of real revenues from the production of all crops in the region. The second and third terms, $P_L^j \sum_i (L_i^j)$ and $P_X^j \sum_i (X_i^j L_i^j)$, represent real land and fertilizer costs in region j which are associated with the other factors of crop production.

The profit function in Equation (25) can be maximized in a variety of ways depending upon the assumed extent of price endogeneity, and the rivalry of the crop producers in the various regions. We consider two somewhat limiting cases below; these correspond to models of perfect competition and "Cournot Monopolistic Competition." In both of these cases the control variables associated with the maximization of Equation (25) are the specific land and fertilizer variables, L_i^j and X_i^j, i = 1, . . . , N.

7.3.4.1 The Competitive Case. In this model all prices associated with the maximization of Equation (25) are taken as given. This model corresponds to the case in which crop production in each region is "small" relative to the world market, and within each region, each producer is a "small" demander of land and fertilizer.

In light of Equations (19) and (20), the first-order conditions for the maximization of π^j can be expressed as

$$\frac{\partial \pi^j}{\partial L_i^j} = [\phi_i^j(T^j, R^j) + \gamma_i^j(X_i^j) + B_i^j] P_i - h^j (\sum_i L_i^j) - P_X^j X_i^j = 0 \qquad (26)$$

$$\frac{\partial \pi^j}{\partial X_i^j} = \frac{d[\gamma_i^j(X_i^j)]}{d[X_i^j]} L_i^j P_i - P_X^j L_i^j = 0, \qquad i = 1, \ldots, N; \quad j = 1, \ldots, M. \qquad (27)$$

The condition shown in Equation (26) states that, in the competitive model, land should be employed for each crop's production up to the point where the marginal revenue of the last unit of land equals the cost of using that last unit. Similarly, Equation (27) states that in each region, fertilizer should be applied to each unit of land used in the production of a particular crop to the point where its contribution to total revenue equals the cost of the last unit of fertilizer (applied to all units of that crop's land). Equation (27) also yields the familiar implication that, in equilibrium, the use of fertilizer in the two regions will be such that the ratio of the marginal products will be equal to the ratio of the fertilizer prices.

Consider, now, another point which may not be so evident. The competitive model generally determines the prices of the commodities, and the <u>total</u> quantities traded. It does not generally determine the production of each of the competitive units. In the present case, the model will not determine the specific land variables L_i^j, nor therefore the distribution over the regions of world crop production. To see this, note that the equations determining the values of the control variables in this model are Equations (23), (24), (26), and (27). The land use variables, L_i^j, appear in this system because they appear in Equation (24), and because their sum, ΣL_i^j, appears in Equation (26). There are N equations in (24), and M total land variables in (26). Therefore, in general, there are not enough equations to uniquely solve for the land variables, and so the variables L_i^j are not determined.

<u>7.3.4.2 Cournot Monopolistic Competition</u>. In this case, crop and land prices are endogenous in the maximization model shown in Equation (25). Specifically, account is taken of the dependence of world crop prices on crop production in the j-th region, and of land prices in the j-th region on the total use of land as an agricultural input in that region. We avoid game-theoretic issues by maximizing π^j for each region j, subject to constant levels of crop production in regions other than j. As in the competitive model, we take fertilizer prices as exogenous. This model corresponds to the case in which regional crop producers are "large" both relative to world crop production, and to the total own-region use of land.

The first-order conditions for the maximization of π^j, for each region j, subject to Equation (19) through (24), can be expressed as

$$\frac{\partial \Pi^j}{\partial L_i^j} = [\phi_i^j(T^j,R^j) + B_i^j + \gamma_i^j(X_i^j)]P_i + \sum_k\{[\phi_k^j(T^j,R^j) + B_k^j + \gamma_k^j(X_k^j)]*$$

$$*L_k^j\left[\frac{\partial P_k}{\partial L_i^j}\right]\} - h^j(\sum_k L_k^j) - \sum_k(L_k^j)\left[\frac{dh^j}{dL_i^j}\right] - P_x^j X_i^j = 0 \qquad (28)$$

$$\frac{\partial \Pi^j}{\partial X_i^j} = \gamma_i^j L_i^j P_i + \sum_k\{[\phi_k^j(T^j,R^j) + B_i^j + \gamma_k^j(X_k^j)]L_k^j\left[\frac{\partial P_k}{\partial X_i^j}\right]\} - P_x^j L_i^j = 0 \qquad (29)$$

for i = 1, . . . , N; j = 1, . . . , M.

The first two terms in Equation (28) represent the marginal revenue of land used for crop i; the first term, expressible as $Y_i^j P_i$, is the value of the marginal product of a unit of that land, and the second term, expressed as $\sum_k [Y_k^j L_k^j (\partial P_k / \partial L_i^j)]$, is the change in total revenue resulting from crop price changes associated with a change in the quantity of land used. The last three terms represent the change in total production cost associated with the use of an additional unit of land on crop i; these costs include the cost of the land itself, P_L^j, the change in the total land bill due to a change in the price of land for all crops in the region, $\sum_k [L_k^j (\partial P_L / \partial L_i^j)]$, and the cost of the fertilizer used on the additional unit of land, $P_X{}^j X_i^j$. Likewise, in Equation (29), the first two terms, which are expressible as $(\partial Y_i^j / \partial X_i^j) L_i^j P_i$) and $\sum_k [Y_k^j L_k^j (\partial P_k / \partial X_i^j)]$, represent the addition to total revenue attributable to an additional unit of fertilizer on each unit of land producing crop i in region j; the last term, $P_X^j L_i^j$, represents the addition to total cost.

In Equation (28), the derivatives dP_L^j / dL_i^j can be evaluated from the land supply functions of Equation (21). The derivatives $(\partial P_k / \partial L_i^j)$ and $(\partial P_k / \partial X_i^j)$, in Equations (28) and (29), can be obtained as described in the five steps below:

1. Aggregate the demand equations in Equation (22) over the regions to obtain the world demand for each crop.

2. Impose the "trade" assumption concerning the price of each crop as described in Equation (23).

3. Invert the resulting system in order to express the price of each crop in terms of the world demands for the N crops.

4. Equate each world demand to the corresponding supply as described by the equilibrium condition of Equation (24).

5. Finally, obtain the derivatives $\partial P_k / \partial X_i^j$ and $\partial P_k / \partial L_i^j$ from the resulting functions of step 4.

This five-step procedure is a direct generalization of the usual textbook procedure for determining monopoly behavior in the one-product case.

7.3.5 The Link with Stratospheric Pollution

In the previous section, we presented a model of monopolistic competition that relates to regional and worldwide demands and supplies of agricultural crops. In this section, we describe how the model's predictions for each crop in each region depend upon the climatic variables--rainfall and temperature--in all of the regions. We also demonstrate the serious shortcomings of one-crop models which purport to explain the economic costs in terms of crop production perturbations as a function of changes in climatic variables.

7.3.5.1 The Solution of the Model of Monopolistic Competition. If the equilibrium trade conditions in Equation (23) are used to equate the price of a crop in one region to its price in another, and the yield, land supply, and demand functions are specified, the equations of our model reduce to the N equilibrium conditions of Equation (24), and the 2MN first-order conditions of Equations (28) and (29). Under reasonable specifications, these 2MN + N conditions will determine the 2MN + N variables (P_i, L_i^j, X_i^j, i = 1, . . . , N; j = 1, . . . , M) in terms of the exogenous variables, which can be viewed as parameters of the system.

For purposes of illustration, denote the solution of the land use variable, L_i^j, as follows:

$$L_i^j = G_i^j(T^1, \ldots, T^M; R^1, \ldots, R^M; P_X^1, \ldots, P_X^M; POP^1, \ldots, POP^M; I^1, \ldots, I^M), \quad i = 1, \ldots, N; \; j = 1, \ldots, M, \tag{30}$$

where G_i^j is the function which relates the solution value of L_i^j to the exogenous variables.

From Equation (30), we see that the amount of land devoted to any crop in any region generally depends upon worldwide climatic conditions, fertilizer prices, population, and income levels. As an illustration, if a given region were to suffer severe and unexpected weather conditions, and, consequently, its wheat crop were to suffer, that region would be expected to import wheat from other regions. This, in turn, would be expected to lead to wheat price changes, which would presumably lead to changes in the acreage devoted to wheat, its supply, and so on. Thus, the climatic conditions present in one region would have economic consequences that extend to other regions.

<u>7.3.5.2 The "Bumping" Process--Why More Than One Crop Must be Considered.</u> Equation (30) relates the land-use variables and the five independent variables in each of the M regions of the world. Consequently, Equation (30) determines the effect that a change in one or all 2M of the climatic variable has on each land use variable in each of the M regions. However, for purposes of illustration, consider the case in which the regional temperature variables $(T^1, \ldots, T^M)$ change in proportion to one another, so that they can all be represented by a global temperature variable, T. We assume that the change in global temperature is caused by changes in the release rates of various pollutants. For purposes of the example, also assume that the values of the remaining independent variables of Equation (30) are fixed. Finally, assume for simplicity that there are only two crops, so that N = 2. Of these two crops, suppose that one is a "cold weather" crop, such as wheat, and the other is a "warmer weather" crop, such as corn. Denote the first crop, i = 1, as the cold weather crop and the second as the warmer weather crop.

The assumptions above suggest that, in region j, the land-use equations reduce to functions of global temperature alone so that:

$$L_1^j = H_1^j(T); \; L_2^j = H_2^j(T) \tag{31}$$

Under our assumptions, Equation (31) describes the quantity of land in region j which will be devoted to crops 1 and 2 for each value of global temperature T. Let T^*_{ij} be the global temperature value which maximizes the land use variable L_i^j and let MIN_{ij} and MAX_{ij} be the minimum and maximum values of global temperatures for which $L_i^j > 0$, i = 1,2. Then, it is reasonable to suppose that:

$$T^*_{1j} < T^*_{2j}; \; MIN_{1j} < MIN_{2j}; \; MAX_{1j} < MAX_{2j} \tag{32}$$

Figure 7-3 illustrates the essentials of the relationships. The basic point of the figure is that the curve describing the land use for the cold-weather crop lies to the left of the curve corresponding to the warmer-weather crop. In terms of figure 7-3, note that if global temperature T is between MIN_{1j} and MIN_{2j}, region j would produce only the cold-weather crop (1); if T is between MIN_{2j} and MAX_{1j}, both crops would be

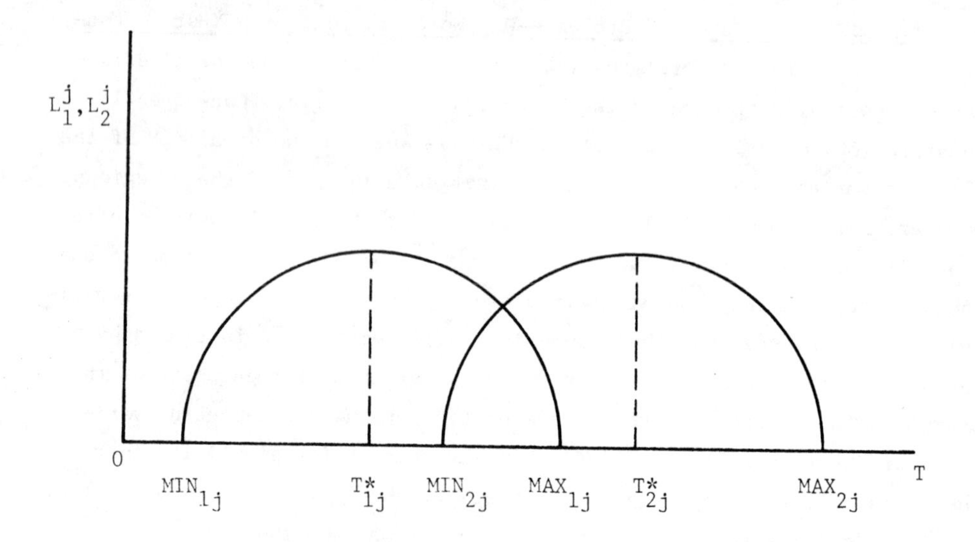

Figure 7-3. Land use functions for two agricultural commodities as a function of global temperature

produced; if T exceeds MAX_{1j}, but is less than MAX_{2j}, only the warmer crop (2) would be produced; and finally, if $T \geq MAX_{2j}$, or $T \leq MIN_{1j}$, neither crop would be produced.

The consequences of an increase in global temperature will depend upon the preexisting temperature in region j. For example, assume the preexisting temperature is between MIN_{1j} and T^*_{1j}, and the global temperature, after the increase, is still less than T^*_{1j}. In this case, region j would produce only the cold-weather crop before and after the temperature increase. However, more land would be devoted to the production of the cold-weather crop after the temperature increase. As another illustration, assume that the preexisting temperature is between MIN_{2j} and MAX_{1j} and that global temperature after the increase is less than MAX_{1j}. In this case, the region would be producing <u>both</u> crops; after the temperature increase, <u>more land</u> would be devoted to the warm weather crop and <u>less</u> to the cold weather crop. Clearly, a very biased view of the economic costs of pollution related to temperature changes would be presented by a study which considers only one crop.

Table 7-7. Effects of Temperature Change on Land Use

	Change in land use, given			
	Temperature increase		Temperature decrease	
Temperature interval	L_1^j	L_2^j	L_1^j	L_2^j
(A) MIN_{1j} to T^*_{1j}	+	0	-	0
(B) T^*_{ij} to MIN_{2j}	-	0	+	0
(C) MIN_{2j} to MAX_{1j}	-	+	+	-
(D) MAX_{1j} to T^*_{2j}	0	+	0	-
(E) T^*_{2j} to MAX_{2j}	0	-	0	+

Note: See figure 7-3, for the underlying relationships.

This discussion is generalized in table 7-7, which shows the effect of both a temperature increase and a temperature decrease on land use in each of the intervals defined by figure 7-3.

7.3.5.3 The "Bumping" Process--Why More Than One Region Must be Considered. Studies which deal with the economic costs of temperature changes as they relate to crop production often point out that, if global temperatures were to decrease, a certain amount of northern lands which are now devoted to wheat production could no longer be used for that purpose. Therefore, wheat production would fall and an economic loss would result. These studies, however, do not point out that if global temperatures were to fall, the southern border of the acreage devoted to the cold-weather wheat crop would also shift to the south. Consequently, the possibility that wheat acreage may extend beyond its previous southern border (thus increasing the land devoted to wheat production) is not considered. Still again, if wheat acreage were to extend, due to a temperature reduction, in a southern direction, the same analysis suggests that the crop it would replace at the southern margin (an example is corn) would itself extend its southern acreage border. In the Southern Hemisphere, this replacement at the margin of one crop for another, caused by a temperature reduction, would take place in mirror-image fashion at the northern border. In terms of table 7-7, we may think of the northernmost region in the Northern Hemisphere devoted to wheat production as corresponding to the interval (A). If only this region is considered, the reduction of wheat acreage would be recognized. However, we see that this single region analysis would not be very meaningful if other geographic crop regions, extending in a southern direction down to the equator, correspond to the regions B, C, D, and E of table 7-7. According to table 7-7, land devoted to wheat production--while decreasing in the northern (colder) region--would actually increase in two other regions, and not change from the zero level in the southernmost regions. Judging from the qualitative results in table 7-7, one possibility is that a global temperature reduction may lead to an increase in the acreage devoted to the cold weather crop, and reduction in the acreage of the warm weather crop.

7.3.6 Measuring the Benefits and Costs to Society of Changes in the Agricultural Markets

Using the methodology introduced in the previous sections, we are able to trace the quantitative effects of pollution release on the world agricultural markets. For purposes of policy concerning the regulation of pollutant release rates, one final link in this chain of reactions should be considered, namely, the consequent social welfare implications.

Ideally, this assessment would result in our being able to determine the change in the net "position" of consumers and producers in each region in response to changes in pollutant release rates. This determination, as it relates to producers, is fairly straightforward in that net profits (producer surplus) can be used as a measure of their net "position," with changes in that net position yielding our evaluation. Analogously, consumer surplus can be used to determine changes in the well-being of consumers. Consumer surplus is a measure of the difference between what consumers would have been willing to pay for the quantity transacted and what they actually pay. Whereas producer well-being is typically measured in monetary terms, the use of consumer surplus requires measurement of consumer well-being, typically measured in utility terms, to be measured in those same monetary terms. It is well-known that consumer surplus is a strictly accurate measure of consumer well-being if the marginal utility of income is constant, or if the demand curve under consideration is an "income-compensated" one. Since, in our particular application, we consider a relatively small number of commodities which represent a small fraction of consumers' budgets, the error in our use of this approximation should not be large.

Assume that P_e and Q_e are the equilibrium values of price and quantity in a given market, so consumers pay a total of P_eQ_e for the purchase of Q_e units of the commodity. Then our measure of total benefits, M, is given by:

$$M = PS + CS \tag{32}$$

where producer surplus, PS, equals P_eQ_e less the cost of producing Q_e units, and consumer surplus, CS, equals A less P_eQ_e, A being the area under the inverse demand curve up to output level Q_e. It is not difficult to show that M can also be interpreted as the difference between the consumer benefits associated with a given level of output, and the cost of producing that level of output.

It should be noted that the consumer surplus measure can be extended to the case of many commodities, using a line integral to measure changes in benefits, as suggested by Hotelling.[19]

7.3.7 Policy Applications

There are two major policy applications of our framework of analysis. The first is that changes in various pollution release rates can be evaluated in terms of the induced changes in the total benefit measure in Equation (32). The second is that advice can be given on the expenditure of research funds. For instance, assume that policy decisions will be taken in response to estimated changes in M with respect to pollutant release rates. Then, if that policy is to be defended and not be misguided, the estimated changes should be "accurate." Clearly, these estimates will depend upon estimates of the model parameters. If existing estimates of certain of these parameters are such that the corresponding estimated changes in M are widely different (not different), further research concerning the values of these parameters would (would not) be suggested.

7.4 Econometric Specification and Estimation of the Economic Model

In order to provide for a more manageable study while illustrating important features of the economic model we consider only two crops and two geographic regions. The crops considered are wheat and corn. These are important food and feed grains and may compete for land at their margin if temperatures and rainfalls change. The two regions are the United States (US) and the World excluding the United States (the rest of the world, ROW). While this geographical aggregation is extreme, it does allow for at least an approximate assessment of the U.S. position vis-à-vis other nations. (Because of data problems, we selected seven countries to represent, in aggregate, the rest of the world. Country data are used for most variables in our estimated yield and demand equations (to follow); however, for climate information we use data for major cities located near the wheat- and corn-growing areas of the respective countries. The seven countries considered, and the cities used for climatic information, include: Argentina (Cordoba), Australia (Canberra), Canada (Edmonston, Winnipeg), France (Paris, Toulouse), India (Indore), Italy (Milano, Messina), and Turkey (Istanbul, Antalya). For the United States, Bismark and Wichita data are used to approximate the relevant climate information.)

Implementation of the economic model requires estimation of its basic yield and demand equations. Our estimation uses annual data for the period 1960 through 1977.

7.4.1 Crop Yield Equations

The yield for each of the two crops in each geographic region is related to the corresponding climate variables of temperature and rainfall, a fertilizer variable, and a time trend which can serve as proxy for the time paths of other factors involved in crop production. Following the theoretical discussion above, the climate and fertilizer variables are entered in quadratic form. Two-stage least squares is used to estimate the yield equations, since fertilizer (an endogenous variable) appears as an independent variable in the equations. (Instruments used in the two stage process include: US and ROW climate variables, entered linearly and squared; U.S. wheat export subsidies, measured in 1970 dollars; transport cost for grain, measured in 1970 dollars; U.S. per capita disposable income measured in 1970 dollars; a time trend; and a constant term.)

Definitions of the variables used in the estimated equations, and the units of these variables, are as follows: crop yields are total crop production divided by total land planted in that crop (metric tons per hectare); average annual temperature (degrees centigrade); average monthly rainfall (millimeters per month); total application of nitrogen, potash, and phosphate fertilizer per unit of land (metric tons per thousand acres of agricultural land for US, and metric tons per hectare of land devoted to wheat and corn for ROW); and an annual time trend (1960=1). The sources used for these data are listed in an appendix to this chapter.

The estimated yield equations appear in table 7-8. With the exception of the climate variables in the ROW wheat-yield equation, all climate and fertilizer variables are estimated to have the anticipated downward-arching (concave) relation to yield, as is indicated by the positive linear terms and negative squared terms. We can calculate estimated "optimal" values of the climate and fertilizer variables as those values that _ceterus parabus_ imply maximal crop yields. (Mathematically, the value of X that maximizes $Y = a + bX + cX^2$, $b>0, c<0$, is calculated from $dY/dX = b + 2cX = 0$; thus $\hat{X} = -b/2c$.) These estimated optimums along with the sample means for the period 1960-1977 appear in table 7-9.

The most striking aspect of table 7-9 is the nearness of the climate optimums to the sample means. Technologies seem to have adjusted to make nearly best use of the existing climatic conditions. The United States could benefit, in terms of crop yields, from a moderate increase in both temperature and rainfall; other nations could, in aggregate, benefit from increased rainfall; however, any change in temperature would be detrimental to corn production. Average fertilizer applications in both the United States and ROW are well below their estimated optimums, indicating the potential benefits to increased use of fertilizer in crop production.

7.4.2 Demand Equations

The data necessary to estimate per-capita demands for wheat and corn are readily available in the case of the United States, but much of the data for ROW countries are unreliable or unavailable. As a result, sev-

Table 7-8. Estimated Yield Equations

Independent variables	Yield for US Wheat	US Corn	ROW Wheat	ROW Corn
Constant	-6.90 (9.88)	-38.2 (23.4)	7.64 (74.2)	-52.3 (109.)
Temperature	1.26 (2.06)	6.42 (4.89)	-0.847 (10.7)	6.76 (15.7)
$(\text{Temperature})^2$	-0.0642 (.108)	-0.329 (0.255)	0.0276 (0.385)	-0.245 (0.568)
Rainfall	0.0567 (0.0548)	0.218 (0.132)	-0.00428 (0.0941)	0.177 (0.137)
$(\text{Rainfall})^2$	-0.00058 (0.00063)	-0.00203 (0.00150)	0.0000387 (0.000788)	-0.0013 (0.00116)
Fertilizer	0.514 (0.576)	2.77 (1.36)	16.1 (47.7)	176. (70.2)
$(\text{Fertilizer})^2$	-0.0182 (0.0772)	-0.0969 (0.184)	-30.5 (302.6)	-1161. (445.)
Time	-0.0525 (0.0769)	-0.348 (0.183)	0.206 (8.73)	0.206 (0.129)
R^2	0.728	0.879	0.936	0.958
Durbin-Watson Statistic	1.53	2.29	2.00	1.69

Note: Standard errors in parentheses. Data are annual for the period 1960-77.

eral assumptions are made which facilitate estimation of demand equations in ROW. Results of our estimations for the US and ROW are presented in table 7-10 and are described as follows.

The U.S. per capita demand function for wheat (corn) is specified in terms of the real price of wheat (corn) relative to the Consumer Price Index; the average real price of grains not including wheat (corn), real per-capita income, a time trend, and a dummy for the period 1964-73. Specific definitions of the variables are as follows: demand is per-capita

Table 7-9. Optimal and Average Climate and Fertilizer Values

	US		ROW	
	Optimal	Sample average	Optimal	Sample average
Wheat yield equation				
Temperature	9.81	9.31	a	13.8
Rainfall	48.9	48.8	a	58.8
Corn yield equation				
Temperature	9.76	9.31	13.8	13.8
Rainfall	53.7	48.8	68.1	58.8
Fertilizer	14.3	4.04	0.076	0.033

[a]Estimated quadratic relations were convex, implying no optimal values.

wheat (corn) disappearances in metric tons per person, scaled by 10^3; wheat (corn) price is dollars per bushel divided by CPI; grain price not including wheat (corn) is the unweighted average price per bushel of corn (wheat), rye, oats, barley, sorghum, and rice, divided by CPI. The 1964-73 dummy was used to account for the effect of marketing certificates on crop demand. These marketing certificates represent payments ($0.75 per bushel) that domestic food processors made to the U.S. Department of Agriculture for each bushel of wheat processed. Because of the endogenous nature of prices and quantities, two-stage lease squares was used to estimate the demand equations. (Instruments used in estimation of the US(ROW) demand equations include: US(ROW) temperature and rainfall, real U.S. wheat export subsidies, lagged real corn or wheat price, lagged real grain price not including wheat or corn, lagged per-capita U.S. corn exports and corn inventories; and the included exogenous variables, real per capita income, the time trend, the constant, the marketing certificate dummy, and additionally for ROW the CPI and transport cost.)

Formulation of per-capita crop demands in ROW that utilize available and accurate data involves making some assumptions. Following the US demand specification for, say, wheat, we specify per-capita ROW demand

Table 7-10. Estimated Demand Equations

Independent variables	Per-capita demands for			
	US		ROW	
	Wheat	Corn	Wheat	Corn
Constant	36.18 (112.7)	-1022. (483.9	75.5 (60.8)	50.9 (13.9)
Real prices				
Wheat	-95.50 (1535.)	--	-2825. (1304.)	--
Grain (excluding wheat)	-1317. (2402.)	--	3323. (2115.)	--
Corn	--	-16290. (14950.)	--	-677. (168.)
Grain (excluding corn)	--	6978. (14350.)	--	867. (515.)
Real income per capita	3114. (5630.)	74440. (24410.)	119900. (45600.)	360. (2580.)
1./CPI	--	--	25200. (12600.)	1.86 (0.548)
Time	-1.090 (3.954)	-44.52 (17.47)	-14.1 (7.11)	--
Marketing certificate	260. (1323.)	-8216. (4830.)	--	--
Grain transport cost	--	--	--	-17.7 (2.60)
R^2	0.679	0.770	0.859	0.926
Durbin-Watson Statistic	1.72	1.43	2.55	3.09

Note: Estimated standard errors in parentheses. Data are annual for the period 1960-77.

(Dw_t^{ROW}/POP_t^{ROW}) as

$$\frac{Dw_t^{ROW}}{POP_t^{ROW}} = a_0 + a_1 \frac{Pw_t^{ROW}}{GPI_t^{ROW}} + a_2 \frac{PG_t^{ROW}}{GPI_t^{ROW}} + a_3 \frac{I_t^{ROW}}{GPI_t^{ROW}} \quad (33)$$

where at time t in region ROW

Pw_t^{ROW} = a price index for wheat

GPI_t^{ROW} = a general price index for all consumer goods

I_t^{ROW} = per capita income

PG_t^{ROW} = a price index for grains, excluding wheat

To reduce the data requirements for estimation we make three assumptions. First, assuming world trade and noting that the United States is a net exporter of wheat, the price of wheat in the US and ROW should be approximately linked as

$$\frac{Pw_t^{ROW}}{GPI_t^{ROW}} = (Pw_t^{US} + TC_t + TAX_t) \frac{K_t^{ROW}}{GPI_t^{ROW}} \quad (34)$$

where P_{wt}^{US} = the US price of wheat at time t

TC_t = the average transportation cost of shipping a unit of grain to ROW

TAX_t = average taxes and tariffs per unit of wheat that ROW imposes on wheat imports

K_t^{ROW} = the exchange factor that converts dollars into the "index currency" of ROW

Essentially this assumption implies that if two regions trade, the price of wheat in the importing sector should differ from that in the exporting sector by the costs of transportation and the taxes and tariffs imposed on the wheat.

The second assumption is similar to the first and relates the grain price index in the US and ROW,

$$\frac{PG_t^{ROW}}{GPI_t^{ROW}} = (PG_t^{US} + TC_t + TAX_{Gt}) \frac{K_t^{ROW}}{GPI_t^{ROW}} \quad (35)$$

where PG_t^{US} = the U.S. grain price index

TAX_{Gt} = the average taxes and tariffs per unit of grain imposed on ROW grain imports.

Finally, we assume that both the price level and real income per capita in the US and ROW increase over time proportionately. That is,

$$\frac{K_t^{ROW}}{GPI_t^{ROW}} = b\left(\frac{1}{CPI_t^{US}}\right), \text{ and } \frac{I_t^{ROW}}{GPI_t^{ROW}} = c\left(\frac{I_t^{US}}{CPI_t^{US}}\right) \qquad (36)$$

where b and c are constants and CPI_t^{US} is the U.S. Consumer Price Index.

Taken together these assumptions imply a demand equation of the form,

$$\frac{D_{wt}^{ROW}}{POP_t^{ROW}} = A_0 + A_1 \frac{P_{wt}^{US}}{CPI_t^{US}} + A_2\left(\frac{TC_t}{CPI_t}\right) + A_3\left(\frac{1}{CPI_t^{US}}\right) + A_4\left(\frac{I_t^{US}}{CPI^{US}}\right) + A_5\left(\frac{PG_t^{US}}{CPI_t^{US}}\right) \qquad (37)$$

This equation and a similar one for ROW corn appear in table 7-10.

Examining the estimated crop demand equations in table 7-10 we see that even at the present level of aggregation the price and income effects (with one exception) are of the signs predicted by microeconomic theory. The average elasticities implied by the price terms are calculated in table 7-11 and are seen to be quite inelastic.

7.5 An Illustrative Synthesis of the Model's Components

7.5.1 A Pilot Study

It is the purpose of this section to demonstrate briefly the use of our model in assessing effects of environmental regulatory policies. Our application can best be described as a pilot study which only suggests the detailed results which the model can yield. With this in mind, we specify scenarios concerning the release of carbon dioxide, chlorofluoromethanes, and nitrous oxide through the year 2000. We then use the relationships of section 7.2.2 to relate these pollution releases to changes in global mean surface temperature and, in turn, to temperature and rainfall changes in the U.S. and ROW. (Specifically, we use the average climate changes in the higher- and lower-mid-latitude belts for both the US and ROW.) The climatic changes are used along with our estimated yield equations to

Table 7-11. Average Price Elasticities of Demand

Demand equation	Own price	Grain price (excluding own price)
US		
Wheat	-0.0155	-0.175
Corn	-0.416	+0.189
ROW		
Wheat	-0.519	+0.501
Corn	-0.242	+0.327

determine percentage changes in wheat and corn yields, under the assumption that other variables in the yield equation do not change. Further, assuming quantities of land used in production do not change, we can interpret this percentage change in yield as the change in production as well.

The above is a partial equilibrium analysis and should lead to overestimates of the change in crop yields and production. This is easily seen for crop yields in terms of the following example. Assume that a given increase in temperatures would lead to a decrease in wheat yield if fertilizer inputs remained constant. Now allow the other variables of the model, including fertilizer, to adjust to the given increase in temperature. If the marginal effect of the temperature change is to decrease wheat yields, the supply of wheat would decrease, and this should lead to an increase in wheat prices. This, in turn, should lead to an increase in fertilizer input because its marginal product would increase in value. The increase in fertilizer input should moderate the initial decrease in yield which was due to the temperature change and which is calculated in our partial equilibrium analysis.

Consider now the calculation of wheat production. If one assumes that the land devoted to wheat production, L_w, does not change, the partial equilibrium estimate of lost wheat production would be $\Delta Y_w \times L_w$, where ΔY_w is the partial equilibrium estimate of the change in wheat yield. This overstatement is compounded, in the same direction, by the assumption that L_w does not change. For example, if wheat supply decreases in response to the initial decrease in wheat yields, the price of wheat would increase and consequently more fertilizer and land would be devoted to wheat production. The increase in wheat land would again moderate the decrease in wheat production. Therefore, $\Delta Y_w \times L_w$ would overstate the loss in wheat production on both counts. Finally, we note that if L_w is assumed to be constant, the estimated percentage change in yield is equal to the estimated change in production since $(\Delta Y_w \times L_w)/(Y_w \times L_w) = \Delta Y_w/Y_w$.

The results of this analysis, although yielding overestimates of the likely changes in production, should still be useful as they give upper limits on the anticipated changes. For example, suppose a given scenario leads in this partial equilibrium framework to an estimated loss in wheat production of 0.1 percent. Then, there would be little motivation to simulate the entire model for we might best discount the loss completely and turn our attention to other matters. On the other hand, if our partial equilibrium analysis were to produce an estimate of the wheat loss of 20 percent, there would be considerable motivation to simulate the model more thoroughly.

7.5.2 Some Illustrative Calculations

Our pilot study had the following features. For each of the three pollutant groups (carbon dioxide, chlorofluorocarbons, and nitrous oxide), we employed the reasonably conservative scenarios for release through 2000. For CO_2, we considered a 5 percent growth through the year 2000. This growth is approximately equal to the historic growth rate of CO_2 since 1950. For CFCs (F-11 and F-12), we considered two possible scenarios: one a 10 percent growth rate in CFC production and the other a zero growth rate, based on the 1975 production level. The 10 percent rate approximates the pre-1973 growth rate which preceded development of voluntary and legislative restrictions, and the zero growth rate approximates the 1973-76 rate under these restrictions. For N_2O, we used a 2.5 percent growth rate, reflecting increased world use of nitrogen fertilizer at a rate of over 10 percent annually, and assuming that all other sources of N_2O will continue at their 1975 levels.

Given these scenarios, table 7-12 shows the resultant changes in atmospheric pollutant concentrations, by the year 2000, and the corresponding changes in global mean surface temperatures. These temperature changes are due to the greenhouse and ozone effects, combined. Table 7-12

Table 7-12. Atmospheric and Climatic Changes Caused by the Production and Release of Various Pollutants

Pollutant group	Production growth rate from 1975 to 2000 (%)	Atmospheric concentration in 2000	Cumulative global ΔT by 2000 (°C)	Cumulative regional ΔT (°C)	Cumulative region change in precipitation (%)
CO_2	5	393.9 ppmv	+1.06	+1.30	+3.2
CFMs[a]	10	2.58 ppbv	+0.61	+0.77	+1.97
	0	0.60 ppbv	+0.13	+0.28	+0.11
N_2O	2.5	397.5 ppmv	+0.18	+0.32	+0.27

Note: Each of these four effects is partial in the sense that it assumes other pollutants have zero effect.

[a]For this pollutant group only, both greenhouse and ozone effects are estimated (see section 7-2).

also presents the regional temperature and rainfall changes that are likely to occur both in the United States and the rest of the world.

In column 4 of table 7-12, we see that the most influential pollutant in terms of climatic variables is CO_2. However, the effects of the other pollutants are by no means inconsequential, especially since their assumed growth rates are somewhat conservative. It should be noted that the climatic effects of these pollutants are additive, implying that the total increase in global mean surface temperature under these scenarios would be on the order of 1.4°C to 1.85°C, by the year 2000, depending upon the assumed scenario for CFMs.

In table 7-13, we see the resultant partial equilibrium effects on wheat and corn yields and production for the United States and the rest of the world, holding fertilizer application and land acreage constant at their 1975 levels. Several observations can be made concerning the results in table 7-13. First, for comparatively small increases in temperature (rows 2, 3, and 4 of table 7-13) the United States is seen to gain in crop productivity, while the rest of the world loses. The nonlinear nature of the yield functions (estimated in section 7-4) make interpretation somewhat difficult. However, the United States, unlike the rest of the world, appears to be at a slightly below-optimal temperature for the production of these crops. With a regional increase in temperature of as little as 1.3°C, however, the United States would find itself facing declining production along with the rest of the world. Moreover, that same 1.3°C temperature increase in the rest of the world could place those countries in a serious situation. From row 1 of table 7-13, we see a dramatic 22 percent decline in corn yields and production in the ROW, were the temperature to increase by the indicated amount. This figure is an overestimate; it therefore only indicates that the situation could be potentially serious.

The calculations in tables 7-12 and 7-13 were made under the assumption that only one of the pollutants is changing at a time. They would, of course, all be changing in the absence of specific environmental regulations. If we simultaneously consider the production of all three pollutants, we can obtain an indication of how these forces interact. Since we will again be relying on the partial equilibrium method, the resulting

Table 7-13. Annual Changes in Crop Yields by 2000 Resulting from Various Pollution Scenarios

Pollutant group	Production growth rate from 1975 to 2000 (%)	Cumulative regional temperature change[b] (°C)	Annual percentage change in yields and/or production: US Wheat	US Corn	ROW Wheat	ROW Corn
CO_2	5	+1.3	-1.42	-5.70	-3.92	-22.5
CFMs	10	+.77	+0.86	+1.60	-3.07	-6.70
	0	+.28	+0.78	+1.80	-0.59	-7.80
N_2O	2.5	+.32	+0.96	+2.30	-1.61	-0.70

Note: The effect of each pollutant is partial in the sense that it assumes the effects of all other pollutants equal zero.

[a]Both the United States and our rest-of-the-world countries are in the Higher Mid- and Lower Mid-Latitude Belts and thus we use the same regional temperatures in each.

estimates will have upward biases. They will therefore indicate whether or not a situation is potentially serious, and the possible extent of the seriousness.

If all three pollutants were produced at the indicated levels (with CFCs growing at 10 percent annually), the total cumulative regional temperature change, by the year 2000, would be 1.30 + 0.77 + 0.32 = +2.39°C. Performing the same calculations as those in table 7-13, we obtain the following changes in crop production and yields corresponding to this assumed scenario. For wheat production and yield in the United States, the estimated decline is 13.1 percent; for corn production and yield in the United States, the estimated decrease is 41.3 percent; for wheat production and yield in the rest of the world, the decrease is 2.7 percent; and for corn production and yield in the rest of the world, the decrease is 82.3 percent. In checking our calculations, the low estimate of 2.7 percent for wheat yield and production in the ROW appears to be caused by offsetting effects of changes in rainfall and temperature. The magnitudes of the remaining three figures indicate that potentially serious food production fluctuations due to climate changes cannot be ruled out. They also indicate that a full simulation of the entire model is needed before more accurate estimates of these environmental impacts can be determined in response to the pollution scenario considered.

7.5.3 Calculations of Model Sensitivity to Parameters and Suggestions for Further Research

Sensitivity of our results to the primary scientific estimates given in Section 7.2 can be exemplified by considering the uncertainty associated with a particular CFC parameter. The original CFC-concentration relation (Equation (6)) related the release of CFCs to their atmospheric concentration in the following manner:

$$\Delta CONC_{CFC} = -.100 + 0.0361\ REL_{CFC} \qquad (38)$$

We noted that there was a significant amount of uncertainty with regard to this relationship. If, in fact, this coefficient were 50 percent greater than our original estimate (1.5 x .0361 = .0542), this would imply an increased global mean surface temperature, by the year 2000, of 1.03°C or a regional increase of 1.25°C, assuming the 10 percent growth in CFC production. This is very nearly equal to our original temperature estimate for CO_2, given in row 1 of tables 7-12 and 7-13. Therefore, in terms of table 7-12, the resulting estimated changes in yields and production would be closer to the figures in row 1 than to those in row 2. This suggests that, instead of increasing, the production of wheat and corn in the United States would decrease. The magnitudes involved also suggest that it would be worthwhile to support research which might lead to a more accurate estimate of the technical parameter of interest. Clearly, similar results could be calculated with respect to the other technical parameters of the model. (The model currently is limited to climate-induced changes, and other possible CFC effects might be incorporated in future work. In particular, if increased ultraviolet radiation in fact affects crop production, that relation should be incorporated in the model.)

We can also demonstate the importance of the degree of uncertainty in the estimates of our model's economic parameters. This can be done by calculating an approximate confidence interval for each of the estimated percentage changes corresponding to wheat and corn production in table 7-13. For example, the estimated change in the value of wheat yield in the United States in response to one of the pollutant scenarios depends upon the estimated values of the parameters in the U.S. wheat-yield equation. Under certain assumptions, the joint distribution of these estimated values can be determined. Then, by linearizing the ratio defining the percentage change in U.S. wheat yield, an approximation to the variance of the estimated percentage change in wheat yield can be determined. As an illustration of the magnitudes involved, the estimated variance corresponding to the 0.86 percent figure in table 7-13 turns out to be 6.4 percent. If we approximate the distribution of the estimated yield change by the normal distribution, a 95 percent confidence interval for the estimated wheat yield change in response to the 10 percent growth scenario for CFC, turns out to range between -11.6 percent to 13.3 percent. These figures indicate that our results may be quite sensitive to the degree of uncertainty in our economic parameter estimates. This, in turn, suggests that more research should be devoted to obtaining improved estimates of the economic parameters.

More detailed and sophisticated sensitivity studies should be applied to the structure of our model, but the evidence thus far seems clear. Scientific and economic research to date has defined many of the important questions relevant to environmental regulatory policy, but the answers to these questions are still speculative. A model of the type developed in this study is essential to the meaningful evaluation of existing and future scientific research in this area.

7.6 Appendix: Sources of Data Used in Economic Model

The following is a list of variables used in crop yield and demand equations and their sources.

Yield, US wheat: U.S. Department of Agriculture, Wheat Situation (Washington, D.C., USDA, May 1973 and August 1975).

Yield, US corn: U.S. Department of Agriculture, Food Grain Statistics Through 1967 (Washington, D.C., USDA, 1968).

Yield, ROW wheat and corn: U.S. Department of Agriculture, Foreign Agricultural Service, Foreign Agricultural Circular (Washington, D.C., May 1976 and June 1977).

Rainfall, US and ROW: National Oceanic and Atmospheric Administration, Environmental Data Services, Monthly Climatic Data for the World, Washington, D.C., GPO, January 1977 and previous issues.

Temperature, US and ROW: National Oceanic and Atmospheric Administration, Environmental Data Services, Monthly Climatic Data for the World, Washington, D.C., GPO, January 1977 and previous issues.

Fertilizer, US and ROW: United Nations, Food and Agricultural Organization, Annual Fertilizer Review 1976 (Rome, United Nations, 1977 and previous issues.)

Demand, US wheat: U.S. Department of Agriculture, Wheat Situation (Washington, D.C., USDA, May 1973 and August 1975).

Demand, US corn: U.S. Department of Agriculture, Economic Research Service, Feed Situation (Washington, D.C., USDA, November 1977 and previous issues).

Demand, ROW wheat and corn: U.S. Department of Agriculture, Foreign Agricultural Service, Foreign Agricultural Circular (Washington, D.C., May 1976 and June 1977).

Population, US: National Bureau of Economic Research, Time Series Data Bank (New York, N.Y., NBER, Inc., 1975).

Population, ROW: United Nations, Department of Economic and Social Affairs, Demographic Yearbook, 1976 (New York, N.Y., United Nations, 1977 and previous issues).

Prices, wheat, corn, grain: U.S. Department of Agriculture, Crop Reporting Board, Agricultural Prices--Annual Summary, 1977 (Washington, D.C., USDA, 1977 and previous issues).

CPI, disposable income: National Bureau of Economic Research, Time Series Data Bank (New York, N.Y., NBER, Inc., 1975).

Marketing certificate: U.S. Department of Agriculture, Supplement to Food Grain Statistics (Washington, D.C., USDA, 1971).

Grain transport cost: International Wheat Council, World Wheat Statistics 1976 (London, IWC, 1977 and previous issues).

Wheat exports, inventories, US: U.S. Department of Agriculture, Wheat Situation (Washington, D.C., USDA, May 1973 and August 1975).

Corn exports, inventories, ROW: U.S. Department of Agriculture, Agricultural Statistics 1976, U.S. Government Printing Office, Washington, 1977.

Wheat export subsidies: U.S. Department of Agriculture, Food Grain Statistics Through 1967 (Washington, D.C., USDA, 1968).

References

1. Unpublished correspondence between the authors and V. Ramanathan of the National Center for Atmospheric Research, April 1978.
2. W.W. Kellogg, "Effect of Human Activities on Global Climate," World Meteorological Technical Note No. 156 (Geneva, World Meteorological Organization, February 1977).
3. National Research Council, National Academy of Sciences, Committee on Impacts of Stratospheric Change, Halocarbons: Environmental Effects of Chlorofluoromethane Release (Washington, D.C., NAS, 1976); National Research Council, National Academy of Sciences, Committee on the Impacts of Stratospheric Change, Responses to the Ozone Protection Sections of the Clean Air Act Amendments of 1977: An Interim Report (Washington, D.C., NAS, 1977); and National Research Council, National Academy of Sciences, Panel on Atmospheric Chemistry, Halocarbons: Effects on Stratospheric Ozone (Washington, D.C., NAS, 1976).

4. Council on Environmental Quality, Federal Council for Science and Technology, Task Force on Inadvertent Modification of the Stratosphere: Fluorocarbons and the Environment (Washington, D.C., GPO, 1975).
5. G.M. Woodwell, "The Carbon Dioxide Question," Scientific American vol. 238 (January 1978), pp. 34-43.
6. Ibid., p. 35.
7. J.M. Mitchell, "Carbon Dioxide and the Climate," Environmental Data Service, U.S. Department of Commerce (Washington, D.C., March 1977).
8. National Research Council, Halocarbons.
9. Ramanathan, unpublished correspondence.
10. National Research Council, Halocarbons, p. 39.
11. Ibid.
12. Ramanathan, unpublished correspondence.
13. National Research Council, Halocarbons, p. 25.
14. Kellogg, "Effect of Human Activities," and Ramanathan, unpublished correspondence.
15. National Research Council, Halocarbons, p. 18.
16. National Research Council, Responses to the Ozone Protection Sections, p. 16.
17. National Defense University, Climatic Change to the Year 2000: A Survey of Expert Opinions (Washington, D.C., Defense Advanced Research Projects Agency, Department of Defense, Fort McNair, 1978).
18. John Firor, personal communication.
19. Harold Hotelling, "The General Welfare in Relation to Problems of Taxation and of Railway and Utility Rates," Econometrica vol. 6 (1938), pp. 242-269.

Chapter 8

CLIMATE, ENERGY USE, AND WAGES

Irving Hoch

8.1 Introduction

This chapter examines the relationship of climate to energy use and to money wage rates in the United States, and applies those relationships in predicting the effects of climate changes on real income or utility. Although energy use and wage rates are related to a number of climate variables, predictions are based solely on temperature change, one of the predicted effects of chlorofluorocarbon (CFC) emissions. Sections 8.2 and 8.3 consider the relation of energy use and climate, respectively employing a sample of metropolitan data on residential electricity use, and samples of state data on residential use of electricity, natural gas, and petroleum products. In both cases, predictions are made of changed expenditures, given temperature changes ranging between -2°F and +2°F (which corresponds to a range of -1.1°C to +1.1°C. Section 8.4 relates specific occupation wage rates to climate variables, and develops predictions of changes in wage rates given temperature changes, employing metropolitan area data for both 1969 and 1975. Changes are interpreted as the amount workers need to be paid in order to compensate for a change in temperature. Section 8.5 integrates predictions and suggests topics for future research. In particular, analysis of individual areas, involving specific patterns of response, seems worth extending.

Predictions can be summarized by these percentage changes in costs for a given temperature change, where a negative value indicates a benefit, and where all consumer costs are estimated by the income compensation necessary to leave the consumer's welfare unchanged:

	Percentage change in costs, given			
	-2°F	-1°F	+1°F	+2°F
All residential energy use	2.33	1.00	-0.67	-1.01
All consumer costs	0.63	0.24	-0.22	-0.41

Energy expenditure changes account for about 10 percent of total cost changes, with the remaining 90 percent presumably reflecting both changes in other consumer expenditures that move in the same fashion and direct changes in utility. In chapter 2, Forziati presents an estimate of a year-2000, CFC-induced temperature increase of about +0.6°C or, roughly, +1°F. This change, per se, implies a reduction in energy costs of about $270 million, and a reduction of all consumer costs of about $2.7 billion. Forziati also shows a projection of a temperature increase of 0.5°C for the year-2000, carbon dioxide (CO_2)-induced greenhouse effect. Hence, the combined CFC and CO_2 effect is 1.1°C, which is almost exactly equal to 2°F. This total change implies a reduction of about $400 million in energy costs, and of roughly $5 billion in total costs. Hence, CFC emission effects yield net benefits for the year 2000 when considered either in isolation or in combination with CO_2 effects. However, there is considerable evidence suggesting that the cost function is U-shaped, so that a high enough temperature increase imposes net costs rather than benefits. If the year 2000 temperature increase is not a long-term equilibrium value, but rather a point on a long-term upward trend, then estimated benefits will eventually be replaced by estimated costs of higher temperature. Further, a temperature increase imposes costs rather than benefits in the U.S. South, suggesting that populations in areas closer to the equator than the United States will also suffer net costs, rather than benefits. Those populations, of course, comprise a substantial share of the world's total.

8.2 Residential Use of Electricity in Metropolitan Areas

This section analyzes residential use of electricity by drawing on metropolitan area data, which yields a larger sample than the state data employed later to analyze use of petroleum products and natural gas, as well as electricity. Annual consumption of electricity by residential consumers in 1970 was related to climate variables, price, income, and location characteristics, including metropolitan area population size and density. A number of regression equations were fitted, reflecting the use of alternative equation forms and measures of key variables, and of a stepwise process employed in obtaining a "best-fitting" equation. This section presents the most meaningful results, and applies them in predicting the impact of temperature change.

8.2.1 Data and Variables

Data series employed included climate and location characteristic information listed by the U.S. Census Bureau;[1] electricity prices and quantities obtained or developed from Federal Power Commission publications;[2] and income data developed from the U.S. Bureau of Economic Analysis income series.[3]

The following list shows variables employed within general categories, their shorthand designations (or mnemonic code), and their units of measure, along with some interspersed discussion of measurement decisions.

DEPENDENT VARIABLE

KWH:	Average annual consumption of kilowatt-hours (kWh) by residential customers (use per customer)

CLIMATE VARIABLES

STEMP	Summer temperature, degrees Fahrenheit
STEMP2	The square of summer temperature
WTEMP	Winter temperature, degrees Fahrenheit
RAIN	Precipitation in inches
WIND	Average wind velocity, miles-per-hour
SUN	Average percentage of possible sunshine (reflecting cloudcover)
DAMP	Interaction term, product of STEMP and RAIN
SLEET	Interaction term, product of WTEMP and RAIN

Degree-days and the squares of winter temperature and of precipitation were used in a number of equations but were never statistically significant, so those cases are not reported.

ECONOMIC VARIABLES

P	Price of electricity in cents per kilowatt-hours, measured as average value on rate schedule (as of January 1, 1971)
PRATE	Price of electricity in dollars per 1,000 kilowatt-hours, or cents per 10 kWh; measured as average price actually paid (1970); alternative to P
P2, PRATE 2	The squares of P and PRATE, respectively
DY	"Deflated" income per capita, accounting for price level differences both by region and population size of metropolitan area (Standard Metropolitan Statistical Area)

Income per capita was deflated by accounting for a 7 percent lower price level in the South than in the rest of the country, and for an estimated increase in price level of 6 percent per order of magnitude of population size (drawing on evidence in Hoch[4]). The deflated variable performed better, in terms of t test, than did per capita income in undeflated dollars. (In practice, DY was measured by lagged income, or 1969 values; since metropolitan area incomes between successive pairs of years have correlations around .99, results for the lagged measure should hardly differ from those for the current measure.)

The alternative measures of price (P versus PRATE) occur because electricity has a declining rate schedule, so that increased use yields lower average prices to the consumer, even if rate schedules are the same. Hence, an apparent causal effect on consumption attributed to price may not be real, so that the use of P, rather than PRATE is usually recommended in the literature, and that recommendation is generally followed in this section. However, in some situations one must make do with PRATE, such being the case in the next section. Given the availability of both measures here, PRATE was sometimes employed in place of P for purposes of comparison. Typically, there was little impact on measured climate effects when this substitution was made.

LOCATION CHARACTERISTIC VARIABLES

LSPOP	Log of Standard Metropolitan Statistical Area (SMSA) population
CDENS	Central city density, in thousands
S	The South (southern region as defined by the Bureau of the Census); other major regions included the Northeast (NE), North Central (NC), and West (W), but results for these variables were never significant, and are not reported.

In addition, there was a set of "specialized" regions:

BOS	Boston, Mass.
NYC	New York City, N.Y.
TVA	TVA area: four SMSAs in Tennessee, one in Alabama
PACNW	Pacific Northwest hydropower area: Seattle, and Tacoma, Wash.; Eugene, Salem, and Portland, Ore.; and Sacramento, Calif.
FLA	Florida SMSAs (nine metropolitan areas)
LOU	Louisiana SMSAs (six metropolitan areas)
TEXOK	Texas-Oklahoma SMSAs (twenty-six metropolitan areas)

These regions are distinguished by special climate, location, or price characteristics that could affect electricity use. Boston and New York have very high electricity prices and also have very high population densities; the TVA area and the Pacific Northwest have mild climates and very low prices; and the three southern areas have hot climates with considerable reliance on electricity for winter heating. Statistically, regions appear as dummy variables taking on values of one or zero (for appearance or absence). Average electricity consumption by region ranged from a low of 3,078 kWh in New York City to a high of 17,535 kWh in the TVA region.

Alternative equations were obtained with the use of P versus PRATE, with the use of an arithmetic versus a logarithmic equation form, and with the use of samples of different size, depending on the availability of data on SUN and WIND. An original sample of 210 observations was reduced to 161 observations because only the latter set contained information on SUN and WIND. For a given case, final results were obtained by a winnowing-out process in which potential explanatory variables were introduced into an equation and then eliminated (one-by-one) if their t ratios were below the 10 percent level of significance. However, a variable eliminated in an initial stage might return at a later stage if its reentry brought with it statistical significance; in short, a great many combinations were examined in deriving a final, best-fitting equation. That equation maximized explained variance subject to the constraint that all variables were significant at the 10 percent level. (An exception was a variable whose appearance caused another variable to become statistically significant; the first variable was retained if its t ratio was above 1.0.) Results across the alternative equations showed general consistency, but it was decided that the best individual equation to use for predictions was the arithmetic, 161-observation case employing P as price measure. For that case, sample means and standard deviations for key variables were:

Variable	Mean	Standard deviation
KWH	7,232.217	3,253.274
STEMP	76.594	5.439
STEMP2	5,896.187	828.582
WTEMP	35.837	11.827
RAIN	35.003	13.142
WIND	9.421	1.725
SUN	61.658	8.167
DAMP	2,692.741	1,079.599
P	2.523	0.375
P2	6.507	1.860
PRATE	20.980	4.543
PRATE2	460.804	187.578
DY	3.613	0.400
LSPOP	2.608	0.444
CDENS	4.681	3.555

Results for the most important equations are now reported.

8.2.2 Fitted Equations

The three linear climate variables, used alone, yield implicit Equation (1):

Dependent variable: kWh, n = 210, $\bar{R}^2$ = .242 (1)

Independent variable	Coefficient	t ratio
CONSTANT	-4865.747	1.785
STEMP	100.451	2.532
WTEMP	72.815	4.035
RAIN	45.706	3.186

In more conventional format, Equation (1) is:

$$\text{kWh} = -4{,}865.75 + 100.45\ \text{STEMP} + 72.82\ \text{WTEMP} + 45.71\ \text{RAIN},$$

with an explained variance of .242. In Equation (1), both summer and winter temperature have positive coefficients; the former seems an obvious air-conditioning effect, while the latter probably is best interpreted as reflecting a shift from oil and natural gas to electricity for home heating as winter temperatures increase. Although electricity is a much more expensive source of heating than the alternatives, its increased cost can be outweighed by reduced investment in heating equipment and by greater convenience.

After the introduction of quadratic climate variables and variables accounting for economic and location characteristics, and use of the stepwise winnowing-out process, Equation (2) was the best-fitting equation:

Dependent variable: kWh, n = 210, $\bar{R}^2 = .746$ (2)

Independent variable	Coefficient	t ratio
CONSTANT	57737.869	2.656
STEMP	-1247.419	2.212
STEMP2	10.223	2.775
WTEMP	-102.329	3.659
RAIN	411.975	2.801
DAMP	-7.238	3.418
SLEET	4.683	5.483
CDENS	-234.712	5.855
DY	739.765	2.561
P	-10744.580	3.405
P2	1937.554	3.185
TVA	9100.157	12.255
PACNW	2737.594	2.556

In Equation (2), most of the climate variables enter the equation (that is, are statistically significant), while most of the specialized regions do not, despite considerable differences in average consumption between regions. This implies that most of the regional differences can be explained by the other variables. The coefficients for summer temperature and its square imply that electricity use first declines and then increases with the variable; however, the minimum level of use is predicted from Equation (2) as occuring at 61°F, which falls below all observed values of STEMP in the sample. Hence, in practice, increases in summer temperature everywhere yield increases in electricity use. Density has a strong negative association with electricity use, but this may merely involve such factors as smaller dwelling units with increased density, and some tendency for residential apartment use to be (mis)classified under the heading of commercial, rather than residential, use. Increased income yields a significant increase in electricity use, with income measured in thousands of dollars. Thus, an increase of $1,000 in deflated income brings with it an annual increase of 740 kWh of electricity use. Income deflated by both region and city-size deflators (DY) had a t ratio of 2.56 in Equation (2); in contrast, income deflated only by the North-South regional deflator had a t ratio of 2.28 when it replaced DY; and undeflated income had a t ratio of 2.15 when employed as the income measure. Thus, use of estimated "real" income improves the statistical fit. Price has a negative and significant coefficient of -760 when it appears only as linear term (that is, when P2 is deleted), implying that a 1-mill increase in price (0.1 cents) causes a decline in use of 76 kWh per year. The introduction of the squared term improves the statistical fit considerably, and implies a tapering off of the effect of prices on quantity as price increases. Within a range of prices, electricity may well become more of a "necessity" at higher prices. Beyond that point, however, a questionable implication occurs, for a price rise then leads to a quantity increase; such might best be interpreted as a limitation of the particular form used.

The analysis was now extended by introducing wind velocity, WIND, and percentage of possible sunshine, SUN, at the cost of a smaller sample size (n=161). It turned out, however, that with SUN and WIND omitted, the use of the same variables yielded results paralleling those of the larger sample. In the smaller sample, both SUN and WIND entered the equation with negative coefficients, though SUN was not statistically significant. Since SUN only measures absence of cloud cover and does not account for variation in length of day, the effect of the amount of sunshine on electricity use will tend to be understated, perhaps accounting for the absence of statistical significance. The wind relation may involve roundabout causation; perhaps increased wind velocity exacerbates the effects of winter temperature (for example, the wind-chill factor), so that there is more reliance on fossil fuels and less on electricity for home heating at any given temperature.

The best-fitting equation for the smaller sample is shown as Equation (3) on the next page.

Dependent variable: kWh, n = 161, $\bar{R}^2$ = .779 (3)

Independent variable	Coefficient	t ratio
CONSTANT	79990.457	3.070
STEMP	-1729.683	2.555
STEMP2	13.619	3.117
WTEMP	-152.190	4.320
RAIN	412.216	2.354
DAMP	-7.763	3.053
SLEET	5.602	5.387
WIND	-258.963	3.076
CDENS	-262.201	5.518
DY	827.197	2.323
P	-11804.340	3.470
P2	2180.320	3.296
TVA	8838.224	11.464
PACNW	2455.746	2.156

Because Equation (3) had the highest $\bar{R}^2$ of all equations employing P, primary reliance was placed on it for predictions. The use of PRATE in place of P gave much the same pattern of results, although there was some increased significance for specialized regions and a reduction of climate effects, with RAIN and DAMP falling out of the equation.

Despite the primary reliance on Equation (3) in arithmetic form, some logarithmic cases seem worth reporting. Notation is extended by use of an L with a mnemonic code to show the log of a variable; thus LSTEMP is log of summer temperature. A number of climate-region interactions were introduced here, each consisting of the log of a climate variable times a dummy variable representing a major census region, but statistical significance held only for LWTEMPS, the log of winter temperature times the dummy variable for the South, S. Equation (4), appearing on the next page, presents the best-fitting equation for that case, with log P as the price measure.

The substitution of log PRATE for log P yields Equation (5), which also is shown on the following page. That equation parallels Equation (4), though the price elasticity increases considerably, and most other elasticities decrease in absolute terms. In both Equations (4) and (5), the winter temperature effect is negative in the North and positive in the South. (The sum of coefficients for LWTEMPS and LWTEMPS yields the southern coefficient.) Perhaps the southern effect involves the substitution of electricity for fossil fuels in home heating in warm climates. The northern effect implies that as winter temperature decreases, electricity use increases; this may involve greater use of electricity for both lighting and heating. Thus, the North-South classification reveals that a specific variable has opposite effects in the two regions. Equations (4) and (5) are as follows:

Dependent variable: LKWH, n = 161, $\bar{R}^2$ = .719 (4)

Independent variable	Coefficient	t ratio
CONSTANT	1.644	2.506
LSTEMP	1.566	4.266
LWTEMP	-0.164	2.030
LWTEMPS	0.658	2.906
LWIND	-0.384	3.639
LP	-0.482	3.376
LDY	0.326	1.817
LCDENS	-0.132	3.052
LSPOP	-0.048	2.116
S	-0.986	2.692
TVA	0.279	6.123
PACNW	0.311	5.886
FLA	0.084	1.777

Dependent variable: LKWH, n = 161, $\bar{R}^2$ = .806 (5)

Independent variable	Coefficient	t ratio
CONSTANT	2.290	4.176
LSTEMP	1.603	5.297
LWTEMP	-0.131	1.979
LWTEMPS	0.459	2.420
LWIND	-0.157	1.722
LPRATE	-0.874	9.087
LDY	0.222	1.482
LCDENS	-0.132	3.718
LSPOP	-0.027	1.410
S	-0.699	2.280
TVA	0.072	1.576
PACNW	0.133	2.782
FLA	0.076	1.949

8.2.3 Predictions

The fitted equations can be directly applied to predict changes in electricity use, given changes in climate measures. In the present study, predictions are limited to the impacts of a temperature increase or decrease of 1° or 2°F; hence, there are four variants, with a change in both summer and winter temperature of: (1) -2°F; (2) -1°F; (3) +1°F; (4) +2°F. Note that changes occur in STEMP and WTEMP, and also in their squares and interactions (DAMP and SLEET). Estimated electricity consumption was found for each observation by multiplying each independent variable by its equation coefficient and then summing the cross-products. A base period prediction was found by using the original temperature measures, and then temperature values were changed, the process was repeated, and predictions for the variants were obtained. Multiplication of predicted kilowatt-hours by actual rate (PRATE) then yielded an estimate of expenditures. Simple averages were obtained for the United States and for regions by summing over the predicted observations and dividing by sample size. This process was carried out, in turn, for each of Equations (2), (3), and (4). Finally, a weighted average was obtained for Equation (3) predictions by weighting the predicted value for each observation by metropolitan population relative to the total population covered by the sample (120 million persons.).

Table 8-1 exhibits individual expenditure predictions based on Equation (3) for selected metropolitan areas, yielding some notion of the variability between individual areas. (For convenience, areas are classified within regions as defined by the U.S. Bureau of Economic Analysis.[5]) In table 8-2, predicted average impacts for both quantity and expenditures are exhibited as a percentage of the base period value for each predictive equation. It seems clear that the patterns are quite similar between equations; predictions from Equation (3) fall between the other two sets, reinforcing the decision to place primary reliance on its results. Weighted averages were obtained only for that case, and were treated as best predictions. In all cases, however, a temperature increase brings greater use of electricity, with increased use and cost occurring somewhat more than proportionately with the temperature increase; a temperature decrease causes a decline in electricity use, but the decline occurs in less than proportionate fashion. Table 8-3 exhibits the weighted average predictions of changes in both kilowatt-hours and expenditures for the United States as a whole and for major Census regions. For both consumption and expen-

Table 8-1. Predicted Effects of Temperature Change on Electricity Expenditures Per Customer, 1970, for Selected Metropolitan Areas, from Equation (3)

Metropolitan area, by region and state	Base period dollars spent	Change in dollars of spending, given:			
		$-2^{o}F$	$-1^{o}F$	$+1^{o}F$	$+2^{o}F$
New England					
Boston, Mass.	$124.93	-0.38	-0.63	1.50	3.87
Burlington, Vt.	108.91	3.49	1.51	-1.04	-1.61
Portland, Me.	164.05	7.59	3.44	-2.74	-4.77
Providence, R.I.	109.82	1.99	0.62	0.13	0.99
Mideast					
Baltimore, Md.	104.08	-4.30	-2.48	3.13	6.91
Buffalo, N.Y.	69.10	3.93	1.65	-1.02	-1.41
New York City, N.Y.	143.92	-7.07	-4.06	5.12	11.28
Philadelphia, Pa.	76.53	-2.97	-1.83	2.52	5.74
Pittsburgh, Pa.	105.96	1.17	0.24	0.44	1.56
Rochester, N.Y.	117.06	1.28	0.32	0.33	1.31
Washington, D.C.	111.85	-5.84	-3.22	3.82	8.24
Wilmington, Del.	165.16	-3.13	-1.90	2.55	5.77
Great Lakes					
Chicago, Ill.	78.74	-4.11	-2.41	3.12	6.96
Cincinnati, Ohio	151.81	-4.86	-2.76	3.42	7.51
Cleveland, Ohio	102.21	1.33	0.33	0.34	1.35
Detroit, Mich.	97.79	-2.35	-1.49	2.13	4.89
Indianapolis, Ind.	129.65	-2.13	-1.33	1.86	4.26
Madison, Wis.	109.33	1.39	0.44	0.06	0.62
Toledo, Ohio	129.18	-0.24	-0.44	1.08	2.79
Plains					
Duluth-Superior, Minn.	160.89	9.51	4.41	-3.72	-6.75
Kansas City, Mo.	190.58	-12.46	-6.59	7.30	15.32
Rochester, Minn.	102.47	0.04	-0.21	0.69	1.84
Saint Louis, Mo.	108.90	-6.81	-3.73	4.39	9.43
Southeast					
Atlanta, Ga.	137.82	-4.55	-2.50	2.95	6.34
Chattanooga, Tenn.	173.18	-3.35	-1.81	2.06	4.39
Jackson, Miss.	153.80	-8.01	-4.25	4.74	9.97
Miami, Fla.	209.57	-7.18	-3.85	4.36	9.24
New Orleans, La.	183.82	-8.02	-4.29	4.84	10.24
Shreveport, La.	198.06	-12.88	-6.76	7.41	15.47
Southwest					
Dallas, Tex.	154.20	-11.81	-6.15	6.64	13.77
Houston, Tex.	169.93	-8.52	-4.51	5.00	10.49
Laredo, Tex.	168.69	-20.34	-10.49	11.13	22.89
Tucson, Ariz.	226.41	-22.27	-11.50	12.22	25.17
Rocky Mountain					
Cheyenne, Wyoming	123.41	1.64	0.50	0.14	0.91
Denver, Colo.	141.58	-2.30	-1.50	2.21	5.12
Great Falls, Mont.	103.86	2.23	0.81	-0.19	0.24
Salt Lake City, Utah	143.68	-6.53	-3.55	4.12	8.82
Far West					
Bakersfield Calif.	178.22	-15.44	-8.00	8.57	17.69
Eugene, Ore.	130.54	3.13	1.45	-1.21	-2.18
Las Vegas, Nev.	147.64	-12.31	-6.31	6.63	13.58
Los Angeles, Calif.	78.51	2.05	0.77	-0.25	0.01
Sacramento, Calif.	128.16	-3.00	-1.69	2.06	4.49
San Diego, Calif	76.22	0.95	0.18	0.40	1.38
San Francisco, Calif.	77.96	10.00	4.72	-4.15	-7.74
Seattle, Wash.	117.40	3.88	1.82	-1.59	-2.94
Spokane, Wash.	95.57	0.65	0.15	0.19	0.72

Table 8-2. Predicted Average Kilowatt-hours and Expenditures as Percentage of Base Values, by Equation

Consumption per customer	Base value (predicted average)	Percentage of base value given temperature changes of			
		-2°F	-1°F	+1°F	+2°F
Kilowatt-hours					
Simple averages					
Equation (2)	100	97.17	98.44	101.85	103.99
Equation (3)	100	97.15	98.39	101.99	104.36
Equation (4)	100	95.33	97.75	102.28	104.60
Weighted averages					
Equation (3)	100	97.63	98.58	101.88	104.23
Expenditures					
Simple averages					
Equation (2)	100	97.05	98.37	101.94	104.17
Equation (3)	100	96.97	98.28	102.12	104.64
Equation (4)	100	95.67	97.83	102.21	104.46
Weighted averages					
Equation (3)	100	97.38	98.45	102.07	104.66

ditures, the effect in the South is substantially above that in the other regions, in absolute terms and as a percentage of base value. The southern base value typically is about one and a third times that for the other regions, but its changes in use and spending are typically two to three times the corresponding changes elsewhere. In the case of the West, a decrease from -1°F to -2°F causes an increase in use and cost. That result might be explained by increased heating costs of electricity in winter outweighing decreased cooling costs in summer places of cool climate. Inspection of individual SMSA predictions, as shown in table 8-1, lends some support to this hypothesis. Note the patterns for Burlington and Portland, Maine, and Duluth-Superior, Minnesota, for example. In addition, the sensitivity of the predictive equation is displayed in the differentiation that occurs for California between hot Central Valley SMSAs (such as Bakersfield) and coastal SMSAs, that is, between hot versus warm (Los

Table 8-3. Weighted Average Predicted Changes in Kilowatt-hours and Expenditures per Customer as a Function of Temperature Change, by Region, Applying Equation (3)

	Base values	Temperature change			
		-2°F	-1°F	+1°F	+2°F
		Changes from base value			
Kilowatt-hours					
United States	5,847	-138.4	-82.8	110.1	247.4
Northeast	3,971	-64.5	-45.9	73.1	173.4
North Central	4,865	-111.9	-69.9	96.8	220.8
South	8,603	-384.5	-205.9	233.1	493.4
West	6,376	48.1	10.4	16.8	60.9
Expenditures, in $					
United States	124.88	-3.27	-1.95	2.59	5.82
Northeast	117.21	-2.47	-1.64	2.46	5.73
North Central	113.77	-2.72	-1.68	2.33	5.30
South	157.87	-7.42	-3.97	4.50	9.52
West	107.80	0.39	-0.06	0.57	1.65

Angeles) and cool (San Francisco) summer temperatures, as shown by the following data from table 8-1.

	Change in Spending Per Customer Given Temperature Change			
	-2°F	-1°F	+1°F	+2°F
Bakersfield, Calif.	-15.44	-8.00	+8.57	+17.69
Los Angeles, Calif.	+2.05	+0.77	+0.25	+0.01
San Francisco, Calif.	+10.00	+4.72	-4.15	-7.74

Further, there are a number of SMSAs (Cleveland, Madison, Cheyenne, and San Diego, for example) where any change appears to be for the worse: predicted costs increase for <u>either</u> a temperature decrease or increase. For these cases, there is likely to be a range between zero and 1° where net benefits occur (cost changes are negative); the results for Great Falls, Montana, shown in table 8-1, are close to this pattern. In general, these cases exhibit a U-shaped pattern of increasing costs with temperature change, and it is plausible that such a pattern holds eventually for all metropolitan areas. Admittedly, for such locales as Tucson, temperatures would probably have to drop considerably before costs turn up.

8.2.4 Conclusions

Results can be summarized as follows: (1) a number of climate variables affect electricity consumption in fairly complex (nonlinear) fashion; (2) despite the complexity, alternative equation specifications yield similar patterns of change in use nationally, for given changes in temperature; (3) considerable variation in pattern occurs between regions and even between SMSAs within regions; and finally, (4) there is evidence of an underlying U-shaped cost function.

8.3 Residential Use of Energy Per Capita, State Data

This section broadens the study of energy use by considering residential energy use per capita for petroleum products and natural gas as well as for electricity. Those three energy sources account for the bulk of residential consumption (97.4 percent), with coal use making up the small remainder (2.6 percent). Data at the state level for 1972 were developed as part of a detailed investigation of energy use,[6] with one application the estimation of residential demand equations for each energy source.[7] The work of that previous application was extended here as follows: (1) additional variables were brought into the equations, including log of precipitation, log of wind velocity, the South, and log of winter temperature times the South (following up on the results in section 4.2); (2) regardless of significance, temperature variables were retained in all equations to facilitate prediction of the impact of temperature change; and (3) additional stepwise regressions were carried out to yield the best-fitting equation for the present extension. Alternative equations, using arithmetic rather than logarithmic measures, gave results consistent with the log form, but the latter seemed preferable, so it was the primary vehicle employed in making predictions here. (In practice, the log form yielded asymetric effects for temperature increases versus decreases.)

8.3.1 Data and Variables

For all fitted equations in this section, there were fifty-one observations (covering the fifty states and the District of Columbia). As noted earlier, a sample of state data is inherently less satisfactory than one with observations on SMSAs, both because of considerably smaller sample size and because of measurement problems. Thus, state averages will involve greater aggregation, and aggregation error, than SMSA averages. Further, state averages are often not available, and then one must make do with an estimate based on a reading for a major city in the state, a procedure followed for much of the climate data employed here. Nevertheless, most of the estimates developed are plausible and, hence, their application to prediction seems defensible. The mnemonic codes for variables are shown below.

Quantities and Prices for Residential Use of Energy

LPFCAP	Log of petroleum products (heating fuels) use per capita
LGASCAP	Log of natural gas use per capita
LELCCAP	Log of electricity use per capita
LPFP	Log of petroleum products price
LGASP	Log of natural gas price
LELCP	Log of electricity price

All energy quantities and prices are in units of million BTU, that is, million BTU per capita for quantities, and dollars per million BTU for prices.

Climate Variables

LSTEMP	Log of summer temperature, degrees F
LWTEMP	Log of winter temperature, degrees F
LWTEMPS	LWTEMP times the South
LRAIN	Log of precipitation in inches
LWIND	Log of wind velocity in miles per hour

Income

LDY	Log of deflated income per capita, including special price level adjustments for Alaska and Hawaii. The adjustments parallel those described in section 4.2 (See Hoch for more detail on measurement.[8])

Population and Location Variables

FSMSA	Fraction of state population located in SMSAs
LGRO760	Log of population in 1970 as a percentage of population in 1960. (Climate, population and location variables are based on Census of Population series cited in Hoch.[9])

Regions

S	The South (Census definition)
NE	New England
MA	Middle Atlantic
ENC	East North Central

WNC	West North Central
SA	South Atlantic
ESC	East South Central
WDC	West South Central
MTN	Mountain
PAC	Pacific

These regions correspond to "divisions" as defined by the Census Bureau. The South consists of SA + ESC + WSC. In addition, there was a special region consisting of eastern coal-producing states, ECOAL: Pennsylvania and West Virginia.

In arithmetic terms, means and standard deviations of key variables were as follows, with deletion of L as first symbol denoting arithmetic value:

Variable	Mean	Standard deviation
PFCAP	22.656	16.644
GASCAP	24.580	13.710
ELCCAP	8.823	3.011
PFP	1.872	0.421
GASP	1.341	0.526
ELCP	6.942	1.448
STEMP	75.249	5.703
WTEMP	33.108	12.373
RAIN	34.666	13.779
WIND	9.465	1.761
DY(IN 000)	3.520	0.320
FSMA	0.540	0.265
GR07070	112.045	10.276

8.3.2 Fitted Equations

The best-fitting equation for petroleum products was obtained using the procedures noted in section 8.2, with LWTEMPS brought into the equation to further differentiate regional climate effects. Equation (6) on the following page presents results. In Equation (6), all the temperature effects are negative, so an increase in temperature is associated with a decrease in petroleum heating fuel use, while WIND has a positive effect, presumably reflecting the wind-chill factor. The own price elasticity is negative and close to unity, while the cross elasticity for natural gas price is positive and significant. (The cross elasticity for electricity price was also positive, but not significant.)

Dependent variable: LPFCAP, n = 51, $\bar{R}^2$ = .773 (6)

Independent variable	Coefficient	t ratio
CONSTANT	4.738	2.773
LSTEMP	-1.570	1.676
LWTEMP	-0.412	1.661
LWTEMPS	-0.654	1.174
LWIND	0.448	1.459
LPFP	-1.096	2.654
LGASP	0.419	1.956
FSMSA	-0.383	3.215
ECOAL	-0.398	2.942
NE	0.334	3.423
MA	0.446	3.950
S	1.145	1.286

Equation (7) was obtained by deriving the best fitting equation for natural gas, and then extending that equation by bringing in LWTEMP and LWTEMPS.

Dependent variable: LGASCAP, n = 51, $\bar{R}^2$ = .862 (7)

Variable	Coefficient	t ratio
CONSTANT	3.512	2.132
LSTEMP	-1.780	2.049
LWTEMP	0.127	0.542
LWTEMPS	-0.533	0.976
LRAIN	0.343	1.466
LGASP	-2.253	10.600
LELCP	0.876	3.481
LPFP	-1.075	2.529
FSMSA	0.407	3.866
NE	-0.195	2.248
SA	-0.156	1.370
WSC	-0.250	2.034
MTN	0.231	1.996
S	0.877	1.045

The effect of a temperature increase is decreased natural gas use, but the causal mechanism is summer temperature. The own price elasticity is negative and high, in absolute terms; the cross elasticity for electricity is plausibly positive; and that for petroleum products is implausibly negative. A difficulty that may be involved is that some areas of the country (particularly, New England) had very high natural gas prices and very low quantities of use, primarily because gas lines did not extend throughout those areas in 1972. Such could lead to distortions in estimating price impacts.

When the best-fitting equation for electricity use was developed, both S and LWTEMPS were statistically significant. The results showed excellent correspondence with Equations (4) and (5) which employed metropolitan area data. To facilitate that comparison, LWIND was brought into the equation here, to yield Equation (8):

Dependent variable: LELCAP, n = 51, $\bar{R}^2$ = .9269 (8)

Variable	Coefficient	t ratio
CONSTANT	-4.143	4.642
LSTEMP	1.093	4.845
LWTEMP	-0.126	2.652
LWTEMPS	0.262	2.034
LWIND	-0.067	0.862
LELCP	-1.204	20.948
LGASP	0.155	3.434
LDY	0.791	4.452
LGR07060	0.757	3.742
FSMSA	-0.165	5.422
WSC	0.055	2.301
MTN	-0.065	3.553
S	-0.440	2.167

The price in Equation (8) is the average payment per unit actually paid, corresponding to PRATE in section 8.2. Hence, a comparison of results is best made by comparing Equation (8) to Equation (5). Considering the coefficients for the climate variables and for price, these comparisons emerge:

	Eq. (5)	Eq. (8)
LSTEMP	1.603	1.093
LWTEMP	-0.131	-0.126
LWTEMPS	0.459	0.262
LWIND	-0.157	-0.067
LPRICE	-0.874	-1.204

where LPRICE refers to LPRATE in Equation (5) and to LELCP in Equation (8).

Consistency of results seems excellent; all variables have the same signs in Equations (5) and in (8), and magnitudes seem similar. (Equality of corresponding magnitudes usually would be accepted by statistical test.) Equation (8) does show a higher own price elasticity, but this might have occurred because Equation (8) also includes a (positive) cross elasticity for natural gas, which does not appear in the specification for Equation (5). The comparison thus suggests that the state data yield fairly good approximations to results for the larger samples of SMSA data, increasing confidence in the reliability of the results for petroleum products and natural gas.

8.3.3 Predictions

Predictions were obtained using the same general procedure employed in section 8.2. A base period prediction was developed for each observation, employing the sample data and corresponding fitted equation. Then, temperature was changed and corresponding predictions of consumption were obtained. Consumption times price gave expenditure predictions. Both simple averages and weighted averages were obtained, but the latter is preferable because some low population states had extreme consumption values. Predictions were obtained for each energy source, and then the sum over the three sources gave a residential total (in terms of dollars per capita). The national weighted averages obtained from the logarithmic equations for all three energy sources, combined, were:

Base period spending	Change in spending with temperature change			
	-2°F	-1°F	+1°F	+2°F
$122.624	$2.826	1.353	-1.242	-2.383

In percentages, alternative predictions for all energy sources combined were:

	Base value spending (predicted average)	Percentage of base value given temperature change of:			
		-2°F	-1°F	+1°F	+2°F
Weighted average, logarithmic	100%	102.30	101.10	98.99	98.06
Simple average, logarithmic	100	102.82	101.34	98.78	97.68
Weighted average, arithmetic	100	101.73	100.86	99.14	98.27
Simple average, arithmetic	100	101.84	100.92	99.08	98.16

Alternative predictive equations and averages yield quite similar results. The 1972 dollar predictions for individual energy sources obtained as weighted averages from the logarithmic equations were:

	Base value spending (predicted average)	Dollar change in spending with temperature change			
		-2°F	-1°F	+1°F	+2°F
Petroleum products	$ 32.396	2.697	1.303	-1.222	-2.369
Natural gas	33.828	1.538	0.756	-0.731	-1.439
Electricity	56.400	-1.409	-0.707	0.711	1.426
Total, all three sources	$122.624	$2.826	1.353	-1.242	-2.383

Thus, changes in spending on petroleum products and natural gas move in opposite direction to those on electricity, and outweigh them, so that a temperature increase causes a decline in spending, and a temperature decrease causes a rise in spending, on average. The relationship is somewhat nonlinear, decelerating with temperature increase and accelerating with temperature decrease. The amount of change in spending is greater for a temperature decrease than for a temperature increase. (Corresponding percentage changes are shown in section 8.5, below.)

Tables 8-4 and 8-5 present area predictions of spending change, with table 8-4 showing predicted effects of temperature change by major census region for each energy source and for the three sources combined, and table 8-5 showing that latter total by state. In table 8-4, there is considerable variation in the impact of temperature change. The level of changes in spending for individual energy sources reflects base levels, but hardly on a one-to-one basis, so that a good deal of the differences in response can be attributed to climate differences between regions. Considering total change for all three energy sources, the Northeast region is most strongly affected, for better or worse, by a change in temperature, followed in descending order by the North Central region, the West, and the South. The South is most heavily reliant on electricity, so that the offset to the changes in the other sources is strongest in the South. However, table 8-5 shows that only in Florida do the changes in electricity outweigh the changes in petroleum products and natural gas, so that energy expenditures decrease as temperature falls, and increase as temperature rises. All energy expenditures also increase for Hawaii with an increase in temperature; however, as temperature decreases, Hawaiian spending first falls and then rises.

Table 8-4. Predicted Changes in Residential Spending on Energy Per Capita as a Function of Temperature Change by Region and Energy Source

(weighted averages from logarithmic equations)

	Base values (dollars) per capita	Temperature change -2°F	-1°F	+1°F	+2°F
		change from base value, in dollars			
Petroleum products					
United States	32.396	2.697	1.303	-1.222	-2.369
Northeast	59.902	4.561	2.211	-2.085	-4.054
North Central	27.643	2.478	1.187	-1.099	-2.120
South	23.087	2.224	1.075	-1.007	-1.953
West	18.821	1.329	0.646	-0.613	-1.195
Natural gas					
United States	33.828	1.538	0.756	-0.731	-1.439
Northeast	39.652	1.594	0.786	-0.764	-1.506
North Central	44.210	1.683	0.833	-0.814	-1.609
South	19.366	1.305	0.637	-0.607	-1.185
West	35.358	1.648	0.809	-0.781	-1.535
Electricity					
United States	56.400	-1.409	-0.707	0.711	1.426
Northeast	54.302	-1.118	-0.561	0.564	1.131
North Central	55.076	-0.988	-0.499	0.509	1.026
South	62.095	-2.068	-1.035	1.037	2.077
West	51.117	-1.290	-0.646	0.646	1.293
Total, three sources					
United States	122.624	2.826	1.353	-1.242	-2.383
Northeast	153.856	5.037	2.436	-2.284	-4.429
North Central	126.929	3.173	1.520	-1.404	-2.703
South	104.548	1.461	0.676	-0.577	-1.061
West	105.295	1.687	0.810	-0.747	-1.436

8.3.4 Comparisons, Applications, and Conclusions

As a check, the electricity predictions obtained here are compared to those of section 8.2 with the caveat that some differences can be attributed to dates of samples (1970 in section 8.2, and 1972 here), and to sample coverage (major SMSAs, accounting for about 60 percent of U.S. population, in section 8.2, and states here, covering the total U.S. population). In section 8.2, weighted average predictions of dollar changes

Table 8-5. Predicted Changes in Spending on Energy (Aggregate of Petroleum Products, Natural Gas, and Electricity), from Logarithmic Equations

	Base period spending in $ per Capita	Change in spending with temperature change			
		-2°F	-1°F	+1°F	+2°F
New England					
Conn.	146.938	4.541	2.179	-2.012	-3.870
Maine	214.000	12.405	5.949	-5.494	-10.580
Mass.	141.515	4.898	2.366	-2.210	-4.275
N.H.	182.099	8.352	3.992	-3.664	-7.031
R.I.	136.235	4.766	2.301	-2.149	-4.158
Vt.	227.592	14.440	6.866	-6.251	-11.961
Mid-Atlantic					
N.J.	176.692	5.519	2.674	-2.516	-4.883
N.Y.	163.014	5.610	2.721	-2.563	-4.980
Pa.	125.007	2.861	1.388	-1.310	-2.549
East North Central					
Ill.	132.946	2.995	1.451	-1.363	-2.645
Ind.	119.925	2.314	1.116	-1.039	-2.006
Mich.	127.290	2.744	1.326	-1.240	-2.401
Ohio	125.054	2.666	1.291	-1.216	-2.360
Wisconsin	124.487	3.956	1.893	-1.738	-3.338
West North Central					
Iowa	126.766	3.580	1.711	-1.570	-3.010
Kans.	123.519	2.020	0.976	-0.913	-1.767
Minn.	138.468	6.259	2.937	-2.619	-4.970
Mo.	114.076	1.972	0.953	-0.890	-1.722
Neb.	123.547	2.587	1.240	-1.142	-2.193
N.D.	158.587	13.260	5.995	-5.053	-9.383
S.D.	136.886	6.107	2.880	-2.583	-4.910
South Atlantic					
Dela.	121.492	4.103	1.937	-1.728	-3.267
D.C.	105.931	2.920	1.384	-1.245	-2.362
Fla.	103.879	-0.674	-0.362	0.405	0.853
Ga.	101.345	1.382	0.638	-0.539	-0.987
Md.	118.221	4.000	1.897	-1.709	-3.246
N.C.	107.010	1.710	0.795	-0.683	-1.265
S.C.	96.263	1.029	0.470	-0.387	-0.696
Va.	109.541	2.043	0.955	-0.831	-1.547
W. Va.	89.743	2.025	0.953	-0.842	-1.583
East South Central					
Ala.	92.462	0.892	0.409	-0.342	-0.619
Ken.	109.717	3.738	1.770	-1.590	-3.017
Miss.	96.711	1.281	0.596	-0.512	-0.945
Tenn.	105.582	1.711	0.797	-0.687	-1.274
West South Central					
Ark.	103.647	1.682	0.783	-0.676	-1.253
La.	99.954	0.725	0.327	-0.263	-0.463
Okla.	114.723	2.870	1.356	-1.212	-2.291
Texas	104.326	0.748	0.336	-0.265	-0.463
Rocky Mountain					
Ariz.	90.597	0.086	0.033	-0.013	-0.007
Colo.	118.763	2.533	1.231	-1.163	-2.265
Idaho	112.432	2.341	1.122	-1.029	-1.974
Mont.	130.701	4.872	2.341	-2.170	-4.183
Nev.	107.880	0.913	0.425	-0.368	-0.681
N. Mex.	97.813	1.845	0.893	-0.838	-1.623
Utah	121.208	2.886	1.405	-1.335	-2.603
Wy.	134.205	4.425	2.138	-2.000	-3.870
Far West					
Alaska	169.186	8.085	3.874	-3.569	-6.861
Calif.	102.323	1.532	0.736	-0.681	-1.310
Hawaii	78.844	0.016	-0.004	0.027	0.076
Oreg.	106.927	1.449	0.689	-0.620	-1.176
Wash.	108.279	1.479	0.700	-0.627	-1.185

in spending on electricity are on a per-customer basis. In 1970, there were 3.23 persons per customer (dividing U.S. population by number of customers); hence, dividing entries by 3.23 puts them on a per capita basis, allowing direct comparisons with predictions here. An alternative comparison consists in considering spending as a percentage of base spending for the alternative predictions. The comparisons are:

	Change in electricity spending for temperature change of			
	-2°F	-1°F	+1°F	+2°F
Dollar amounts per capita				
Section 8.2 (1970)	-1.012	-0.604	0.802	1.802
Section 8.3 (1972)	-1.409	-0.707	0.711	1.426
Percentage of base spending (100)				
Section 8.2 (1970)	97.38	98.45	102.07	104.66
Section 8.3 (1972)	97.50	98.75	101.26	102.53

The comparisons show a fair degree of consistency, though section 8.2 predictions are much more asymmetric than those of section 8.3, showing more response to a temperature increase than to a decrease. This may reflect better accounting for a curvilinear relationship because of larger sample size. There was also some indication of a U-shaped pattern of costs, the deceleration of benefits with temperature increase suggesting an eventual turning-up of costs, given large enough temperature increases. A U-shaped pattern actually occurred, however, only for Hawaii, probably reflecting both the relatively narrow range of temperature changes employed, and the limited size of sample.

8.4 Wage Rates in Metropolitan Areas

This section relates wage rates to climate. It begins with a short presentation of a theoretical structure for wage rate analysis and then presents results for several empirical applications.

8.4.1 Theoretical Structure

There is considerable evidence that money wage rates for the same work increase with metropolitan area population size, and it is a reasonable hypothesis that such represents a compensatory payment for an excess of costs over benefits that occur with urban size. With increased size there are increases in rents, and, by way of multiplier effects, in many components of the market basket for the conventional cost-of-living index; there are also increased congestion and pollution costs; and the cost increases appear to outweigh the benefits of increased size, which usually reflect increased specialization, for example, specialized medical services.[10] It is also a plausible hypothesis, supported by empirical evidence, that climate affects money wage rates in similar fashion, with "good" climate tending to drive down wage rates, and "bad" climate tending to drive them up.[11] Good and bad climate involves more than direct utility effects (personal comfort or discomfort), since a variety of costs should depend on climate, including clothing, transportation and medical care, as well as residential energy use.

The population size and climate hypotheses have implications for the form of the long-run supply schedule for all workers, as shown in figure 8-1. Here money wage rates are related to population size, viewed as a multiple of labor in all employment, and climate, in an equation of the form: $Y = k + aX + bZ$, where Y is money wage rate, X is log of SMSA population, and Z measures climate; a, b, and k are parameters that are to be estimated in empirical applications. In practice, the log of population size has more explanatory power than the corresponding arithmetic measure; Z represents a vector of climate characteristics; and a number of other variables enter the equation. In particular, regional dummy variables are employed to avoid ascribing effects to climate that are properly attributed to other variables related to geography.

Single equation estimation is employed assuming that both climate variables and population size are independent of disturbances in the wage rate. Certainly, climate can be treated as exogenous, while population is viewed as essentially a predetermined variable, since migration and natural increase generally comprise only a small portion of total population. Typically, total population has been determined by a decades-long or centuries-long process. Empirically, for a large sample of metropolitan areas, the correlation between current and recently past levels of population is about .99, even for periods some years apart. In effect, popula-

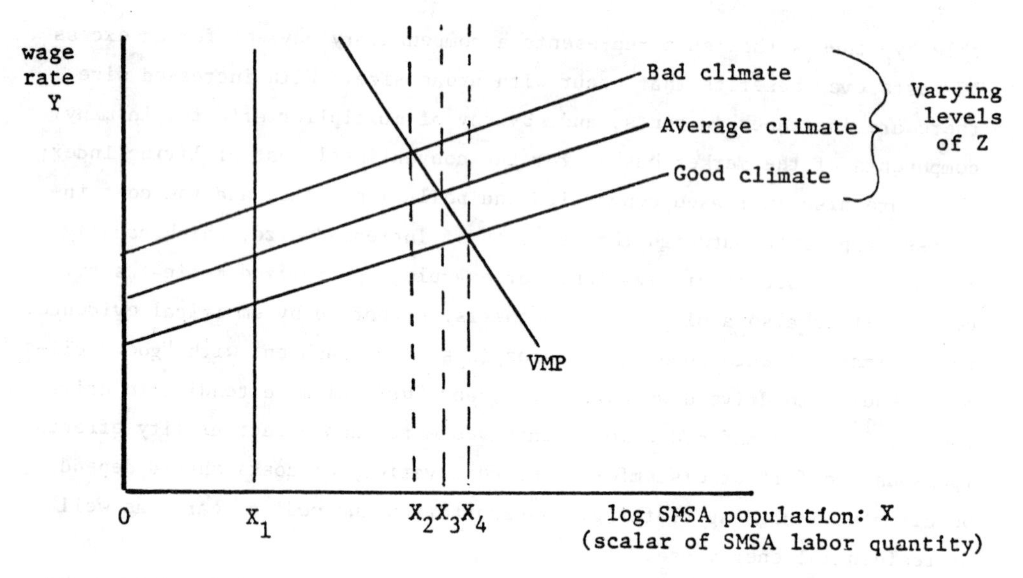

Figure 8-1. Representation of labor market exhibiting size and climate effects

tion (or supply of all labor) can reasonably be viewed as perfectly inelastic in the current period, as shown at X_1 in figure 8-1. Of course, if we envision urban population size as determined by productivity relations, the intersection of value of marginal product (VMP) and labor supply will determine population size as well as wage rate, as shown by X_2, X_3 and X_4 in figure 8-1. But it seems reasonable to view that equilibrating process as expressing longer-run relationships than hold in the current period; hence, the dashed lines at X_2, X_3 and X_4 are viewed as short-run, fixed levels of population, corresponding to perfectly inelastic supplies of labor at those production levels, as well as at X_1. In this model, a shift in VMP will cause wage rates to temporarily rise above or below the equilibrium level (that is, non-zero disturbances do occur), but population shifts in response occur in future periods, rather than in the current period.

Some additional points are worth noting. First, each urban area can be viewed as having its own VMP curve, so that a large family of curves in effect trace out points on the long-run supply schedules of figure 8-1. Second, climate and population may well be correlated, but this poses no problem statistically, for independent variables are generally correlated with one another.

Finally, if a climate change occurs for a given metropolitan area, its population should also change in the long run. In figure 8-1, consider an urban area at X_3 with average climate, and on the specific VMP curve shown in that figure. Say that climate improves, so the relevant labor supply curve shifts downward to the "good climate" line. Then population should expand from X_3 to X_4, in the long run. That expansion should raise rents, in line with the intuition that "better" climate is capitalized into land values. But higher rents in turn correspond to higher wage rates at X_4, relative to wage rates paid in an urban area on the good climate supply curve at X_3 (on a lower VMP curve). Thus, wage rates can be viewed as encompassing rents, with both related to population size (for given climate). These considerations are also relevant to the question of measuring the benefits or costs of climate change. Figure 8-2 exhibits measures of short-run versus long-run benefits of the climate change discussed in the present example. In the short run, those benefits correspond to the rectangle ABCD, equal to the increase in real wage rates (CD) times initial labor quantity (DB or OG). Workers continue to receive wage rate OC in the short run, but would now accept OD, so CD is the gain in real income per worker. In the long run, population expands from G to H, and benefits now correspond to figure AGHE, in turn equal to the increment in total product, the increased area under the VMP curve. This area consists of increased wage payments, or rectangle FEGH, plus increased payments to all other factors of production (including rent to land), triangle AFE. In the present study, empirical estimates of benefits (or costs) are limited to the short-run case. It is not at all obvious that the corresponding long-run measure can be estimated, in practice. It is also worth noting that a fixed standard of living, or real income, is implicitly assumed in figure 8-2, so all increases in total wages occur through population increase. If we think of the given metropolitan area as representative of all settlements, then a national population increase would "sop up" the gains in welfare generated by better climate, with real wage rate unchanged over time. Alternatively, people may decide to increase their standard of living, given the higher real productivity of the system, and this decision

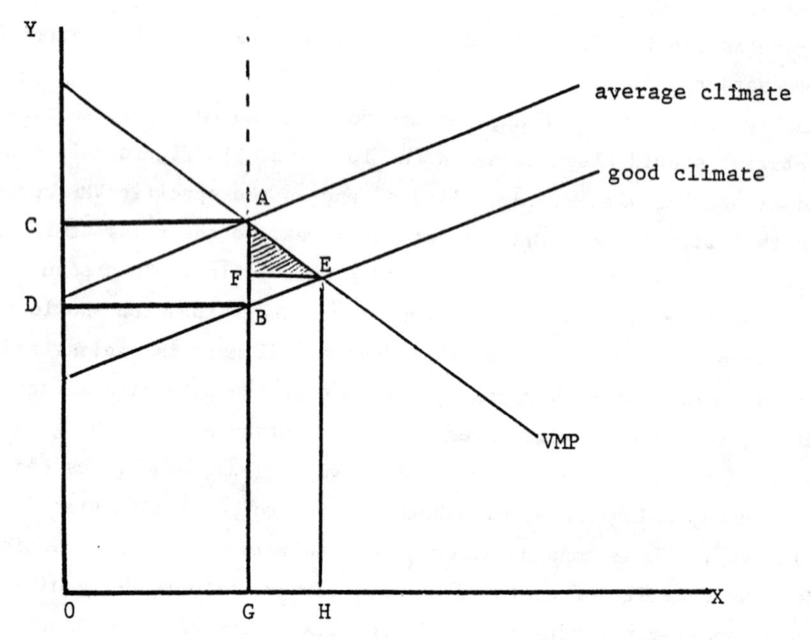

Figure 8-2. Measures of short-run versus long-run benefits of climate change

would entail shifting the labor supply curve up. An additional complication is the transition cost of induced migration, given climate changes that benefit some places and harm others, although a long-run adjustment process may primarily involve changes in direction of migration streams, without much increase in total migration.

These long-run questions are important subjects for future research, but there is a case for treating the short-run estimate as more-or-less equivalent to the long-run estimate. The rationale is that an initial short-run net "surplus" will eventually be distributed between an increase in population and an increase in living standards, with the amount to be distributed ex ante, equivalent to the amount actually distributed, ex post. Of course, parallel arguments hold for a climate change for the worse, with a net loss in per capita income in the short run, and losses in per capita income or population, in the long run.

8.4.2 Data, Variables, and Predictive Equations

This section extends some analyses of impacts of climate change by wage-rate equations, the earlier analyses having been carried out for the Climatic Impact Assessment Program (CIAP). For an overview of the economic and social analysis for CIAP, see d'Arge;[12] for reports on the wage-rate equations, see Hoch and Drake.[13] In that previous effort, wage rates for specific occupations were related to climate variables and a number of other explanatory variables, including region, scale of urban settlement (both log of population size and population density) and racial composition (percentage black). The data on wage rates were published in several series by the Bureau of Labor Statistics[14] (also see Hoch and Drake[15]), and had the important feature of homogeneity in coverage: for a given occupation, essentially the same work was performed, and the same level of skill was involved for each observation. Each sample covered U.S. metropolitan areas over a one-year period in the late 1960s and in 1970, with sample size ranging from forty-three to eighty-six. Twenty-five occupations were analyzed, and in most of the corresponding twenty-five fitted equations, at least some climate variables were statistically significant. On the criterion of statistical significance, summer temperature was the strongest climate variable, but winter temperature, precipitation, wind velocity, and the interaction of summer temperature and precipitation all showed explanatory power. A temperature decrease of as much as -5°F turned out to be neutral or beneficial in relatively hot and humid areas, but was adverse in areas above 30° latitude, in which most of the U.S. population is located. An increase in cost corresponded to a predicted higher wage rate which would be necessary to keep utility, or real income, constant. However, nonlinearities in the equations yielded some suggestion of an optimal summer temperature, estimated as 74°F for average precipitation, with a higher optimum as precipitation decreased, and a lower optimum as precipitation increased, in line with the familiar observation, "It's not the heat, it's the humidity." These results were obtained in the face of considerable "noise" in the estimating equations and so can only be viewed as suggestive. Nevertheless, there is the implication that a large enough temperature increase in any locale would have an adverse impact, yielding a U-shaped cost function.

Data for eight of the equations came from a 1969 series containing the maximum number of observations available, and this was probably a factor in the success obtained with those cases, for they had the highest explained variance and the greatest number of statistically significant climate coefficients of all the equations. For the present study, it was decided to draw on those eight particularly successful cases, both by extending the analyses using the original data, and by replicating the work using 1975 data (which were again obtained from the Bureau of Labor Statistics[16]). In contrast to the eighty-six observations available for the 1969 cases, coverage for the 1975 cases ran from seventy-three to seventy-seven observations. Wage rates for the occupations of interest were designated by these mnemonic codes:

JAN	Janitor, male
LABR	Laborer, male
MECH	Auto mechanic, male
ACTAF	Accounting clerk, class A, female
KP	Key punch operator, class B, female
STENO	Stenographer, general, female
REC	Switchboard operator-receptionist, female.
TYP	Typist, class B, female

Classes refer to skill levels, with class A the highest level. Only seven of the occupations were employed in the 1975 analyses because accounting clerk, class-A wage rates are no longer disaggregated by sex. Since equal pay by sex is not a reality, and hence, differences in proportions by sex between metropolitan areas can affect observed average wage rates, the occupation was deleted. Note that all occupations are sex-specific.

In practice, wage rates were standardized by dividing each observation by the respective sample mean wage rate for the occupation and then multiplying by 100; the resultant reading was then a percentage of a base value of 100.

For both the 1969 and 1975 cases, alternative predictive approaches were employed. In the first approach, individual best-fitting equations were estimated for each occupation, and then an overall predictive equation was obtained as an average of the individual equations, simply by summing corresponding coefficients and dividing by number of cases. In the second approach, the observations on the individual occupations were combined into one large pooled sample, with standardized wage rate as the dependent variable, and a single best-fitting equation obtained for that pooled sample. Several interaction terms in the pooled sample accounted for differences between occupations.

The independent variables employed in the individual equations, and their mnemonic codes, were:

Climate variables

STEMP	Summer temperature, July average, °F
STEMP2/100	Square of STEMP divided by 100
WTEMP	Winter temperature, January average, °F
WTEMP2/100	Square of WTEMP, divided by 100
RAIN	Precipitation, annual average, inches
DAMP/100	STEMP times RAIN, divided by 100
SLEET/100	WTEMP times RAIN, divided by 100
WIND	Average wind velocity, miles per hour
SUN	Average percentage of possible sunshine

Nonclimate variables

Regional dummies

NE	Northeast
NC	North Central
S	South
W	West
CONF	Confederacy (region encompassing Confederate states during the Civil War)
PAC	Pacific Southwest (Arizona and California)
CALIF	California only

Socioeconomic variables

% BLACK	Percentage Black
CB	Confederacy times percentage Black
LSPOP	Log SMSA population, in 000
CDENS	Central city density, in 000 (population per square mile)
HDENS	High central city density, in 000 (variable set at zero for density below 10,000, and at CDENS for density above 10,000)
GROW	Percent growth in SMSA population 1960-1970 (1975 cases only)

Average values for key independent variables were:

STEMP	75.530
WTEMP	34.436
RAIN	36.216
WIND	9.624
% BLACK	10.662
SPOP	660.600 (antilog of LSPOP = 2.820)

Table 8-6 presents the best-fitting individual equations obtained for the 1975 cases, designated as Equations (9.1) through (9.7). Corresponding equations for the 1969 cases appear in Hoch.[17] The two sets of equations show generally good correspondence, both across occupations for a given year, and for pairs of occupations between years.

Table 8-7 presents the predictive equations obtained by averaging the individual occupation equations for 1969 and 1975, with the respective average equations designated as Equations (10.1) and (10.2).

In both Equations (10.1) and (10.2), a small temperature increase, occurring near the means of summer and winter temperatures, yields a predicted wage decrease (or utility increase). The interaction of temperature and precipitation implies that near the mean of summer temperature, increased precipitation has little effect on wage rates. However, for relatively low summer temperatures, increased precipitation has a negative effect on wage rates, interpreted as an increase in utility; and for relatively high summer temperatures, the reverse effects occur. The population effect is of similar magnitude for the two cases, as are the effects for % BLACK and CB. The latter results imply that wage rates increase as the percentage of black population increases, but the effect is much more pronounced in the states outside than inside the Confederacy. A variety of interpretations for this result are possible, including discrimination, white antipathy to blacks or higher social costs because of the relatively greater poverty of blacks.

In the alternative approach to prediction, in which all observations were combined into one large pooled sample, some additional variables were employed. Those variables were:

SEX	Sex of occupation; male occupation set at 1, female at 0.
AW	Average money wage for specific occupation
SSEX	SEX times South (dummy variable for Southern region)
SAW	South times average wage
SEXAW	SEX times average wage
LSPAW	Log of SMSA population times average wage

The average wage was constant over the observations on a given occupation, with variability relative to that average measured by the standardized dependent variable.

Some additional regional dummy variables were introduced to investigate the effect of special climate conditions in the Southwest and Pacific

Table 8-6. Best-Fitting Equations for Individual Occupations, 1975 Data

Independent variable	Equation number and dependent variable						
	(9.1) JAN	(9.2) LABR	(9.3) MECH	(9.4) KP	(9.5) STENO	(9.6) REC	(9.7) TYP
	Coefficients						
Constant	199.644	126.027	73.528	114.174	216.489	128.577	84.915
NE	-	-	-5.626	-	-	-	-
NC	12.590	18.276	-	3.873	-	-	-
S	-13.514	-	-14.149	-5.850	-	-6.258	-
W	-	32.780	16.586	-	8.098	-	-
CALIF	-	-	-	-	-	6.371	19.185
% BLACK	-	0.432	0.372	0.564	0.531	0.416	-
CB	-	-0.502	-	-0.414	-0.523	-0.321	-
LSPOP	-	14.983	8.816	8.626	-	8.321	9.147
CDENS	0.936	-	-	-	-	-	-
STEMP	-1.506	-1.056	-	-0.744	-1.780	-0.766	-
WTEMP	0.272	-0.508	-0.428	0.380	-	-	-0.701
RAIN	-	-	-	-	-3.595	-1.404	-
DAMP/100	-	0.585	-	-	4.865	1.966	-
SLEET/100	-	-	0.399	-	-	-	0.830
WIND	-	-	1.029	-	0.999	-	-
SUN	-	-	-	-	-	-	-
$\bar{R}^2$	0.667	0.752	0.730	0.628	0.288	0.626	0.227
	t ratios						
Constant	7.807	4.389	10.171	6.938	4.672	3.546	11.325
NE	-	-	2.793	1.960	-	-	-
NC	3.712	5.533	-	-	-	-	-
S	3.097	-	4.918	1.947	-	2.743	-
W	-	4.825	5.062	-	1.989	-	-
CALIF	-	-	-	-	-	2.011	3.054
% BLACK	-	1.649	3.053	3.522	3.384	3.229	-
CB	-	2.088	-	2.788	3.416	2.581	-
LSPOP	-	4.311	4.422	3.990	-	4.723	3.841
CDENS	3.308	-	-	-	-	-	-
STEMP	4.223	3.065	-	3.544	3.029	1.693	-
WTEMP	2.023	3.125	3.285	4.519	-	-	3.351
RAIN	-	-	-	-	2.701	1.338	-
DAMP/100	-	2.232	-	-	2.805	1.438	-
SLEET/100	-	-	1.950	-	-	-	2.633
WIND	-	-	2.035	-	1.657	-	-
SUN	-	-	-	-	-	-	-

Table 8-7. Comparison of Predictive Equations Obtained by Averaging Individual Equations

Independent variable	Equation number and average value of coefficients	
	(10.1) 1969 (8 cases)	(10.2) 1975 (7 cases)
CONSTANT	319.001	134.765
NE	-1.916	-0.804
NC	2.388	4.963
S[a]	-6.758	-4.788
W	0.794	8.209
California, etc.[b]	5.976	3.651
% BLACK	0.299	0.331
CB	-0.284	-0.251
LSPOP	5.888	7.128
CDENS	--	0.134
HDENS	0.110	--
STEMP	-5.387	-0.836
STEMP2/100	2.877	--
WTEMP	0.169	-0.141
WTEMP2/100	-0.271	--
RAIN	-1.495	-0.714
DAMP/100	2.050	1.059
SLEET/100	--	0.176
WIND	0.123	0.290
SUN	--	--

[a] Includes Confederacy in 1969, that is, coefficients for the South and Confederacy were added together.

[b] Includes Phoenix, Arizona, in 1969, and only California SMSAs in 1975 (CALIF).

Coast. The availability of a larger sample through the pooling of data facilitated that investigation, but the occurrence of only eight SMSAs in the region of interest limited the effort. The regional dummies were:

PHOENIX	Phoenix, Arizona
SCAL	Southern California, including Los Angeles, San Bernardino, San Diego, and San Jose
CST	The "Coast": southern California plus San Francisco, Portland, Oregon, and Seattle, Washington
CSTST	CST times summer temperature
CSTWT	CST times winter temperature

Table 8-8 presents three variations of the pooled-data predictive equation using the 1969 observations, listed as Equations (11.1) through (11.3). The variations occur through first including the dummy variables for Phoenix and southern California (11.1) and then deleting first Phoenix (11.2) and then southern California (11.3). In Equation (11.1), the explicit recognition of an effect for Phoenix caused the square of summer temperatures to lose significance. Now, the large estimated coefficient for Phoenix in (11.1) could legitimately account for disequilibrium, since Phoenix has been a fast-growing area, but it is also possible that differences really attributable to climate may well be incorporated into that (small) region dummy. Similar qualms hold for the use of the southern California dummy. In both cases, accounting for very specific regional differences may be questioned if a hypothesis for such differences is not obvious, a priori. Given these qualms, it seemed best to consider all three variations listed in table 8-8. The coefficients for the climate variables are the most sensitive to those variations, while, at the other extreme, the average wage and sex interactions are unaffected. Those latter interactions imply that: (1) not surprisingly, males receive higher wage rates than females; (2) the North-South wage differential is greater for males than for females; (3) as average wage increases, the North-South wage differential narrows; and (4) as average wage increases, the wage differential between large and small places narrows.

Two variations were developed for the pooled date predictive equation using the 1975 observations, and appear as Equations (12.1) and (12.2) in table 8-9. Squared temperature terms appear in the first of those equations, but not in the second. In Equation (12.1), results generally parallel the corresponding 1969 Equation (11.1); however, a sign reversal

Table 8-8. Predictive Equations for 1969 Data Obtained by Pooling All Observations

Independent variable	Coefficients			t Ratios		
	Eq.			Eq.		
	(11.1)	(11.2)	(11.3)	(11.1)	(11.2)	(11.3)
CONSTANT	339.562	480.524	381.708	3.919	5.854	5.827
STEMP	-4.916	-9.555	-7.079	2.213	4.773	4.406
STEMP2/100	1.769	5.524	4.056	1.209	4.508	4.046
WTEMP	0.398	0.213	--	2.169	1.173	--
WTEMP2/100	-0.718	-0.450	-0.183	3.115	1.992	2.330
RAIN	-4.501	-2.462	-2.048	6.009	4.046	4.076
DAMP/100	6.047	3.348	2.791	6.210	4.276	4.247
WIND	0.704	0.367	0.333	3.406	1.877	1.735
LSPOP	8.163	8.494	8.484	6.970	7.162	7.157
% BLACK	0.394	0.450	0.456	5.223	5.954	6.585
CB	-0.451	-0.444	-0.431	6.335	6.145	6.195
NE	-1.925	-2.238	-2.054	2.006	2.401	2.350
S	-19.197	-20.774	-20.545	5.513	5.909	5.876
SSEX	-16.372	-16.372	-16.372	11.897	11.726	11.711
SAW	8.567	8.567	8.567	5.143	5.069	5.062
SEXAW	2.318	2.318	2.318	6.373	6.282	6.273
LSPAW	-1.518	-1.518	-1.518	4.078	4.019	4.014
CSTST	0.500	0.312	0.125	2.653	1.672	4.661
CSTWT	-0.635	-0.384	--	2.072	1.255	--
PHOENIX	28.552	--	--	4.544	--	--
SCAL	5.592	6.576	--	1.785	2.073	--
$\bar{R}^2$	.581	.570	.568			

occurs for summer temperature and for its square. With the deletion of the squared term, STEMP becomes negative in Equation (12.2), and the explained variance in that equation approximately equals that in Equation (12.1). It is possible to downplay or dismiss the sign reversal as simply expressing sampling variability in a situation that is close to colinearity, since 0.999 of the variance in STEMP2 is explained by other climate variables. But it is also possible that some structural change occurred between 1969 and 1975. The sample includes two low-skilled occupations--laborer and janitor; perhaps illegal Mexican immigration was concentrated in some border SMSAs and, hence, tended to force down wages in the two low-skilled occupations in areas which also happened to have high summer temperature. In any event, caution must be exercised in interpreting results.

Table 8-9. Predictive Equations for 1975 Data Obtained by Pooling All Observations

Independent variable	Coefficients Eq. (12.1)	Eq. (12.2)	t ratios Eq. (12.1)	Eq. (12.2)
CONSTANT	-103.837	288.817	0.756	7.494
STEMP	7.609	-3.076	2.139	5.502
STEMP2/100	-7.066	--	2.968	--
WTEMP	--	0.272	--	1.514
WTEMP2/100	0.899	--	2.803	--
RAIN	-5.585	-5.472	4.708	4.794
DAMP/100	8.165	7.676	4.643	4.651
SLEET/100	-1.911	-0.984	2.900	2.175
SUN	--	0.180	--	1.984
LSPOP	11.741	11.477	8.039	7.846
% BLACK	0.360	0.397	3.835	4.302
CB	-0.312	-0.386	3.792	4.928
NE	-3.614	-3.915	2.783	2.913
S	-17.596	-17.825	5.109	5.143
SSEX	-22.566	-22.532	12.560	12.480
SAW	4.298	4.313	4.878	4.871
SEXAW	2.104	2.103	7.141	7.101
LSPAW	-1.091	-1.092	4.676	4.657
CST	230.506	91.033	4.271	5.617
CSTST	-1.881	--	2.368	--
CSTWT	-2.274	-2.018	4.445	5.319
$\bar{R}^2$	.601	.597		

8.4.3 Predictions

For predictive Equations (10.1) and (10.2), which were obtained by averaging individual equations, predictions in turn were obtained as single point estimates. For a given change in temperature, all variables which included temperature were changed accordingly from their sample mean values, and inserted in the predictive equation to yield a corresponding wage rate. The difference from the mean wage rate yielded the percentage change in wages; for the temperature changes of interest, the following percentage changes in wage rates were obtained:

	-2°F	-1°F	+1°F	+2°F
1969, eq. (10.1)	+0.738	+0.343	-0.286	-0.529
1975, eq. (10.2)	+1.048	+0.524	-0.524	-1.048

The predictions are interpreted to mean that for a decrease in temperature of 2°F, workers need roughly 1 percent more in wages to keep their real income at the same level, while for an increase of 2°F, workers are better off by roughly 0.5 percent of wages (1969 prediction) to 1 percent of wages (1975 prediction).

For the predictive equations obtained from pooled samples, temperature changes were used to yield predictions for each individual metropolitan area. Those individual predictions were then averaged, using both simple and weighted averages. Weighted averages were found by weighting by the populations of each SMSA, and then by adjusting for differences between the population distribution of the sample and that of the United States, in effect accounting for the nonmetropolitan population. In particular, the population distributions by region were distinguished as follows:

	Sample SMSAs	United States
Northeast	.314	.241
North Central	.287	.278
South	.209	.309
West	.190	.172
	1.000	1.000

For the 1969 cases, the following percentage changes were predicted in wage rates, given a base wage of 100.000 percent:

	-2°F	-1°F	+1°F	+2°F
Simple averages				
Eq. (11.1)	0.364	0.172	-0.151	-0.280
Eq. (11.2)	0.404	0.151	-0.050	0.002
Eq. (11.3)	0.269	0.096	-0.018	0.041
Weighted averages				
Eq. (11.1)	0.663	0.238	-0.217	-0.413
Eq. (11.2)	0.429	0.164	-0.062	-0.023
Eq. (11.3)	0.254	0.088	-0.011	0.056

The weighted average predictions for the 1975 pooled sample equations were the following percentages, on a base of 100.000 percent:

	-2°F	-1°F	+1°F	+2°F
Eq. (12.1) - Squared terms enter	1.382	0.752	-0.876	-1.875
Eq. (12.2) - Linear terms only	1.068	0.534	-0.534	-1.068

Predictions for major Census regions obtained from Equations (11.1), (11.2), and (11.3) are compared in table 8-10. For all three equations,

Table 8-10. Predicted Effects of Temperature Change on Wage Rates by Region, 1969 Equations (Weighted Averages)

Predictive equation and region		Percentage change in wage rates, Base = 100.000% given temperature change of			
		$-2^{o}F$	$-1^{o}F$	$+1^{o}F$	$+2^{o}F$
(11.1)	United States	0.663	0.238	-0.217	-0.413
	Northeast	-0.343	-0.182	0.203	0.427
	North Central	0.526	0.252	-0.231	-0.442
	South	-0.681	-0.351	0.372	0.765
	West	3.745	1.862	-1.842	-3.662
(11.2)	United States	0.429	0.164	-0.062	-0.023
	Northeast	0.147	0.023	0.078	0.258
	North Central	0.596	0.247	-0.146	-0.190
	South	-1.086	-0.594	0.695	1.492
	West	3.276	1.587	-1.485	-2.870
(11.3)	United States	0.254	0.088	-0.011	0.056
	Northeast	0.097	0.010	0.068	0.213
	North Central	0.506	0.214	-0.137	-0.196
	South	-0.917	-0.497	0.575	1.227
	West	2.170	1.047	-0.969	-1.861

interpretations of results for the South and the North Central regions seem relatively straightforward; with an opposite pattern of signs between those regions, a temperature decrease is beneficial in the South and costly in the North Central region, and a temperature increase has the reverse effect. Interpretations for the Northeast and the West seem less obvious. In the Northeast, the pattern of signs is the same as that of the South in Equation (11.1), while wage changes are always positive in the other two cases, indicating that any change is for the worse; the magnitudes of change are typically below those of the other regions. The sign pattern for the West is the same as that of the North Central region, but its magnitudes are much larger.

Table 8-11 shows some of the underlying predictions for the national and regional results in the form of predictions for selected individual metropolitan areas obtained from Equation (11.1). (The regional classifi-

Table 8-11. Predictions of Percentage Changes in 1969 Wage Rates (Base = 100%) for Selected Metropolitan Areas, from Equation (12.1)

Metropolitan area, by region and state	Percentage change in wage rate, given: -2°F	-1°F	+1°F	+2°F
New England				
Boston, Mass.	-0.451	-0.236	0.257	0.535
Manchester, N.H.	-0.340	-0.181	0.202	0.424
Portland, Me.	-0.297	-0.159	0.180	0.381
Providence, R.I.	-0.281	-0.151	0.171	0.364
Mideast				
Baltimore, Md.	-0.605	-0.313	0.334	0.689
Buffalo, N.Y.	0.503	0.241	-0.219	-0.418
New York City, N.Y.	-0.527	-0.274	0.295	0.611
Philadelphia, Pa.	-0.481	-0.251	0.272	0.565
Pittsburgh, Pa.	0.435	0.207	-0.186	-0.352
Rochester, N.Y.	0.925	0.452	-0.432	-0.842
Washington, D.C.	-0.328	-0.174	0.195	0.412
York, Pa.	-0.416	-0.219	0.239	0.500
Great Lakes				
Chicago, Ill.	0.458	0.219	-0.197	-0.374
Cincinnati, Ohio	-0.174	-0.097	0.119	0.258
Cleveland, Ohio	0.572	0.276	-0.254	-0.487
Detroit, Mich.	0.878	0.428	-0.408	-0.794
Indianapolis, Ind.	-0.154	-0.088	0.109	0.238
Milwaukee, Wis.	1.239	0.609	-0.588	-1.155
Toledo, Ohio	1.001	0.490	-0.468	-0.916
Plains				
Kansas City, Mo.	0.108	0.043	-0.023	-0.024
Minneapolis, Minn.	1.272	0.626	-0.604	-1.187
St. Louis, Mo.	0.091	0.035	-0.014	-0.008
Sioux Falls, S.D.	1.214	0.597	-0.575	-1.130
Southeast				
Atlanta, Ga.	-0.912	-0.466	0.488	0.997
Chattanooga, Tenn.	-1.613	-0.817	0.838	1.697
Jackson, Miss.	-1.283	-0.652	0.673	1.368
Miami, Fla.	-2.016	-1.019	1.040	2.100
New Orleans, La.	-1.626	-0.823	0.844	1.710
Savannah, Ga.	-1.105	-0.563	0.584	1.189
Southwest				
Beaumont, Tex.	-1.900	-0.960	0.982	1.985
Dallas, Tex.	0.210	0.095	-0.074	-0.126
Houston, Tex.	-0.855	-0.438	0.459	0.939
Phoenix, Ariz.	3.203	1.591	-1.570	-3.119
Rocky Mountain				
Boise City, Ida.	3.231	1.605	-1.584	-3.147
Denver, Colo.	2.955	1.467	-1.446	-2.871
Salt Lake City, Utah	2.801	1.390	-1.369	-2.717
Far West				
Los Angeles, Calif.	4.593	2.286	-2.265	-4.509
Portland, Ore.	1.199	0.589	-0.569	-1.116
San Bernardino, Calif.	3.191	1.585	-1.564	-3.108
San Diego, Calif.	4.759	2.369	-2.348	-4.676
San Francisco, Calif.	4.047	2.013	-1.992	-3.963
Seattle, Wash.	1.766	0.873	-0.851	-1.681
Spokane, Wash.	1.039	0.509	-0.487	-0.954

cation employed here is that of table 8-1, above.) Patterns noted in table 8-10 are sharpened and refined here. Thus, Cincinnati and Indianapolis are far enough South to exhibit the southern rather than the North Central pattern of wage increases and decreases. Buffalo and Pittsburgh are northern in pattern, while Baltimore and Washington are southern; somewhat surprisingly, New York City and Philadelphia also fit the southern pattern. The magnitudes of wage changes are pronounced for most of the western metropolitan areas, and there must be some question of overstatement, perhaps reflecting some specific features of the estimating equation. Equation (11.3) deleted the dummy variables for Phoenix and southern California, and consequently yielded more plausible predictions for those areas. Thus, these percentage changes from Equation (11.3) should be contrasted with the corresponding entries in table 8-11:

	-2°F	-1°F	+1°F	+2°F
Phoenix, Ariz.	-0.531	-0.304	0.381	0.840
Los Angeles, Calif.	2.509	1.216	-1.139	-2.200
San Bernardino, Calif.	0.860	0.391	-0.315	-0.551

A temperature decrease now becomes beneficial to Phoenix, while the magnitude of impacts in the California cases is considerably reduced. These results suggest that a sensitivity of predictions to relatively small changes occurs generally for the western region. Because the western region accounts for only 17 percent of total population, that sensitivity is diluted in the national predictions.

8.4.4 Applications, Conclusions and Suggestions for Future Research

Collecting the major predictions by predictive process, the following list summarizes U.S. results, in terms of percentage changes:

	-2F°	-1F°	+1F°	+2F°	Equation source
Average of individual equations					
1975	1.048	0.524	-0.524	-1.048	(10.2)
1969	0.738	0.343	-0.286	-0.529	(10.1)
Pooled sample					
1975 (linear)	1.068	0.534	-0.534	-1.068	(12.2)
1975 (squared terms)	1.382	0.752	-0.876	-1.875	(12.1)
1969 (all regions in)	0.663	0.238	-0.217	-0.413	(11.1)
1969 (small regions out)	0.254	0.088	-0.011	0.056	(11.3)

There seems a rough consistency to these results: (1) magnitudes are fairly close across cases, (2) signs tend to be consistent with negative association between temperature change and wage rate change, and (3) there seems at least some suggestion of a U-shaped relation, with a temperature decrease imposing costs, a small increase yielding benefits, and a large enough increase again yielding costs. Figure 8-3 exhibits the general form of the U-shaped relationship, while the results for equation (11.3) clearly conform to such a relationship.

In practice, the results for Equation (11.1) were employed in making the ultimate predictions utilized here. The Equation (11.1) results were chosen because they fell between the extreme values obtained, and because the procedures employed in the pooled sample approach seemed more comprehensive than those used in the average of individual equations approach. Further, the results for Equation (11.1) are roughly the average of those for Equations (12.1) and (11.3), both of which have some appeal. Equation (12.1) is from data closest to the present, while results for Equation (11.3) are based on the omission of small region dummy variables whose use is questionable, and whose omission yields more plausible predictions for the western region, and they most clearly exhibit the U-shaped pattern of figure 8-3.

In future research, a larger sample with broader geographic coverage should be of help in testing for the existence of the U-shaped pattern, and if such holds, in examining the specific form it takes. In particular, if the U-shaped pattern holds, it will be important to find where the cost-curve crosses the axis, given a temperature increase. In the work reported above, plots of the weighted average results for Equations (11.1) through (11.3) suggest that the cost curve cuts the axis at 1.5°F for Equation (11.1), at 2.3°F for Equation (11.2) and at roughly 5°F for Equation (11.3). The point of intersection becomes quite important if the temperature increase continues beyond the year 2000.

The wage studies here have focused on wage rates, or wages per unit of time (wages per hour). A preliminary analysis carried out here suggests that climate also has some impact on hours of work, and consequently on total wages. That analysis drew on data developed to extend one of the samples employed in the CIAP study noted earlier.[17] Weekly hours of work

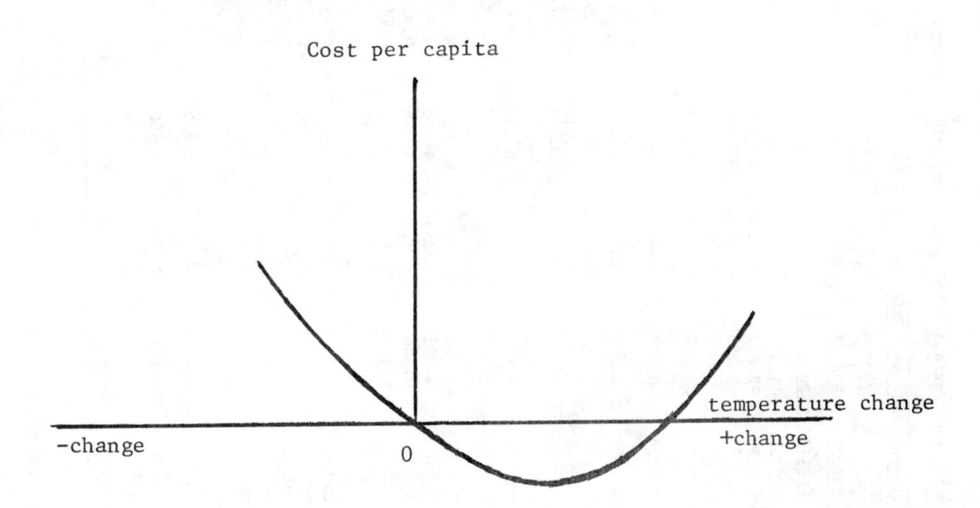

Figure 8-3. Inferred U-shaped relationship between costs and temperature change

for each of eight occupations were derived from the data, to form these dependent variables:

HCMPB	Computer operator, class B
HPROGB	Computer programmer, business, class B
HDRAFTA	Draftsman, class A
HDRAFTB	Draftsman, class B
HDRAFTC	Draftsman, class C
HACTAM	Accounting clerk, class A, male
HBOY	Office boy
HNURSE	Industrial nurse, registered, female

Here, H denotes hours of work and the remainder of the mnemonic code indicates occupation. Sample readings were individual metropolitan area averages, and number of observations ranged from 43 to 82. Table 8-12 presents the best-fitting equation for each dependent variable, in turn [Equations (13.1) through (13.8)], where the independent variables are as previously defined, save for the addition of NYC, a dummy variable accounting for New York City.

In these best-fitting equations, central city density turned out to be the dominant measure of urban scale, and, indeed, the dominant variable in the equation. Its use eliminated log of population size and a set of population-size dummies, save for the dummy for New York City. It

Table 8-12. Weekly Hours of Work Related to Significant Variables Only, for Eight Occupations, 1970 Data

Independent variable	Equation number and dependent variable by occupation							
	(Eq. 13.1) HCMPB	(Eq. 13.2) HPROGB	(Eq. 13.3) HDRAFTA	(Eq. 13.4) HDRAFTB	(Eq. 13.5) HDRAFTC	(Eq. 13.6) HACTAM	(Eq. 13.7) HBOY	(Eq. 13.8) HNURSE
				Coefficients				
CONSTANT	40.3595	41.2819	41.2409	40.9741	41.4512	40.3415	39.2194	71.1771
CDENS	-0.0685	-0.0537	-0.0268	-0.0235	-0.0375	-0.0768	-0.0824	-0.0504
NYC	-1.4933	-2.6887	-1.2800	-0.9479	-1.1368	-1.2412	-1.1485	-1.4439
S	—	—	0.1591	—	—	—	—	-0.5957
NE	—	—	—	—	—	-0.3690	-0.5993	—
W	—	-0.8568	—	—	—	—	—	—
STEMP	—	—	-0.0115	-0.0118	-0.0173	—	—	-0.7637
STEMP2/100	—	—	—	—	—	—	—	0.4510
WTEMP	—	—	-0.0028[a]	—	—	—	—	0.0611
WTEMP2/100	—	—	—	—	—	-0.0190	—	-0.0981
RAIN	-0.0132	-0.0546	-0.0067	—	—	—	-0.0168	-0.2248
DAMP/100	—	—	—	—	—	—	—	0.2989
SLEET/100	—	0.0452	—	—	—	—	—	—
WIND	—	—	—	—	—	—	0.1166	—
				t Ratios				
CONSTANT	184.564	88.169	95.628	128.910	98.314	250.203	86.611	6.130
CDENS	4.234	3.144	4.446	4.302	5.393	4.250	4.800	3.699
NYC	2.585	4.831	5.865	4.311	4.164	1.952	1.986	2.928
S	—	—	1.861	—	—	—	—	2.713
NE	—	—	—	—	—	1.999	3.595	—
W	—	2.429	—	—	—	—	—	—
STEMP	—	—	2.062	2.906	3.220	—	—	2.630
STEMP2/100	—	—	—	—	—	—	—	2.485
WTEMP	—	—	1.067[a]	—	—	—	—	2.050
WTEMP2/100	—	—	—	—	—	2.533	—	2.535
RAIN	2.308	3.026	2.693	—	—	—	2.964	2.849
DAMP/100	—	—	—	—	—	—	—	2.855
SLEET/100	—	2.003	—	—	—	—	—	—
WIND	—	—	—	—	—	—	2.802	—
$\bar{R}^2$	.571	.706	.670	.490	0.564	0.478	0.660	0.446
Number of Observations	48	43	71	82	76	74	63	66

[a] Causes another variable to become significant.

— Indicates variable was not significant and did not enter equation.

seems clear that hours of work decline with density, beyond a special decline attributable to New York City, and such may well be an equilibrating response to traffic congestion, and therefore to time lost in the journey to work.

It also turned out that at least one climate variable was statistically significant in every equation. The dominant pattern was a negative impact on hours for both precipitation and summer temperature. A key conclusion established earlier was that hourly wage rates are negatively related to summer temperature, and we now see that this occurs *despite* a negative relation between hours of work and summer temperature, for hours of work appears in the denominator of the definition for the wage rate, and hence, the two should tend to be inversely related. The reverse argument applies to precipitation in areas of above average summer temperature, for in those locales, precipitation was estimated as positively related to wage rate; the relationship established here might be a factor in that earlier result. Presumably, these relationships express the effect of physical conditions on work schedules, but the causal process is not obvious and merits investigation. It must be added, however, that the summer temperature impact on hours of work appears considerably smaller, in percentage terms, than its impact on wage rates. Thus, the *largest* linear coefficient for STEMP occurs in Equation (13.5); here, a change in temperature of 2°F causes a percentage change of 0.083 percent in hours worked, in contrast to a corresponding change in wage rates of around 0.5 percent.

8.5 Conclusion

8.5.1 Integration of Predictions

The "best" predictions made in the previous sections are now assembled, compared, and integrated. The following list shows those best predictions in terms of percentage changes from a base level of 100 percent, for given changes in temperature:

Variable predicted	Source of prediction		Percentage change in variable for temperature change of:			
	Section	Equation	-2°F	-1°F	+1°F	+2°F
Residential energy expenditures (per capita)						
Electricity	8.2	(3)	-2.62	-1.55	2.07	4.66
Petroleum products	8.3	(6)	8.33	4.02	-3.07	-7.31
Natural gas	8.3	(7)	4.55	2.24	-2.16	-4.25
Wage rates	8.4	(11.1)	0.663	0.238	-0.217	-0.413

Data for those predictions came from different years: 1970 for electricity, 1972 for petroleum products and natural gas, and 1969 for wage rates (though similar results were obtained using a 1975 sample). To combine the individual predictions, it was decided to employ 1975 as a common base year, under the reasonable assumption that percentage changes predicted for the initial base would be applicable to a base shifted forward a few years.

Base levels for predicted variables in 1975 were estimated from data obtained from the American Gas Association, American Petroleum Institute, Edison Electric Institute, and the Department of Commerce.[18] It was assumed that the predicted percentage changes in wage rates would also hold for per capita income, under the argument that the wage rate shift involved a change in utility that would affect all forms of consumption, and hence income. (Wages comprised 64.2 percent of income in 1975.) Multiplication of base year levels by the percentage shown in the preceding list yielded the following dollar levels per capita:

Variable predicted	Base level per capita, 1975	Predicted change in dollars per capita for temperature change of			
		-2°F	-1°F	+1°F	+2°F
Residential energy expenditures					
Electricity	$88.379	-2.316	-1.370	+1.829	+4.118
Petroleum products	56.144	+4.677	+2.257	-2.117	-4.104
Natural gas	45.587	+2.074	+1.021	-0.985	-1.937
All energy uses	190.111	+4.435	+1.908	-1.273	-1.923

Variable predicted	Base level per capita, 1975	Predicted change in dollars per capita for temperature change of			
		-2°F	-1°F	+1°F	+2°F
Wage rates (costs)	3,757.0	+24.90	+8.94	-8.15	-15.52
Income (costs)	5,852.0	+38.80	+13.92	-12.70	-24.17
Energy cost as fraction of income	.032	0.115	0.136	0.100	0.080

Energy expenditures made up about 3 percent of consumer income in 1975,[19] but accounted for about 10 percent of the predicted changes in income for the given changes in temperature.

Given the 1975 population of 213.6 million,[20] we can scale per capita changes by that amount to yield predicted total dollar changes per year:

	-2°F	-1°F	+1°F	+2°F
All energy use, in million dollars	+947.32	+407.55	-271.91	-410.75
Income ("needed") in million dollars	+8,287.68	+2,973.31	-2,713.72	-5,162.71

These are estimated annual costs and benefits; for both energy use and income, costs are positive and benefits are negative (that is, negative costs). A rough approximation to capitalized value can be obtained by dividing by the appropriate discount rate, say .10. Of course, there has been no accounting for population growth, nor any concern about the "proper" discount rate.

There is reassurance in the consistency of sign between changes in energy use and income, and in the relative magnitudes of the estimates. It seems plausible that other consumer expenditures vary in similar fashion with temperature, accounting for much of the 90 percent of income change not explained by change in energy cost. Presumably, "psychic" benefits and costs cover the remainder. The shift of population to warmer climates as industry and employment have become more "footloose" (less tied to specific locale) supports the conclusion that temperature increases generally yield net benefits.

8.5.2 Suggestions for Future Research

This study has developed estimates of residential expenditures on energy as a function of climate and other explanatory variables, and has demonstrated that there are significant climate effects in all cases. It would be useful to carry the residential analyses forward in time, to account for possible structural changes, including the spread of air conditioning, and impacts of energy price changes and policies not fully accounted for by the equations developed here. Further, it would be useful to extend the analyses to commercial, industrial, and even transportation use. Greene, for example, finds a small but significant negative impact on gasoline consumption of number of days in the year with subfreezing temperatures.[21]

Some CIAP studies estimated energy use for aggregate categories of use, and their estimates diverge considerably from those developed here. Crocker estimated impacts of temperature change using data on all-electric office buildings,[22] but his results have been applied to both residential and commercial use; and Nelson estimated corresponding impacts for fossil fuels--petroleum and natural gas used as fuels for the aggregate of residential, commercial and industrial uses.[23] (See d'Arge in the CIAP report,[24] and table 5-1 in this volume for summary information on the Crocker and Nelson studies.) On the basis of several rough scalings of estimates to make coverages comparable at a change of -2°F, with base year now 1972, and covering nonresidential as well as residential use, these comparisons emerge, in millions of 1972 dollars of annual costs:

	Electricity	Fossil fuels
CIAP studies	-1,300	350
This study	-525	1,045

These discrepancies seem relevant in comparing results here to those in table 5-5 of this volume. More generally, they reinforce the suggestion that additional analyses are needed.

Turning to wage rates, it would be helpful to work with a much larger sample to extend the analysis. A larger sample might facilitate use in the wage equations of other environmental variables, such as measures of air pollution. Colinearity problems have occurred when such measures were employed in metropolitan samples (see Fisher and Peterson[25]).

Section 8.4.4 established a relationship between hours of work and climate, and a deeper probing of that relationship holds promise, particularly in the prediction of total wages.

A feature of the results developed here both for energy use and wage rates was that of specific predictions for individual areas, and it turned out that a given climate change could have very different impacts on individual metropolitan areas, even when they were located in the same region. Those results seem of considerable consequence, both for analysis and policy, for small area variation tends to be neglected because we are generally concerned with overall impacts. Hence, further examination and clarification of small area differences is strongly recommended.

It was hypothesized that other forms of consumption (clothing, medical care, and so forth) might move in the same fashion with temperature as energy expenditures and, hence, help confirm the wage inferences drawn here. Future empirical work could aim at testing that hypothesis by relating specific forms of consumption to climate.

There is some uncertainty about the impact of CFC emissions on climate, first because climate effects other than temperature may be involved (see chapter 7) and second, because expected temperature change itself is subject to considerable error. In chapter 2, Forziati noted a year-2000 projection of CFC-caused temperature change on the order of 0.6°C (with +0.7°C attributed to the greenhouse effect, and -0.1°C to ozone depletion). He also noted a year-2000 projected increase of 0.5°C attributable to CO_2. Thus, the total increase becomes 1.1°C or 1.98°F, essentially the upper limit of 2°F considered here. A temperature increase of 1° to 2°F generally yielded net benefits for the United States and for most of its regions in the estimating equations employed here. However, the U.S. South generally was adversely affected by that change, and it is reasonable to infer that all areas of the globe as close or closer to the equator than the U.S. South will also be adversely affected. Again, there was evidence of a U-shaped relationship between costs and temperature change, so that for high enough temperature increases, costs rather than benefits should occur for the United States. In particular, if the year-2000 projections presented by Forziati are not "steady-state" (or ultimate) values, but rather trend line values on the way to considerably higher temperatures, then costs rather than benefits should eventually ensue. It becomes important to develop information on trend versus steady-state values, to investigate possible interactions of CFC and CO_2 temperature effects, and to determine at what point costs rather than benefits are engendered.

References

1. U.S. Bureau of the Census, Census of Population: 1970, Number of Inhabitants, Final Report (Washington, D.C., 1971); and County and City Data Book (Washington, D.C., 1972).
2. Federal Power Commission, Statistics of Private Electric Utilities (Washington, D.C., 1970); Statistics of Public Electric Utilities (Washington, D.C., 1970); and Typical Electric Bills (Washington, D.C., 1971).
3. U.S. Department of Commerce, Survey of Current Business (May 1971).
4. Irving Hoch, "City Size Effects, Trends and Policies," Science (Sept. 3, 1976); and Irving Hoch, "Climate, Wages and Urban Scale," in Terry A. Ferrar, ed., The Urban Costs of Climate Modification (New York, John Wiley, 1976).
5. Classification used in Regional Accounts, Survey of Current Business, various issues.
6. Irving Hoch, Energy Use in the United States by State and Region (Washington, D.C., Resources for the Future, 1978).
7. Ibid., pp. 322-329.
8. Ibid., pp. 716-717.
9. Ibid.
10. See Hoch, "City Size Effects"; Irving Hoch, "Interurban Differences in the Quality of Life," in J.B. Rothenberg and Ian G. Heggie, eds., Transport and the Urban Environment (London, Macmillan, 1974); and Irving Hoch, "Urban Scale and Environmental Quality," in Ronald G. Ridker, ed., Population, Resources and the Environment, Commission on Population Growth and the American Future, Research Reports, vol. III (Washington, D.C., GPO, 1972).
11. Hoch, "Climate, Wages and Urban Scale," Irving Hoch and Judith Drake, "Estimation of Climatic Impact by Wage Relationships," in Economic and Social Measures of Biologic and Climatic Change, Monograph no. 6 (Washington, D.C., U.S. Dept. of Transportation, Climatic Impact Assessment Program, 1975); and Irving Hoch and Judith Drake, "Wages, Climate and the Quality of Life," Journal of Environmental Economics and Management (December 1974).

12. Ralph C. d'Arge, "Introduction and Overview," in *Economic and Social Measures of Biologic and Climatic Change*, U.S. Dept. of Transportation, Climatic Impact Assessment Program, Monograph no. 6 (Washington, D.C., DOT, 1975).

13. Hoch and Drake, "Estimation of Climate Impact" and Hoch and Drake, "Wages, Climate and the Quality of Life."

14. U.S. Bureau of Labor Statistics, *Area Wage Survey, Specified SMSA*, Bulletins 1625-1 to 1625-90 (Washington, D.C., BLS, 1970); and U.S. Bureau of Labor Statistics, *Area Wage Surveys, Selected Metropolitan Areas, 1960-70*, Bulletin 1660-91 (Washington, D.C., BLS, 1971).

15. Hoch and Drake, "Wages, Climate and the Quality of Life," pp. 294-295.

16. U.S. Bureau of Labor Statistics, *Area Wage Surveys, Selected Metropolitan Areas, 1975*, Bulletin 1850-88 (Washington, D.C., BLS, 1977).

17. Hoch, "Climate, Wages, and Urban Scale," pp. 184-185.

18. American Gas Association, *Gas Facts '75* (Arlington, Va., AGA, 1975); American Gas Association, *Gas Facts '76* (Arlington, Va., AGA, 1976); American Petroleum Institute, *Basic Petroleum Data Book* (Washington, D.C., API, 1978); Edison Electric Institute, *1975 Yearbook* (Washington, D.C., EEI, 1975); Edison Electric Institute, *1976 Yearbook* (Washington, D.C., EEI, 1976); U.S. Department of Commerce, Survey of Current Business (August 1976; August 1977; and July 1978).

19. Hoch, *Energy Use in the United States*.

20. U.S. Bureau of the Census, *Current Population Reports* (Washington, D.C., June 1978).

21. David L. Greene, *An Investigation of the Variability of Gasoline Consumption Among States*, ORNL-5391 (Oak Ridge, Tenn., Oak Ridge National Laboratory, 1978).

22. Thomas Crocker, "Electricity Demand in All-Electric Commercial Buildings: The Effect of Climate," in *Economic and Social Measures of Biologic and Climatic Change*, Monograph no. 6 (U.S. Department of Transportation, Climatic Impact Assessment Program, September 1975); and Thomas Crocker, "Electricity Demand in All-Electric Commercial Buildings: The Effect of Climate," in Terry A. Ferrar, ed., *The Urban Costs of Climate Modification* (New York, N.Y., John Wiley, 1976).

23. Jon Nelson, "Econometric Analysis of the Effects of Climate on Energy Demand in the Residential and Commercial Sector: Fossil Fuels," in *Economic and Social Measures of Biologic and Climatic Change*, Monograph no. 6 (U.S. Department of Transportation, Climatic Impact Assessment Program, September, 1975); and Jon Nelson, "Climate and Energy Demand: Fossil Fuels," in Terry A. Ferrar, ed., *The Urban Costs of Climate Modification* (New York, N.Y., John Wiley, 1976)

24. d'Arge, "Introduction and Overview."

25. A. Fisher and F. Peterson, "The Environment in Economics: A Survey," *Journal of Economic Literature* (March 1976).

Chapter 9

IMPACTS OF OZONE REDUCTION ON NATIONAL AND INTERNATIONAL COMMERCIAL FISHERIES

Ivar E. Strand, Jr.

9.1 Introduction

Because UV-B radiation generally penetrates only a few meters into water,[1] its biological effects on marine organisms have generally been discounted. Very few commercially valuable species spend a great portion of their life history that close to the surface. However, recent research suggests that UV-B can have pronounced adverse effects on marine microorganisms and on eggs and larvae of a number of marine species.

Bell accepted the then-prevailing view of beneficial biological effects, and his results were incorporated in d'Arge's and coauthors' benefit-cost analysis of fluorocarbon emissions.[2] But Bell's results now seem questionable, casting some doubt on the conclusion that a complete ban of certain fluorocarbons would be justified only in special circumstances.[3] Bell's key assumptions that are open to question include:

1. UV-B radiation associated with ozone reduction will not affect marine biological organisms.
2. The distribution of world fish landings will be very similar, in 1990, to the current distribution.
3. Economic impacts will arise from new competitive equilibria given changing environmental conditions.

The development of detailed information on the validity of these arguments, and their possible (or likely) replacement by other propositions, awaits a considerable investment of scientific resources. As an interim approach, this paper develops an alternative set of propositions:

1. UV-B has an adverse effect on fisheries production.
2. The distribution of fisheries landings will change in response to widely accepted "extended" fisheries jurisdiction.
3. U.S. fisheries management will improve and create increased benefits from U.S. fishery production.

Although these propositions are not universally accepted, they offer a reasonable alternative and, simultaneously, establish the "worst" scenario that could result from UV-B increases. In conjunction with Bell's "best" scenario propositions, the alternative propositions yield a range of estimates which policymakers can use given their judgment on the likely physical effects of UV-B.

Section 9.2 summarizes available biological research findings on ozone-reduction impacts. Section 9.3 presents a qualitative statement on current and future international fisheries production. A discrete-time optimization model, which presents the problem of obtaining maximum social revenues from a fishery which is experiencing increasing natural mortality, is developed in section 9.4. This is the first step in estimating social net benefits from a regulated fisheries sector. Because of enormous requirements for biological data, a simplified version of the model is developed and applied in section 9.5. Finally, section 9.6 summarizes the major findings of the study and makes some recommendations on future research.

9.2 Biological Information

Recent scientific research indicates that UV-B can have pronounced adverse effects on the development of numerous marine organisms. Studies of phytoplankton, fish (mackerel and anchovy), and shellfish (crab and shrimp) larvae have shown convincingly that high rates of UV-B exposure increase mortality rates, retard growth, and reduce fertility.

At the base of the marine food chain, microorganisms have reacted strongly to increased UV-B radiation. Van Dyke and Worrest conclude that, with increased UV-B exposure,[4] Oregon diatoms (Melosira nummuliodes) have retarded growth, and Oregon copepods (Arcartia clausi) have increased mortality rates and decreased fertility. The authors conclude that "current natural levels of UV-B radiation, as well as enhanced levels, inhibit the development of some marine species of marine primary producers."[5]

Fish eggs and larvae, because of buoyancy, are also susceptible to UV-B radiation. Eggs and larvae often stay within the first meter of the water column from two to four weeks. At the egg and larvae stages, increased UV-B has been shown to increase mortality rates and inhibit growth. Hunter, Taylor, and Moser examined UV-B exposure effects on northern anchovy (*Engravilus mordax*) and Pacific mackerel (*Scomber japanicus*).[6] The mackerel and anchovy larvae were irradiated for four days with high exposure rates (19 percent to 50 percent reduced ozone), and the results indicated reduced survival, retarded growth, and lesions to the brain and eggs. The natural variation in the samples allows the inference that anchovy populations would be significantly affected if ozone depletion were on the order of 15 percent. (This corresponds to the most "likely" current estimate of long-term ozone depletion from CFCs; see chapter 2). The mackerel larvae were much more resistant to the exposure, even though adverse effects were prevalent at high dosage rates. For mackerel, the lethal dosage for 50 percent of the larvae population was approximately a third higher than that for anchovy.

Studies of the response of shellfish larvae to increased UV-B exposure have been recently undertaken by Damkaer.[7] West Coast shrimp (*Pandalus platyceros*, *P. danae*, *P. hypsinotus*), and Dungeness crab (*Cancer magister*) reacted noticeably to UV-B exposure increases. The study, although severely handicapped by lack of adequate instrumentation, showed that shrimp larvae had substantially increased mortality rates with increased UV-B radiation. Dungeness crab larvae reacted strongly to near ambient levels of UV-B exposure, whereas the Oregon crab larvae (*C. oregonensis*) required very extreme exposure to produce high mortality rates. Since Dungeness crabs are spring spawners (low UV-B), and Oregon crabs are summer spawners, Damkaer suggested that many species are currently living at their UV-B threshold, and the observed seasonal differences associated with spawning are the result of natural selection on the basis of UV-B sensitivity. The crabs also showed reduced growth rates and inabilities to molt successfully. Shrimp were generally more sensitive than crabs to UV-B dosage increases.

A summary of the present biological information must conclude that:

1. There is evidence that the projected decrease of ozone will impair survival and development of commercially valuable fish and shellfish.
2. There is the possibility of irreversible changes (extinction of some species) in the marine ecological system with projected increased UV-B.
3. There appears to be some indication that faster-growing organisms react more strongly to UV-B exposure. (But this inference is based on only two observations, indicative of the limited data now available.)

9.3 International Effects

The most important nutrient in fisheries products is protein. Approximately 5 percent of the world's protein consumption comes from fish (table 9-1). Moreover, the characteristics of the fish protein (that is, composition of amino acids) makes fish over 50 percent more productive than soybeans in producing additional body weight. Fish products rank second only to chicken eggs in terms of productivity. Improved amounts of protein in diets are particularly important in developing nations because of their current protein deficiencies and projected population growth rates, with about a 25 percent greater population in 1990. Additional food supplies will be necessary to feed this population.

One possible area of increased food production is in the fisheries of developing nations. Recent claims by coastal nations extending jurisdiction over adjacent waters via expanded territorial seas or economic zones indicate a trend toward a substantial alteration in the national and international distribution and time-paths of landings. The movement from open access to coastal-nation-controlled marine resources suggests that major producers of fishery products, such as Japan and the Soviet Union, will be forced to relinquish this dominant position as coastal developing nations increase their control over adjacent seas and build their own fish harvesting and processing industry. Thus, one would not expect the distribution of international landings in 1990 to be similar to the 1975 distribution.

Table 9-1. Protein Production and Availability by Source, 1975

Amounts of protein produced and available	United States			World		
	Fed to livestock	Available to man	Total	Fed to livestock	Available to man	Total
Million metric tons						
Vegetable protein-cereals, legumes, other vegetables	24.6	2.5	27.1	45	86	131
Livestock protein	0.7	5.3	6.0	3	30	33
Fish protein	0.8	0.2	1.0	3	6	9
Protein from all sources	26.1	8.0	34.0	51	122	173
Percent						
Vegetable protein-cereals, legumes, other vegetables	94.25	31.25	79.47	88.23	70.49	75.72
Livestock protein	2.68	66.25	17.60	5.88	24.59	19.08
Fish protein	3.07	2.50	2.93	5.88	4.92	5.20
Protein from all sources	100.00	100.00	100.00	100.00	100.00	100.00

Source: Adapted from David Pimental, William Dritschilo, John Krummel, and John Kutzman, "Energy and Land Constraints in Food Protein Production," Science (November 21, 1975) pp 754-761.

Under previous open-access ocean regimes, the developed nations have used large capital investments to form fleets which travel easily between continents and harvest the majority of fisheries production (table 9-2). The international acceptance of extended fisheries zones affords many developing nations the opportunity to develop marine fisheries in their adjacent seas. This creates employment (through direct, forward, and backward linkages) for the excess work force and provides a relatively inexpensive protein source.

The recent experience of Peru and Brazil helps illustrate how the distribution of world fisheries production is likely to change by the year 2000. Prior to World War II, the fisheries of Peru were underutilized because the extent of these marine resources was not known, and the fish were considered best used for bird feed. (Birds feeding on anchovy excreted guano, a manure used for export. Chemical fertilizer later replaced guano.) World War II and a collapse in California's pilchard fisheries created the economic incentives for development of Peru's fisheries, primarily through U.S. investments. By the end of 1951, Peru's fish industry was valued at $17 million, employed 60,000 Peruvians, and had exports valued at $2 million. Circumstances changed when the United States increased tariffs on imported canned tuna, U.S. ships began offloading elsewhere to avoid a Peruvian landings tax, and Japan's new cannery industry offered stiff international competition. These events, along with the Truman Declaration on the Continental Shelf,[8] led Peru, in concert with Chile and Equador, to declare a 200-mile territorial sea on August 8, 1952. The intention of the countries was to increase their control over the anchovy and tuna resources along the Pacific Coast.

Between 1952 and 1970, Peru benefited from the discovery of the unique attributes of fish meal for producing growth in chickens, and from the collapse of the California pilchard resources, which raised the demand for Peruvian anchovy. Foreign investments quickly developed the fish meal industry and were nationalized during the 1960s. Total landings rose from less than 1 million metric tons (MMT), in the 1950s to over 10 MMT, in the 1970s. Nearly all of the landings were converted to exports, and the value of those exports rose from about $2 million to around $200 million (in undeflated values), accounting for about 5 percent of all exports at

Table 9-2. Fishery Landings in Developed and Developing Nations (in thousands of metric tons)

Areas	Landings
Developed nations	
United States	2,743
Canada	1,029
Europe	12,625
Japan	10,524
USSR	9,936
Oceania	254
Subtotal	37,111
Developing nations	
Africa	4,462
South America	5,942
Peru (alone)	(3,447)
Asia	20,409
Central America and Caribbean	102
Subtotal	31,834
Total	68,945

Source: United Nations, Food and Agriculture Organization, Yearbook of Fisheries Statistics, vol. 42 (Rome, FAO, 1975).

the start, and over 25 percent of all exports at the end of the period. However, much of the advantage was later dissipated because of serious overfishing and a consequent serious drawing down of stocks, with the catch dropping to 10 percent of its previous high level. The overdrawing of stocks can be attributed to limited understanding of fishery biology and conservation economics.

Not all developing nations will have opportunities as fortuitous as Peru, nor necessarily dissipate their opportunities, but it seems likely that most will follow the basic course of extending the controlled fisheries zone and nationalizing or excluding foreign investment. Brazil represents a recent example. Backed by a Food and Agricultural Organization program, Brazil's fishery has slowly developed into a large industry. Brazil's fisheries production expanded from 24,000 metric tons in 1948 to 905,000 metric tons in 1976. In 1970, Brazil expanded her fisheries zone to increase the control of shrimp harvests. The extension of the fisheries zone in 1970 was followed by issuance of permits to foreign vessels. By reducing the number of permits and expanding her own fleet, Brazilian shrimp landings have nearly doubled since 1970. The Brazilian experience confirms the developing nations' ability and will to use a 200-mile extended jurisdiction to increase food supplies and revenue.

The degree to which developing nations' landings may be different from the 1975 level, and the possible impacts of increased UV-B radiation on those landings is considered next. The analysis is confined to those species groups for which some evidence of adverse UV-B damage has been shown. The three major groupings are anchoveta (including herring and sardines), mackerel (including snoeks) and shrimp (prawns and others). The analysis proceeds by determining the geographic and national distribution of 1975 landings in areas adjacent to developing nations. The national distributions are categorized either as developing or developed. The additional amount available to developing nations is represented by the developed nations' landings, and the total amount is represented by the sum of the developing and developed nations.

Table 9-3 shows that the African coast offers developing countries the largest possibilities for increased yields from the species selected. The three fish groups could add 867,000 metric tons of fishery output annually to Africa's developing nations. The tonnage would have been worth about $50 million in 1975 as processed fish meal, the lowest-valued processed form. The next largest gains in the developing regions could be made by South and Central America, about 250,000 metric tons in 1975. It is quite possible that this production has already been appropriated by the South and Central Americas, as the figures are based on 1975 world landings. Thus, it seems very likely that increased UV-B radiation could eliminate some of the developing nations' potential sources of protein.

Table 9-3. Distribution of 1975 Catch of Selected Species in Waters Adjacent to Developing Areas

(million metric tons)

Species		Landings	
		Developed nations	Developing nations
Shrimp[a]	South and Central American coasts[b]	92.8	195.2
	African coasts[c]	28.5	25.8
	Asian coasts[d]	0.0	238.9
Mackerel[e]	South and Central American coasts	0.0	
	African coasts	191.4	108.0
	Asian coasts	0.0	147.4
Anchovy[f]	South and Central Americas	173.2	4269.2
	African coasts	648.0	1669.4
	Asian coasts	0.8	798.0

Source: United Nations, Food and Agriculture Organization, Yearbook of Fisheries Statistics, vol. 42 (Rome, FAO, 1975).

[a]Shrimps, prawns, and others

[b]Includes East Central and Southeast Pacific and West Central and Southwest Atlantic.

[c]Includes East Central and Southeast Atlantic and West Indian Ocean and African Catch in Mediterranean.

[d]Includes East Indian Ocean and Asian Catch in Mediterranean and West Indian Ocean.

[e]Mackerels, snoeks, cutlass fishes, and others.

[f]Herrings, sardines, anchovies, and others.

Table 9-4. Net Exports of Fisheries Products by Continent, 1975 (in $ millions)

Continent	Imports	Exports	Net exports[a]
Africa	273	374	101
North America	1,619	1,060	-559
South America	95	396	301
Asia	1,637	1,586	-51
Europe	3,125	2,390	-735
Oceania	129	198	69
USSR	35	212	177

Source: United Nations, FAO, Yearbook of Fisheries Statistics, vol. 43, 1975.

[a]Net exports do not add to zero because of data error and excluded countries.

An inspection of 1975 fishery import and export data (table 9-4) suggests that the developing nations do obtain export earnings from fisheries products. Continents comprised of mostly developed nations, such as North America and Europe, are extremely large importers of fisheries products, whereas developing continents, such as Africa and South America, have substantial net exports. These trade positions should become even more dramatic with greater coastal nation control. Ozone depletion threatens both the actual and potential export earnings.

A more serious threat, however, appears to be the possible impacts on specific developing nations. In particular, Peru and Chile could be seriously affected if the anchovy resource was substantially damaged. A 30 percent reduction in Peru's anchovy landings (which is within the limits implied by the biological and geographic analyses presented above) would reduce export earnings by approximately $100 million annually. The situation is of particular concern because Peru and other developing nations are already heavily indebted to developed nations (via international banks).

Some U.S. impacts would occur with changes in foreign production, though these would be of significance primarily to domestic fisheries, rather than to domestic food production and consumption as a whole. Almost 60 percent of total U.S. finfish supply and over 50 percent of shellfish are imported, suggesting considerable impacts on U.S. fishery markets given marked fluctuations in foreign supply. On the other hand, the impact on the U.S. consumer should be relatively minor given the relatively small share of fish in total protein consumption (see table 9-1 above).

9.4 National Impacts: Conceptual Framework

Given the high probability that some degree of ozone depletion will occur in the next fifty years (see chapter 2) and, consequently, that the natural mortality of some commercial fish species will be augmented in a systematic manner, it is useful to explore how possible systematic mortality changes might influence the strategies of fisheries management agencies. A general model incorporating environmental alterations should be valuable to fisheries management agencies in suggesting how ozone reduction could affect present and future decisions, and in providing an approach to the estimation of social losses (reduced consumers surplus and government and producers rent) due to uncontrolled chlorofluorocarbon (CFC) emissions. A preliminary version of such a model is presented here, as follows.

A baseline situation is first developed for the marine resources (fishery) sector. The baseline information includes a schedule of use rates over time (landings for each year), total consumers surplus, and government and producers rents generated by the use rates, and price information for the period of analysis. These are based on the assumption of uncontrolled CFC emissions. The simulation serves to reflect the principles expressed in U.S. legislation (PL94-265), which states that fisheries policy will be developed to achieve maximum sustainable yield as modified by relevant social, economic, or biological factors. The model modifies maximum sustainable yield so as to attain maximum economic return from the fishery. A discrete-time optimization process is used to achieve the necessary solution; then the baseline solution is compared to the solution obtained by controlling emissions from F-11 and F-12 over the period 1976-2025, and net benefits are estimated.

The model might be generalized as follows. Variations in the ozone layer could be introduced into the growth functions of individual fish and in the recruitment functions for the stock. Reduction of the ozone layer raises the earth's seawater temperature and this might result in a more rapidly growing fish. UV-B-induced mortality rates in eggs and larvae could easily be incorporated in the stock-recruitment relationship.[9] Using predicted variations in UV-B radiation, one could determine optimal use rates for a variety of environmental situations.

The solutions containing UV-B variations could then be compared to the baseline solution. The comparison would show the present value of losses resulting from various conditions of ozone depletion, and also would present information on fishery management strategy under conditions of increasing natural fish mortality. Additional information could be generated on whether the hypothesized UV-B changes should alter present decisions being made by Regional Fisheries Councils. One must recall, however, that the relationship between ozone reduction and increased natural mortality is truly preliminary, and substantial biological research is necessary before precision could be expected of the model.

9.4.1 The Bioeconomic Model

A discrete-time optimization model is suggested as the basis for the analysis. The model consists of an economic objective function, transition equations for the fish and capital in the fishery, and nonnegativity constraints on fish and capital stocks. For convenience, an infinite time horizon is specified. In practice, of course, the researcher would have to employ one of numerous techniques (for example, Burt and Cumming's iterative terminal value function[10]) to constrain the problem size.

9.4.1.1 The Transition Equations. The dynamics of the system operate through the population equations for species and vessels. The biological system follows the approach taken by Walters,[11] which is diagrammed in figure 9-1. The number of fish in the i-th exploited age class in the t-th year ($n_{k,t}$) proceeds to the next age class ($n_{1+1,\ t+1}$) in the next year via the following relationship:

$$n_{i+1,\ t+1} = g(n_{1,t},\ F_t, M_t) = n_{i,t}\, e^{-F_t - M_t}, \qquad (1)$$

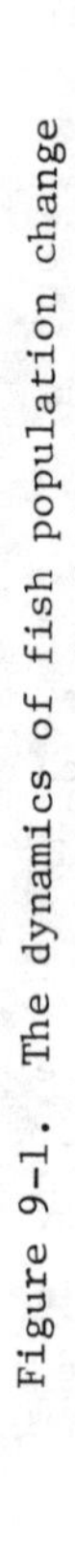

Figure 9-1. The dynamics of fish population change

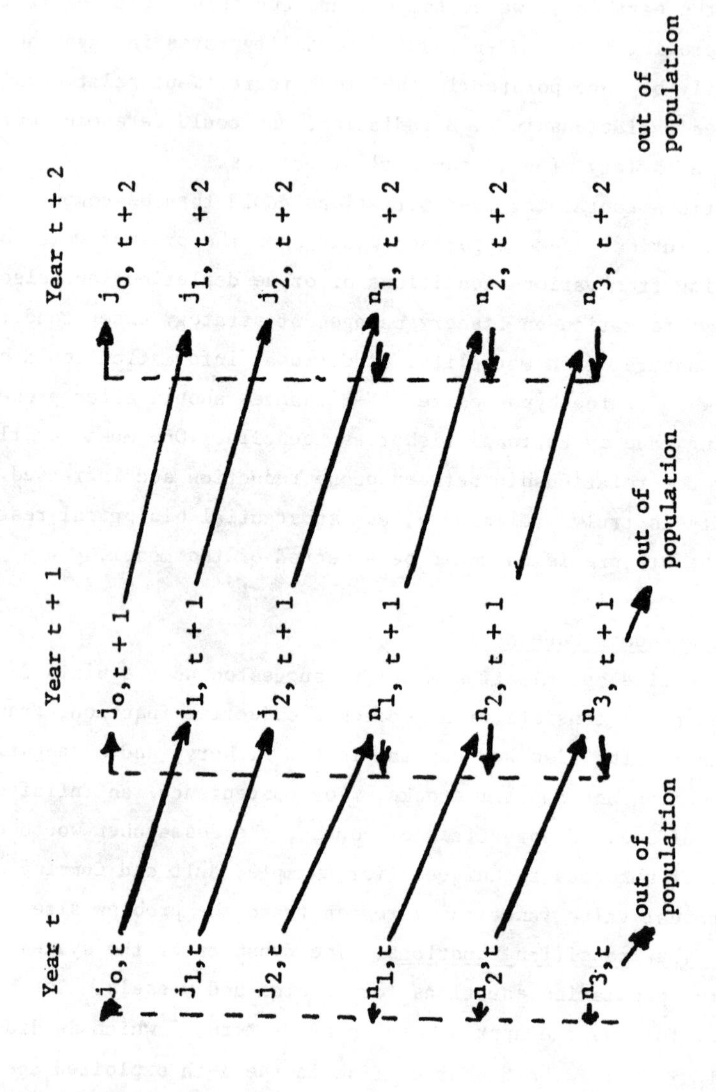

where F_t is the instantaneous fishing mortality rate, and M_t is the instantaneous natural mortality rate. For all of the exploited age classes, the expression is:

$$N_{t+1} = (e^{-F_t - M_t})N_t, \tag{2}$$

where N_t is a (column) vector of numbers of fish in the exploitable age classes.

Juvenile and unexploited age classes are denoted by J_t. Because of the extreme measurement problems associated with egg, larval, and juvenile stages, data on natural mortality during these stages are generally not available. Instead, relationships are estimated between the first exploited age-class (n_o) and total mature population (Σn_i). In computer programs, this relationship is moved sequentially through holding states (j_i) in the following manner:

$$j_{ot} = \Gamma N_t e^{-\beta N_t} \tag{3}$$

$$j_{i,\ t+1} = j_{i,t} \tag{4}$$

$$n_{i,\ t+1} = j_{i-1,t} \tag{5}$$

Equation (3) is the stock-recruitment relationship which has recruits into the exploited population as a fraction of adult population ($\Gamma_t N_t$). This figure is then adjusted to reflect the fact that adults tend to increase the mortality of the juveniles they have produced ($e^{-\beta N_t}$). The latter factor, density dependence, can result from cannibalism or simply lower food resources with greater adult populations. Equations (4) and (5) show the juveniles passing through the juvenile stage into the first exploited age class.

Equation (3) is very important to the analysis of biological impacts of increased UV-B radiation. The adult fish are not greatly affected by UV-B exposure so that the egg and larval stage, represented in Equation (3), is the critical period with respect to UV-B. The density-independent coefficient, Γ, might be used to capture the UV-B-induced mortality increases and fertility decreases. Making Γ a function of the amount of stratospheric ozone, Ω_t, would create the necessary linkage. This can be represented as:

$$\Gamma_t = f(\Omega_t) \tag{6}$$

The capital-transition equations are quite simple in comparison to the biological equations. Certain assumptions, however, must be made about the reversibility of investment or the ability of vessels to enter and exit freely from the fishery. For the purpose of this paper, it is assumed investment is irreversible. Since there is reason to believe normal profits can be returned even with greater-than-efficient levels of landings, the rational entrepreneur will not leave the industry. Returns to the unpriced factor of production, the fish, accrue to the vessel owner, and, hence, produce a great reluctance to exit. Moreover, the political realities of fisheries regulation are such that forced exit is unlikely. The second assumption is that one-year lapses between the decision to invest and the time when new capital is available to land fish. The normal delay in receiving vessels makes this assumption tenable.

The capital transition equation is expressed as:

$$C_{t+1} = C_t - D(C_t, V_t) = dC_t + V_t, \tag{7}$$

where C_t is a (column) vector showing the number of vessels by size class; D(...) is a (column) vector of depreciation functions; d is a row vector of rates of depreciation; and V_t is a (column) vector of investments in vessel according to size class.

9.4.1.2 The Objective Function. The objective of the optimization is to attain a Pareto efficient level of output by maximizing the discounted sum of consumer surplus, and producer and resource rent.[12] The consumer surplus represents the additional value that the fishery output provides over the value of the consumers' next best alternative expenditure. For the producers and government, the rent represents the additional value that the inputs provide over the value they could make in the next best alternative enterprise. The maximization provides a solution where the value of the last unit produced is precisely equal to the cost of producing it. For this level of output to be a Pareto-efficient level, it is necessary that all other economic production be at a level where marginal cost is equal to price,[13] or that fishery is separable from the rest of the economy. The objective disregards distributional considerations of the economic activity, treating dollars received by each individual in a similar manner. Discounting intertemporal returns is necessary in order to recognize the opportunity cost of capital.

The producer and consumer surplus, during a year t, results from use rates or landings (L_t) of fish. The magnitude of the surplus in year t is expressed as:

$$A_t = \int_0^{L_t} [D^{-1}(L_t) - S^{-1}(L_t)]dL_t , \qquad (8)$$

where $D^{-1}(L_t)$ and $S^{-1}(L_t)$ are the inverse demand and supply functions, respectively.

Use rates will be determined from the amount of capital (C_t, vessels) in the fishery, the amount of exploitable resource stock (N_t), and the natural mortality rate (M_t) of the species. (The use rate-capital specification involves the implicit assumption that the labor/capital ratio remains constant over the planning horizon.) The functional form for determining use rates is:

$$L_t = z_t(1-e^{-M_t-PC_t})N_t'W_t, \qquad (9)$$

where Z_t is the percentage of total mortality attributable to fishing; P is the (row) vector of catchability coefficients for each size class of vessels; and W_t is a (column) vector of average weights of fish in the various age classes. Increasing water temperatures from ozone depletion would probably increase the average weight vector, W_t, causing increased harvests. Thus, one could introduce the relationship:

$$W_t = W(\Omega_t) \qquad (10)$$

If the functional form used for the inverse-demand equations is linear, the total willingness-to-pay is:

$$WTP = \alpha L_t + \beta L_t^2 \qquad (11)$$

To obtain total costs of production, a (row) vector of annual operating costs (γ, including returns to capital) per vessel is multiplied by the number of vessels in operation, and added to the total investment costs for the period. Total investment costs are found by multiplying a (row) vector of investment costs (ϕ) times the number of new vessels. Thus, the sum of consumers' surplus, and producers' and government rent is:

$$A_t' = \alpha L_t + \beta L_t^2 - \gamma C_t - \phi V_t \qquad (12)$$

To complete the problem specification, nonnegativity constraints are required for the capital and resource stocks and the investment:

$$\text{a. } C_t \geq 0 \qquad \text{b. } N_t \geq 0 \qquad \text{c. } V_t \geq 0 \tag{13}$$

9.4.1.3 Ozone Reduction Effects on Benefits from Fisheries. A generalization of the problem described by Equations (1) through (12) can be presented so that some of the issues can be seen without unnecessary clutter. Obtaining maximum net economic gain reduces to the problem of choosing investment patterns (V) so that the following expression is maximized:

$$R = \sum_{t=0}^{T} \left[B_t(v_t, \Omega_t)\Delta^t\right] + \lambda_{t+1}\Delta^{t+1}\left[X_{t+1} - g(X_t, C_t, \Omega_t)\right] + \eta_{t+1}\Delta^{t+1}\left[C_{t+1} - C_t + D(C_t, V_t)\right] \tag{14}$$

where T is the terminal period; B_t is the objective function for period t; Ω_t is a measure of ozone layer thickness; Δ is the discount factor; and λ_{t+1} and η_{t+1} are the marginal scarcity values of resource and capital stocks, respectively. Equation (14) is necessarily simplified to concentrate on the issue at hand: initial and terminal constraints are disregarded; recruitment comes in the subsequent period; and vectors of capital and investment in t are reduced to scalars. The expression can be reduced, for convenience, to

$$R = B + \sum_{t=0}^{T} [\lambda_{t+1} \Delta^{t+1}[X_{t+1} - g(X_t, C_t, \Omega_t)] + \eta_{t+1}\Delta^{t+1}[C_{t+1} - C_t + D(C_t, V_t)] \tag{15}$$

Solution of Equation (15) yields an optimum stream of investment (V*) which can be substituted into the benefit function to obtain an indirect objective function:

$$B^*(V^*(\Omega_t), \Omega_t) = \xi(\Omega_t) \tag{16}$$

This indirect function merely represents the value of benefits at the optimum investment level as a function of the ozone parameter.

Establishment of the "primal-dual" problem gives quick insight on how ozone changes would affect net benefits to fisheries under management using optimal economic returns as an objective.[14] The primal-dual problem is to minimize the difference $B^*(\Omega) - B(V, \Omega)$ subject to the afore-mentioned constraints. A first-order condition for the minimization with a marginal change in ozone thickness, at time t, is:

$$\frac{\partial B^*}{\partial \Omega_t} = \frac{\partial B}{\partial \Omega_t} - \lambda_{t+1}\Delta^{t+1}\frac{\partial g}{\partial \Omega_t} \qquad (17)$$

This equation shows that the changes in net benefits to the fishery as the ozone layer marginally changes are equal to its marginal impact on the net benefit function, plus the product of the discounted marginal scarcity value of stock in the next period times the effect on the stock's current growth that the ozone change implies. For instance, a marginal decrease in ozone thickness in period t would marginally increase net benefits in period t because a rise in seawater temperature creates more rapid individual weight growth, but would also decrease future benefits by raising the mortality rates in future recruitment.

Although it is not easily shown using Equation (14), one would expect that the knowledge of ozone-depletion effects would alter the resource stock even before the ozone-depletion effects were prevalent. In the case where seawater temperature change was dominant, one expects the optimal resource stock to become larger so that it will propagate the soon-to-be larger fish. On the other hand, the resource stock falls as a result of knowledge of UV-B effects. Essentially, the user costs of current production is lowered (that is, the subsequent generation of fish will die anyway), and greater current production is undertaken with a concommittant decline in resource stock.

9.5 National Impacts on Marine Resources

Without a sufficient range and depth of biological information, it is difficult to accurately estimate national net benefits to the entire marine resources sector with the proposed model. The U.S. landings include more than one hundred commercial species which Bell collapsed into fourteen major categories in his study.[15] Limited UV-B-biological information is available on only about four of the species. Thus, a difficult information gap exists in the linkage between UV-B radiation and net benefits estimation of marine resources.

One approach to fill the gap would be to await complete biological information; another is to suggest a range within which the net benefits may fall. Awaiting the necessary biological research could take decades, whereas many policy questions will require responses in short order; hence, the second alternative is chosen here. A range requires a lower bound, established by d'Arge under the assumption there were no UV-B impacts, and an upper bound, which we proceed to estimate.

The upper bound is viewed as a worst situation, and is estimated by treating shrimp as a surrogate for all species in the marine resources sector, since shrimp is the most prevalent U.S. species shown to be sensitive to UV-B. Shrimp is assumed to have no response to heat; although there are questions on this assumption,[17] it appears to yield the "worst" results. Using available information, estimates of losses to shrimp producers and consumers from changes in UV-B radiation levels can be obtained. Then, based on shrimp's share of the total value of U.S. fisheries landings, the estimates will be adjusted upward to yield estimated losses to the entire fishing industry. Since shrimp is one of the more sensitive species to UV-B radiation, and is considered to have high income elasticity of demand and low price elasticity of demand,[18] the total U.S. industry estimate should represent the worst conceivable loss to marine fisheries from uncontrolled chlorofluoromethane emissions.

9.5.1 Shrimp Gains

The procedure for estimating shrimp gains from reduced ozone pollution will also provide a methodological illustration of the use of regulated rather than competitive equilibria. Equilibria associated with maximum economic yield will be used to derive net benefit estimates. This requires establishment of that output level where marginal costs equal average revenue for each time period. Intertemporal user costs are not relevant because shrimp are short-lived (usually no more than eighteen months), and seasonal closures are thought to be sufficient for providing future populations. (Externalities in current-period production, however, do exist in a competitive shrimp fishery. Since no one restricts entry by charging a use rate for landed resources, entry persists until the marginal product of the resource is reduced to zero.[19]

A shrimp production function for the Gulf and South Atlantic region has been estimated.[20] When fresh water discharges from the Mississippi River are assumed at their average value, a Spillman-type yield function for shrimp reduces to:

$$L_t = \alpha(1.0 - .9947^{E_t}) \ , \qquad (18)$$

where L_t is landings (millions of pounds), and E_t is effort (thousands of units). The coefficient α represents the maximum average yield from the fishery. It will be assumed that increases in radiation reduce the yields by reducing this coefficient.

To determine how the coefficient would vary with ozone reduction, the biological studies on shrimp were drawn upon. Data in Damkaer were employed to estimate a regression equation relating steady-state larval survival rates (SR_t) to exposure rate (e_t),[21] where an ambient level of e is specified to equal 0.4. The result is

$$SR_t = 102.2 - 77.9e_t \qquad (19)$$

The relationship between ozone change and exposure rate was based on d'Arge and coauthors,[22] and was estimated as:

$$e_t = .4*(1 + \%_t) \qquad (20)$$

where % is the estimated percentage increase in UV-B exposure.

Finally, it was necessary to link UV-B to yield changes by assuming that a change in survival rate proportionally influences maximum yield or

$$\alpha_t = \alpha_o(\frac{SR_t}{SR_o}) = 1.72 \ SR_t \qquad (21)$$

This development, in addition to Griffen's analysis of shrimp costs, produces a total industry cost curve:

$$TC_t = 208.9(41869400 + .16P_t)(\ln(\alpha_t - L_t) - \ln\alpha_t), \qquad (22)$$

where TC_t is total cost (in 1972 dollars) for the t-th year, and P_t is the price of shrimp (in 1972 dollars per pound).

The price for Gulf and South Atlantic shrimp was obtained using Doll's ex-vessel price equation:[23]

$$P_t = \phi_t - .0031L_t \qquad R^2 = .91$$
$$(4.6) \qquad D.W. = 2.04 \qquad (23)$$

where ϕ is a value derived with cold holdings (stocks) of shrimp (assumed constant at the 1976 levels), imports of shrimp (assumed to vary in proportion to α_t), and total personal income for each t-th year.

Doll's price equation was also used to obtain the total benefits of shrimp landings. Consumers surplus (measured in 1972 dollars) associated with Gulf and South Atlantic shrimp was then:

$$CS_t = \phi_t L_t - .0015L_t^2 \qquad (24)$$

Equations (18) through (24), in conjunction with projections of UV-B from d'Arge, and data on total disposal income, make it possible to derive net social gains from unilateral control of ozone-reducing CFM emissions by the U.S. Income projections through 1990 were based on the University of Maryland's INFORUM forecast of June 28, 1978. Personal income was assumed to rise at $20 billion per year from 1990 and $10 billion per year from 2000 to 2025. Assumed controls involved the reduction of F-11 and F-12 by 50 percent by 1977, 70 percent by 1979, and 90 percent thereafter. Landings where net social gains are maximum were derived for each year through 2025 with a controlled and uncontrolled emission policy. Differences in net social benefits between the controlled and uncontrolled cases produce the estimated gains from controlled emission.

The results for a 5 percent discount rate are presented in table 9-5. Net benefits through the year 2025 amount to $43.64 million in 1976 dollars or approximately $1 million annually. If the entire amount went to the shrimp producers, it would represent a gain of $29 thousand per vessel for the usual 1,500 vessels participating in the fishery. In fact, the percentage of gains captured by the producer and consumer vary substantially over the time horizon. The producers initially capture all of the gains, then capture but 60 percent in year 1990, 75 percent in year 2000, 85 percent in year 2010, and finally about 85 percent in year 2015. There are gains from 2025 to 2050 which amount to $14.3 million.

Table 9-5. Estimated Discounted (5%) Annual Net Benefits to the Shrimp Industry from Controlled Emissions of F-11 and F-12, 1976-2025

Year	Controlled emissions		Uncontrolled emissions		Discounted net benefits
	Price (1972 $/lb)	Landings (lb x 10^6)	Price (1972 $/lb)	Landings (lb x 10^6)	(1976 $ x 10^6)
1976	1.30	62	1.30	62	0.128
1977	1.34	64	1.34	64	0.164
1978	1.49	70	1.49	69	0.231
1979	1.59	73	1.59	72	0.295
1980	1.69	75	1.69	75	0.374
1981	1.76	77	1.76	77	0.379
1982	1.83	79	1.84	78	0.456
1983	1.91	80	1.91	80	0.518
1984	1.98	82	1.99	81	0.541
1985	2.06	83	2.07	82	0.605
1986	2.13	84	2.14	83	0.666
1987	2.19	85	2.20	84	0.735
1988	2.26	86	2.27	85	0.742
1989	2.33	87	2.34	86	0.798
1990	2.40	88	2.40	87	0.807
1991	2.46	89	2.48	87	0.897
1992	2.54	89	2.55	88	0.979
1993	2.60	90	2.62	89	0.719
1994	2.67	91	2.69	89	0.999
1995	2.74	91	2.76	90	1.07
1996	2.81	92	2.83	90	1.062
1997	2.89	92	2.90	91	1.121
1998	2.95	93	2.97	91	1.159
1999	3.02	94	3.04	92	1.203
2000	3.09	94	3.11	92	1.188
2001	3.13	94	3.15	92	1.12
2002	3.16	94	3.18	92	1.08
2003	3.19	95	3.22	93	1.08
2004	3.23	95	3.25	93	1.32
2005	3.27	95	3.29	93	1.09
2006	3.31	95	3.33	93	1.07
2007	3.35	95	3.36	93	1.20
2008	3.37	96	3.40	93	1.10
2009	3.41	96	3.44	93	1.08
2010	3.44	96	3.46	94	1.06
2011	3.48	96	3.50	94	1.15
2012	3.52	96	3.55	94	1.07
2013	3.55	96	3.57	94	1.05
2014	3.59	97	3.61	94	1.04
2015	3.63	97	3.66	94	1.0
2016	3.66	97	3.68	94	1.0
2017	3.70	97	3.72	94	0.97
2018	3.73	97	3.75	94	0.89
2019	3.77	97	3.79	94	0.94

Table 9-5 (continued)

Year	Controlled emissions		Uncontrolled emissions		Discounted net benefits
	Price (1972 $/lb)	Landings (lb x 10^6)	Price (1972 $/lb)	Landings (lb x 10^6)	(1976 $ x 10^6)
2020	3.81	97	3.83	95	0.90
2021	3.83	98	3.86	95	0.94
2022	3.87	98	3.90	95	0.88
2023	3.91	98	3.94	95	0.88
2024	3.94	98	3.97	95	0.86
2025	3.98	98	4.01	95	0.84

As expected, the results for a 10 percent discount rate produced markedly smaller benefits. Net benefits through year 2025 amount to $13.5 million (1976 dollars) or $270,000 annually. The value per vessel amounts to $9,000. The distribution of benefits between producers and consumers remains the same. The higher discount rate lowers net benefits to a negligible level by year 2025.

9.5.2 National Gains

The final step in estimating the upper bound on national marine resource gains from emission controls involved extrapolating the shrimp results to a national fisheries total. There are substantial possibilities for error in the extrapolation. For example, decreased yellowtail flounder availability is liable to provide a niche for a species that is not exposed to UV-B during its life history. Dogfish, a species that bears live young, might expand its availability. Large-scale marketing and processing of dogfish, especially in a situation of generally low seafood availability, might make up for losses from reduced cod and haddock production. There are also numerous possible demand and supply interactions which make the extrapolation tenuous. Most of the problems, however, would reduce net gains, and since the objective is to establish an upper bound, the extrapolation was made without further modification.

Gulf and South Atlantic shrimp represented 85 percent of the value of U.S. shrimp production in 1974.[24] With the assumption that surpluses for all shrimp production will vary proportionally with the Gulf and

South Atlantic shrimp estimates, gains to the entire shrimp industry are computed to be $68.12 million at a 5-percent discount rate and $15.88 million at 10 percent.

To extrapolate from the entire shrimp industry to a fisheries total from emission controls, the shrimp figures were expanded in proportion to shrimp's share of the value of total U.S. fisheries production. Shrimp production was $355 million in 1977, or 23 percent of the $1.5 billion ex-vessel value of U.S. fisheries. Based on these figures, the total gain to the marine resources sector from emission control was extrapolated to $300 million at a 5 percent discount of $69.8 million at 10 percent.

The gain of $300 million differs markedly from d'Arge's $661-million marine resource loss if all F-11 and F-12 emissions ceased; d'Arge's estimate was based on the assumption of no UV-B effects, while the estimate here started with such effects for shrimp, and extrapolated using "worst case" assumptions. The substantial range in estimated net social gains from emission control policies underscores the lack of basic scientific information concerning UV-B impacts on marine resources. However, the results here can be viewed as adding at least some support to the control of CFC emissions.

9.6 Summary

Three assumptions underlying previous analysis of impacts on the fishing industry from controlling CFC emissions were changed in the development of this paper. The new assumptions were:

1. UV-B radiation has lethal and sublethal effects on marine organisms.
2. The world distribution of fisheries landings will change in response to philosophical changes in international law.
3. National fisheries will become regulated to provide maximum economic benefits.

International distributions of three UV-B-sensitive species (anchovy, mackerel, and shrimp) were examined and it was concluded that ozone depletion might eliminate some of the developing nations' potential sources of protein. However, since fisheries generally are not large industries in developing nations, it was concluded that grave international impacts of uncontrolled chlorofluoromethane emissions would be specific to certain nations (for example, Peru) rather than being generic to developing nations.

A general conceptual model to analyze national impacts from regulated equilibria was developed, but required a great deal of biological information to implement. A simplified version illustrated the use of regulated rather than competitive equilibria. In that application, gains to the Gulf and South Atlantic shrimp industry were estimated and then extrapolated to cover all fisheries. This produced an upper bound of nearly $300 million (at a 5 percent discount) for national gains from unilateral emissions control of CFCs (F-11 and F-12). The discrepancy between this estimate and a previous estimate of -$661 million from controls suggest that much more scientific evidence is necessary to assure a reasonable degree of accuracy in estimating the net benefits to fisheries from controlled CFC use. However, the present estimate suggests that net benefits to fisheries are liable to strengthen the case for the control of ozone depletion, and hence, for the reduction of CFC emissions.

References

1. National Academy of Sciences, Panel on Atmospheric Chemistry, Halocarbons: Environmental Effects of Fluorocarbon Release (Washington, D.C., NAS, 1976).

2. F.W. Bell. "Economic Consequences of Stratospheric Flight on Living Marine Resources," in Economic and Social Measures of Biologic and Climatic Change, CIAP Monograph E, Part II, DOT-TST-75-56 (Washington, D.C., Department of Transportation, September 1975), pp. 5-198-5-293; and Ralph C. d'Arge, L. Eubanks and J. Burrington, "Benefit-Cost Analysis for Regulating Emissions of Fluorocarbons 11 and 12," Report on Contract 68-01-1918 (Washington, D.C., EPA, December 1976) mimeo.

3. Bell, "Economic Consequences."

4. H. Van Dyke and R.C. Worrest, Assessment of the Impact of Increased Solar Ultraviolet Radiation Upon Marine Ecosystems, Annual Progress Report, NAS 9-14860 (Corvallis, Department of General Science, Oregon State University, December 1977).

5. Ibid. p. 33.

6. J. Hunter, J. Taylor and H.G. Moser, "Effect of Ultraviolet Radiation on Eggs and Larvae of the Northern Anchovy and the Pacific Mackerel During the Embryonic Stage," in Annual Report (La Jolla, Calif.: Southwest Fisheries Center, December 1977).

7. D.M. Damkaer, "Effects of uvB on the Near Surface Zooplankton of Puget Sound," Preliminary Report (Seattle, Washington, Pacific Marine Environmental Laboratory, September 1977) mimeo.

8. Marjorie M. Whiteman, Digest of Internal Law, vol. 4, Department of State, Pub. 7825 (April 1965).

9. D.H. Cushing, "The Dependence of Recruitment of Parent Stock in Different Groups of Fishes," Du Conseil Permanent International L'Exploration de la Mer vol. 33 (May 1971) pp. 340-350.

10. O.R. Burt and R.G. Cummings, "Production and Investment in Natural Resource Industries," American Economic Review, vol. 110 (1970).

11. C.J. Walters, "Adaptive Control of Fishing Systems," Journal of the Fisheries Research Board of Canada, vol. 33 (1976) pp. 145-159.

12. P. Copes, "Factor Rents, Sole Ownership and the Optimal Level of Fisheries Exploitation," Manchester School of Social and Economic Studies, vol. 40 (1972).

13. R.G. Lipsey and K. Lancaster, "The General Theory of the Second Best," Review of Economic Studies, vol. 24 (1956-1957) pp. 11-32.

14. E. Silberberg, "A Revision of Comparative Statics Methodology in Economics, or How To Do Comparative Statics on the Back of an Envelope," Journal of Economic Theory, vol. 7 (1974) pp. 159-172.

15. Bell, "Economic Consequences."

16. Ibid.

17. F.W. Bell, D.A. Nash, E.W. Carlson, F.V. Waugh, R.K. Kinoshita and R.F. Fullenbaum, "A World Model of Living Marine Resources," in Walter C. Labys, ed., Quantitative Models of Commodity Markets (Cambridge, Mass., Ballinger, 1975).

18. Ibid.

19. W.L. Griffen, R.D. Lacewell and J.P. Nichols, "Optimum Effort and Rent Distribution in the Gulf of Mexico Shrimp Fishery," American Journal of Agricultural Economics, vol. 58 (November 1972) pp. 644-653.

20. Ibid.

21. Damhaer, "Effects of uvB."

22. R.C. d'Arge, L. Eubanks and J. Burrington, "Benefit-Cost Analysis for Regulating Emissions of Fluorocarbons 11 and 12."

23. J.P. Doll, "An Econometric Analysis of Shrimp Ex-Vessel Prices," American Journal of Agricultural Economics, vol. 54 (August 1972) pp. 431-441.

24. U.S. Department of Commerce, Fishery Statistics of the United States, 1974, Statistical Digest #68 (Washington, D.C., GPO, 1974).

25. U.S. Department of Commerce, Fisheries of the United States, 1977, Current Fisheries Statistics No. 7500 (Washington, D.C., April 1978).

Chapter 10

INDIVISIBILITIES AND INFORMATION COSTS: A CONCEPTUAL FRAMEWORK FOR ANALYSIS OF POLICY ON CFC EMISSIONS, AND FOR SIMILAR PROBLEMS

Mancur L. Olson

10.1 Introduction

Although theoretical, conceptual, and methodological analyses are sometimes alleged to have little relevance for current decisions about public policy, a moment's reflection makes it obvious that this is by no means necessarily the case. A change in the way in which we *conceive* of a public policy problem can clearly have dramatic effects on the policies that are chosen and on the detailed policy research that is proposed. In the early nineteenth century, when "bleeding" was often used as treatment for a variety of diseases, it might have seemed appropriate to do quantitative research on what amount of blood should optimally be drawn from each type of patient in each diagnostic category. Now we would have grave doubts about such suggestions; if the data were poor, or the sample small, or the preconceptions strong, even the quantitative research might fail to show the futility of bleeding. By contrast, apparently rather academic speculation on the spread of various yeasts and other microorganisms led, as we now know, to the "germ theory of disease," with the most portentous practical consequences.

More recently, some legislators and officials have been anxious to determine the characteristics and costs of "best practical" pollution-control technology. Purely theoretical work in economics does, of course, show that, in general, a higher level of environmental quality can be obtained at the same or lower cost by policies that do *not* stipulate best practical technology. Best practical requirements may give many coal users an incentive to use coal which generates more pollution, even with best practical technology, than would a "cleaner" fuel without the use of such technology.

This chapter contends that the policy debates and research agendas about the impact of chlorofluorocarbon (CFC) emissions on the stratospheric ozone layer and the climate would benefit greatly if the problem were understood with the aid of a better conceptual framework. A better formulation of these problems would have important practical consequences, both for the actual policies chosen and also for the detailed research that is undertaken. If the present paper is even close to correct, the issue of CFC control is a leading example of a large class of problems, and this large class of problems is in an important respect generally misconceived. If certain neglected features of problems in the class are understood, the whole class of problems will often be debated and researched in a somewhat different way and somewhat better policy choices will eventually be made. The class of problems at issue includes not only most ecological and environmental problems, but also many problems of policy choice and resource allocation far outside the environmental area.

One major purpose of the paper will be to show that the cost and availability of information about cause-and-effect relationships and production functions will vary systematically from one class of problem to another, and that this variation can, in large part, be explained in terms of a concept of "indivisibilities," and largely derivative notions about "scant sets" and "multitudinous sets." Though these concepts grew out of discussions in Climatic Impact Assessment Program (CIAP) research on the ozone layer,[1] it is nonetheless vital to see this problem in the context of the whole class of analogous problems, including those that have no ecological or environmental aspect.

It is no secret that there is a great deal of uncertainty and disagreement among physical and natural scientists about the effect of human activities on the ozone layer and the climate. The CIAP study came about because various scientists alleged that jet flights in the stratosphere, such as supersonic planes regularly make, would have catastrophic consequences for the habitability of our planet. More recently, however, it has been claimed that, if the continued operation of the Concorde were to pass a cost-benefit test, the reason would be that it had a favorable effect on the ozone layer! This is presumably an extreme example, but it is clear that there is an extraordinary lack of scientific consensus with respect to the ozone layer and climate change. On the latter, there has been no consensus about whether our planet will be getting hotter or colder in coming generations, or about what policies should be adopted in the light of these divergent possibilities. Though the frontier of knowledge is, of course, pushed farther ahead as time goes on, the rate of progress is disappointingly slow. One student of the fate and consequences of carbon dioxide (CO_2) produced by burning fossil fuels says that "it turns out that our confidence in existing models of the carbon cycle is considerably less now than it was ten years ago."[2] Even less attention has been paid to the possible cumulative effects of CFC emissions and CO_2 on climate (see chapters 2 and 7). One possible explanation for the slow progress is that scientists who specialize in the study of the stratosphere are intellectually inferior to scientists who specialize in areas where there is greater consensus and more rapid progress, so that a plausible public policy would be to subsidize some of the most talented people in the most advanced specialties of science to study the stratosphere. This paper will argue that the problem is _not_ any difference in the degree of talent across specialties, but that, in turn, implies the need to search for the source of the informational disparities across the areas of concern.

When there is a special paucity of information, special questions arise about the utility of cost-benefit analysis and about how such analysis ought to be done in these circumstances. This paper will accordingly suggest some changes in the format used for the cost-benefit analysis of various stratospheric and environmental problems.

Where ignorance is very great, decisions are more than usually likely to be wrong. This makes it especially important for a society to be able to change course quickly and focuses our attention on the _sequence_ of events and on investments in information. This paper, therefore, endeavors to develop a conception of "social and environmental hysteresis" and some general suggestions about research strategy.

10.2 Indivisibilities and Scant Sets

Though deduction can tell us the logical implications of our axioms or assumptions, only empirical observation can provide knowledge of the causal connections in the world around us. This is almost universally accepted in economics as well as in other sciences and has been widely understood at least since the time of David Hume. We can make inferences about cause and effect only through statistical and historical studies of natural variations or through controlled experiments. This epistemological position, or some more sophisticated version of it, is again widely accepted as applying to all fields of scientific inquiry. It is also supposed to apply to the estimation of input-output relationships or production functions, for whatever industry or sector is at issue.

Though the epistemology is much the same, the amount of knowledge about cause-and-effect relationships varies dramatically from one area to another. Not only are there great differences in how much is known in different disciplines, but there are also great differences at the level of the industry and the sector: there is often definite knowledge and general consensus about what variety of an agricultural commodity will provide the highest yield in given conditions or about what general shape an airplane designed to fly at any given range of speeds should have, but there is little definite knowledge and much ideological dispute about how best to control crime or to obtain security *vis-á-vis* the Soviet Union. Why do we know more about how to produce some "goods" than others? And more in some disciplines than others?

One extraordinarily important and heretofore neglected explanation of the different degrees of knowledge in different areas of production functions and cause-and-effect relationships generally can best be seen by considering extreme cases. Let us start with the type of situation in which unusually large amounts of knowledge can be generated. Suppose that the resources best suited to produce a given commodity are highly divisible, and that the loss of control by the manager or entrepreneur as the establishment increases in size is, in combination with this divisibility, sufficient to ensure that the lowest average cost of production will occur in productive units that are so small that each of them provides only a miniscule share of the total output of the commodity in question. In addition, the output or commodity at issue must be so divisible into small units, or be produced in such large quantities, that it constitutes what we shall here call a "multitudinous set." That is, when the commodity is divided into its minimum size unit (if there is a minimum size unit), the number of units becomes, at the extreme, infinitely large. The output must also be homogeneous in the sense that any differences from one unit of output to another, or from the output of one producer to another, are not of interest; that is, all units in any free market would sell at the same price.

The reader may object that the assumptions that have been made, albeit with an unfamiliar and awkward terminology, are essentially those that define that most familiar abstraction in all of economics, the perfectly competitive industry. He may reasonably further object that, outside of agriculture, few such industries exist. The argument here will not, however, even touch upon any of the familiar implications and virtues of perfect competition. Indeed, it will not even suppose that there are privately owned firms or a market, for neither of these features is necessary here.

The assumptions that have been made have unfamiliar and important implications for the cost of information about the production function. They entail that the largest amount of resources needed to obtain an experimental or statistical-historical observation that provides information about the production function will be of negligible cost in relation to the value of the output of the industry. This follows trivially from the assumption that the amount of resources at which minimum costs are attained is small in relation to industry output and from the assumption that the output is divisible into measurable units which are, for all practical purposes, homogeneous: in other words, with an expenditure that is trivial in relation to the value of aggregate output, we have *quantitative* information on both the inputs and outputs used in any experiment by virtue of their *divisible* and, therefore, countable character.

In keeping with the bucolic character of most perfectly competitive industries, let us consider how information about what regimen for crop rotation, forage and feed production, and livestock management might be appropriate for livestock farms that provide some of their own feedstuffs. Each such farm provides a pertinent observation about the regime it uses, and a single, controlled experiment would cost no more than it costs to operate an optimally sized farm of this nature. Yet, the information could have pertinence to hundreds of thousands or even millions of farms. Even a number of experimental farms large enough to provide statistically significant results will not cost much in relation to the value of total output. Statistical and historical study of the natural variation among such farms can also provide a great deal of information at low cost. With so many thousands of units capable of choosing different regimes, there is likely to be a good deal of variation in what is chosen. If more efficient farmers are rewarded by market forces, or in other ways, the sufficiently venturesome entrepreneur will have an incentive to try new regimes that offer the prospect of being more efficient, so that we can then be sure there will be some variability even in the absence of controlled experiments.

Note that the example that has been chosen is one which involves a great deal of interdependence among different parts of the farm, so that the *whole* of a farm must be observed to get a good experimental or statistical observation on the production function. This is, however, a special case that puts the argument being made in the least favorable light. Normally, when we have the divisible, and thus countable, outputs and inputs that have been assumed, an experiment can be conducted on a far smaller scale. Consider the experimental plots in any agricultural experiment station which are used to test the yields of different varieties or their response to different levels and types of fertilizer. Each of these plots is normally about the size of a room in an average home and a few hundred of them (or less) will provide all the information anyone wants about the alternative varieties or fertilizer applications. In the absence of other sources of experimental information, the individual farmer will have some incentive to try different options in different fields or years, so that the amount of natural variability the statistician may observe should then be considerable.

What has just been said can best be summarized in two simple ratios. Suppose a socially optimal level of expenditure for obtaining information on production functions, so that the marginal cost of the last experiment (or observation of natural variation) will just equal its expected value. Since the marginal cost is so low, this will imply that almost any experiment with a chance of producing significant results should be performed. At this optimal level of expenditure for obtaining information on the production function, the total cost of such information, E_c, will then be negligible in relation to the total value of output, V_o, of the pertinent commodity in the period during which the experimentation is conducted, so E_c/V_o in the limit approaches zero. The value of the information obtained from the experiment will, of course, depend upon the magnitude of the increase in output or reduction in costs that results from it in each period and on the number of periods in which this information is worth using. The fact that the costs of experimentation are negligible in relation to the value of the output that is produced with the aid of the information they generate suggests that normally E_c will also approach zero over the value of this information, V_i. This is not necessarily always the case, however. It is possible that the experiments will produce very little useful information. This will be most likely in circumstances where experiments in prior periods have already provided most of the information that experiments of the pertinent kind can bring. Another possibility is that any information provided will almost immediately be rendered useless by still better information. Nonetheless, if we take a sufficiently long time period, we will often (but not necessarily always) find that the E_c/V_i ratio in the limit also approaches zero. This theoretical conjecture is, in a rough way, consistent with the work of Zvi Grilliches and others on the cost-benefit ratio of the research on the development of hybrid corn and some other agricultural innovations.[3]

Now, we must consider the other extreme where information about production functions, or cause and effect generally, is relatively most costly and scarce. This is not monopoly in the casual sense of that term, but rather a situation where there can be no market at all.

Consider a good which has "shared indivisibility;" that is, the good is indivisible in that it is not possible to divide or partition the consumption among different individuals. If any individual in some group obtains any amount of this good, everyone in the group obtains the good.

Of course, a public good or externality from which nonpurchasers cannot be excluded has shared indivisibility; there would be no need for a new term were it not for the connection between this type of indivisibility and others that will be considered.

Though it is not part of the definition of "shared indivisibility," assume also that the indivisibility at issue is shared by the population of a major country, or even by the whole world. The "goods" produced by a country's defense forces and its diplomatic machinery would be examples of this, as would any nationwide public goods from which "nonpurchasers" could not be excluded. The ozone layer in the stratosphere that protects us from ultraviolet radiation is an indivisible good that is shared on a worldwide basis, as is any international organization or machinery that reduces the chances of world war.

All of these nationwide or worldwide public goods are either unique, in the sense that there is only one ozone layer or United Nations on our planet, or else members of a class which contains only a small number of others, as is sometimes the case for the public good of defense for a major power. The indivisibility that gives the goods their worldwide or nationwide significance limits the number of other goods in the same class, simply because we are constrained to one planet and because the number of large countries is inevitably limited. This means that each such public good falls in what we shall here call a "scant set," in contrast to the "multitudinous sets" we dealt with in the perfectly competitive industry. A "scant set" includes only one or a few goods or objects, though the number of people affected or the mass of the objects could be very large because of indivisibilities.

A set may be scant not only because of indivisibilities, but also because the number in the set is scant for other reasons. Suppose we consider the number of active volcanoes at any one time, or the number of people with Legionnaire's disease. A volcano may affect only an isolated community, or none at all, and may have little or no importance as a public good or bad. Similarly, Legionnaire's disease may be so rare that our interest in discovery of a cure is minimal. Still, for reasons that shall become clear directly, from the perspective of the researcher who is specializing on volcanoes or on Legionnaire's disease, the scant character of the set of objects to which he can direct his research, and to which his research is pertinent, is of decisive significance.

We are now in a position to see the two elements that together make for the most extreme cases of costly and inadequate information about production functions or cause-and-effect relations generally. The first is that the production function at issue involves a good in a scant set, and, at the extreme, a set with only one member. In this extreme case, there is obviously no subset on which there may be an experiment, or no subset that varies independently for "natural" reasons, that is smaller than the set itself. This means that the ratio of the cost of the experiment over the value of output in the experimental period, E_c/V_o, which approached zero in our previous extreme, here becomes equal to one. To provide the worldwide or unique public good in a different way in a given period, in order to obtain information about the production function, will cost, on average, about as much as it does to produce the normal output, especially when we take into account both resource costs and the gains and losses that result because the experimental allocation is either better or worse than the original allocation. When there is only one element in the set, there can no longer be any distinction between the experiment and the actual choice or policy. In these situations, we are forced, as historians are, to make inferences from samples of one. The ozone layer and the climate are clear-cut examples of such a situation.

The second element that must also be present, if the situation is, indeed, at the very extreme in terms of paucity of information on production functions, is _shared_ indivisibility. The fact that, by its nature, the good must be shared, that is, it is such that no one in the relevant group can be excluded from getting the good if any one in the group gets it, means that it cannot be straightforwardly counted or measured. If the good were divisible into separate units or portions that could be counted, it would not have to be consumed collectively. Thus, the public good of police protection, or any security provided by defense forces, or the degree of environmental improvement obtained by an effluent fee cannot be directly measured.

One may object that the achievement of the ultimate objectives can often be measured. The public good of police protection is desired primarily to prevent crime, and statistics on crime rates may be gathered; defense forces (if really defensive) are sought to reduce the probability that the citizens of the country or alliance at issue will be attacked or, if attacked, defeated, and judgments of this probability may, perhaps, be made; effluent fees are sought to clean up the environment, and approximate measures of the degree of air or water pollution may be obtained. Unfortunately, the fact that social and environmental indicators (that is, measures of characteristics of social states that are of normative or welfare interest) can often be obtained does not do away with the special information difficulty when there is shared indivisibility. There are three reasons why this is so.

First, none of the nonexcludable public goods of general importance is the only causal variable affecting the ultimate social state that we attempt to measure with the social or environmental indicator. The crime rate depends not only on the police, but also on a great many other variables, including the level of unemployment, the degree of family solidarity, the quality of street lighting, and the strength of the locks on doors of dwellings and offices. The level of pollution depends not only on the severity of the antipollution taxes or regulations, but also on the level and type of economic activity, the movement of population, the amount and timing of rainfall, the prevailing winds and other atmospheric conditions. Since at least the major types of shared public goods combine with other public and private goods and diverse uncontrollable variables to determine the value of the social indicators in which we are ultimately interested, the measures of the quantity of output that can be obtained are inevitably both indirect and approximate.

A second factor that makes information about the quantity of output of public goods harder to obtain than it usually is in the case of private goods is that shared public goods, by virtue of their shared indivisibility, are always intangible, or always "public services" rather than material goods, and this distinguishes them at crucial stages of the production process from much of what is available for purchase on the market. Much of what we buy in the market is tangible and countable, and can be measured quantitatively even if we have no information about its price, whereas none of our public services or nonexcludable public goods can be, again by virtue of the indivisibility that makes us consume them collectively. To be sure, most goods are ultimately desired because of the services they provide; we want cars because they provide transportation services, washing machines because they help us provide ourselves with laundry services, etc. In Lancastrian terms, the "attributes" we value in the real final good are normally not measurable in purely physical terms. The public good that is consumed collectively is, nonetheless, a service at an earlier stage in the productive process than most of that which is available in the market. The consumer who buys food or an automobile is buying a tangible and measureable good, which becomes a service, if at all, only at the stage of the household production function. Producers at all earlier stages are producing a physical good whose output can be counted in some way. The police protection or environmental regulation, or any other nonexcludable public good that the consumer receives is always a service, even before it enters into any household production function. Some of the factors of production used at the earliest stages of production to produce public goods, such as police cars, tanks, and smoke scrubbers, are, of course, private goods that can be straightforwardly counted, but that is also true in the production of private goods and does not contradict the point that is being made.

The third reason why the quantity of output of nonexcludable public goods is harder to measure than the quantity of private outputs is derived indirectly from the fact that individuals do not normally have an incentive to reveal their demands for collectively consumed public goods. Consider the case of a firm that sells janitorial services in the market and which employs many different groups of janitors working under foremen in separate locations to clean up the offices of different firms. Though the services provided by each group of janitors cannot practically be measured in any single physical unit, the firm can, nonetheless, get some information, over the long run, on how much each group of janitors is actually getting done. The reason is that although the total revenue received as a result of the activities of each group of janitors cannot practically be measured in any single physical unit, the firm can, nonetheless, get some information, over the long run, on how much each group of janitors is actually getting done. This is because the total revenue received as a result of the activities of each group of janitors can be known, and this total revenue is, of course, some product of the price the users of the services of each group of janitors are willing to pay for "cleanliness" and the quantity of "cleanliness" produced. This product will be positively correlated with each of the two multiplicands that comprise it. Though, in a given case, one group of janitors may bring in more revenue for a time than another simply because it happens to be assigned to firms that are willing to pay a lot for cleanliness, in the long run, the groups of janitors that bring in more revenue will be those that get more work done, that is, those who have a larger quantity of output. There can be no comparable source of information about the quantity of output of the collectively consumed public good. This third reason for the fact that producers of public goods subject to nonexclusion have less information about quantity of output than producers of private goods, unlike the other arguments, applies only if there is, in fact, a market; the good must not only be marketable but actually marketed.

As a result of the foregoing three reasons, we see that goods with shared indivisibility are not only goods for which experiments are disproportionately costly and their history unusually uninformative, but they are also goods whose output is unusually hard to measure, so that the outcome of such little experimentation or historical variation as occurs is harder to observe. In those cases where both elements of the information problem are present in the fullest degree, there is not only less data, but the data tend to be poorer as well.

The difficulty of the information problem that has just been described will diminish somewhat if there is a "discrete," as opposed to a "shared," indivisibility. An indivisibility is "discrete" if it results from some discontinuity in the production set that causes the average cost curve to be downward-sloping at a Pareto-optimal level of output. When public goods are under discussion, such an indivisibility is usually described as "nonrivalness" or "jointness." At the extreme, when a public good of this type is a "pure" public good, the marginal social cost of an additional consumer is zero, but it may still be perfectly feasible to exclude any potential consumers. Though it is not socially efficient to exclude anyone from the consumption of such a good, the feasibility of exclusion opens up the possibility that insight into the output level can be obtained by observing consumer demand. This means that when there is an experiment or a natural variation, the results of that experiment or variation can be monitored more closely. This makes the problem distinctly less severe than it is when we have both elements of the public good information problem.

If the public good is unique or worldwide, however, we still have the scantest possible set, and E_c/V_o still equals one. It makes no difference to this ratio whether the indivisibility is on the input side (as here), or on the consumption side (as with the nonexcludable public good), or whether there is only one item in the set for some other reason. Looking at the matter from the point of view of the epistemology of science, rather than the economics of the production function, we can say that scant sets make experiments relatively more costly and historical and spatial variation less informative, whether the independent or the dependent variable (or both) are in the scant set. If the causal connections run both ways, so that the system is best described by a set of simultaneous equations, the information problem will exist so long as any one of the interdependent variables is in the scant set.

If the public good is not pure even in its nonrivalness (that is, if the cost of additional consumption is positive), this will somewhat lower the E_c/V_o ratio. A public good of this sort becomes analytically the same thing as the "decreasing-cost industry;" that is, the marginal cost is less than average cost when it equals price, so that a firm that provides a socially efficient level of output is bound to run at a loss, thereby creating a case for public provision of the good at issue. In cases like this, experiments are relatively less costly. The reason is that it is not normally necessary to produce at the socially optimal level to get the needed information about the production function; the experimental activity can take place at the lowest level of output consistent with statistically significant results, and thereby keep the <u>variable</u> costs of the experiment low. Of course, the discrete indivisibility that gave rise to the decreasing costs in the first place entails that the experiment will still be relatively expensive.

A private firm with economies of scale that are sufficiently great can also have an E_c/V_o ratio that is significantly different from zero, though the problems here are normally very much less serious. Nonetheless, students of industrial organization must now add our E_c/V_o ratio to their analyses of market structure, efficiency, and innovation. If other things were equal, we would see better knowledge of production functions in purely competitive industries than in industries with differentiated products, and better knowledge of production functions in industries which can, in the long run, sustain several firms than in those that can sustain only one or a few.

The information problem we have described, obviously, also becomes less serious in the case of nonexclusive public goods as the number of public goods of a given type rises. Other things equal, we should know more about how to provide police and fire protection, for example, than about diplomacy and national defense. Unfortunately, the distinctive features of particular communities, cities, and countries often are important enough to make their production functions different. Big cities have entirely different crime and fire protection problems than small towns, as do cities in warm and cold climates. Cultural and geographic differences, as well as differences in size, must often also make production functions for various national governments also very different. The uncertainty and much of the ideological conflict over foreign and military policy is due, it is hypothesized, to the information problems set out here.

We now have the beginnings of an answer to why knowledge is harder to attain in some areas than others. Scientists know a lot about molecules and even atoms, difficult as such small entities are to observe. A great deal is also known about astrophysics, even though direct experiments with the heavenly bodies have obviously not been feasible, and much of the evidence is light years away. But atoms, molecules, and even stars and galaxies are multitudinous sets. By contrast, when it comes to ecological systems, the scientist confronts an indivisibility which creates a scant set. Thus it should be no surprise that ecologists, whatever other sciences they may also have mastered, agree about very little. Much of the chemistry and physics of the meteorological system is understood, but the climate forecasts leave much to be desired since there is, in essence, only one indivisible climate system for the planet. The difficulties of cosmology and plate tectonics can also be understood in terms of the scantiness of the sets with which the relevant scientists deal.

Similarly, in the social sciences, we can now see why economists can be sure of the effects of setting a price ceiling below the equilibrium level in almost any industry, or about the effect of a tax on marginal costs in a firm, yet disagree so much about macroeconomic and monetary policy and public goods or government. The political scientist, of necessity, deals in the scant sets of institutions which produce, in large part, public goods subject to nonexclusion, so the difficulty of making progress in that vital field need not be puzzling. The "macro-sociological problem" raises the same set of difficulties, as do several other sociological problems.

10.3 Uncertainty-Dominated Cost-Benefit Analysis

If the foregoing line of reasoning has any significant degree of validity, it follows that uncertainty is at the very essence of many cost-benefit calculations about ecological problems and other public goods, and especially those that involve a planetary asset like the ozone layer or climate. An intuitive perception of this exceptional degree of uncertainty may be the hidden source of many of the objections to cost-benefit analyses of major environmental and social policies. Some people have argued that cost-benefit analysis is of little use, or of no use at all, in making decisions that involve the natural environment in an important way. They contend that cost-benefit analysis inherently leaves out, underemphasizes, or in other ways is biased against certain environmental, social, or esthetic objectives, and therefore should be used only in conjunction with completely different concepts, or even left out of account altogether, when major decisions about the environment are made. A major proportion of those who belittle cost-benefit analysis for environmental problems are not economists and do not properly understand this tool of thought. Their criticisms are mistaken in familiar ways and call for nothing more than an increase in economic education.

More recently, however, some distinguished economists, whose arguments must be taken seriously, have offered fresh criticisms of cost-benefit analysis for certain environmental problems. Allen Kneese, for example, has argued that decisions about nuclear power should be made by means other than cost-benefit analysis, and apparently even without the aid of such an analysis.[4] Kneese contends that a decision whether to proceed with nuclear power raises distributional and, thus, moral problems which should be decided through the democratic political system, and apparently without much assistance from the cost-benefit analyst.

In fact, as any economist, skillful and experienced as Kneese, would presumably concede, <u>every</u> cost-benefit analysis, in situations where lump-sum compensation is not possible, has distributional and, therefore, moral implications. Moreover, it has been clear ever since Tibor Scitovsky exposed the fatal flaw in the "new welfare economics" that the outcome of a cost-benefit analysis can be affected by the distribution of income. Thus, it does not seem very likely that major environmental or social decisions, such as those that involve the impact of CFCs on stratospheric ozone or climate, pose distributional or moral problems that are altogether without precedent in cost-benefit analysis.

It may be that some uneasiness among economists about cost-benefit analyses of major ecological alternatives arises, instead, out of a visceral sense of the exceptional magnitude of the uncertainties about the cause-and-effect relationships. Conceivably, also, laymen who object to cost-benefit analysis in certain cases because of the "noneconomic" factors do so, in part, because they implicitly define the "economic" as the "calculable," or that which can be fairly well understood and measured to an approximation.

Though there is no inherent or fundamental contradiction whatever between cost-benefit analysis and uncertainty, the "standard operating procedure" for cost-benefit analysis has evidently not given uncertainty a sufficient role. Uncertainty is the subject of a chapter or two near the end (this is significant) of almost every text on cost-benefit analysis, but these manuals do not give the problems of ignorance and of uncertainty the central place that this paper argues they should have. Nor do the existing manuals provide an adequate agenda for a cost-benefit analysis of problems in which uncertainty is particularly great.

The insufficiency of the standard treatment of uncertainty in cost-benefit analysis can perhaps best be illustrated by confronting the cost-benefit analyst with the problem of allocating resources among alternative "pure science" projects. The output of "pure" or "basic" science--the discovery, say, of a new and inherently unpatentable law of nature--is recognized to be a public good. Most modern governments, accordingly, devote some of their funds to pure science. But the production function by which the public good of the benefits of successful pure research is produced is not well known. Pure research is conventionally defined as research on basic questions that are so difficult, or on which there has been so little advance in the past, that there is no way of predicting whether the research will make progress, or if it does, what the nature of the progress will be, or what practical usefulness, if any, whatever progress occurs would have, or who the major beneficiaries will be. If the research has a distinct likelihood of producing a given practical result, the research is, by definition, applied research or development. Though the term "pure science" seems to come from the physical and natural sciences, much of the research in the humanities, the social sciences, and even economics, if it has any social justification at all, has only whatever justification can be given to pure science.

At the aggregate level, the uncertainty about how much to spend on pure research need not _necessarily_ be utterly overwhelming. As Larry Sjaastad has argued,[5] it may be fruitful to regard all "development" and practical innovations as explicitly or implicitly some kind or form of pure research. Practical innovations are, as it were, fish that must inevitably come out of the sea of pure science or disinterested speculation. If this assumption is accepted, and there is some basis for judgments about the interest lost through the lag between pure research and its practical application, and a notion exists as to what costs are, there could be some basis for supposing that the royalties or other returns that would accrue on all applied inventions, if they were _perfectly_ patentable, would provide a guide to the optimal level on pure science.

Even if this effort to cut the Gordian knot should somehow be acceptable at the aggregate level, it cannot be used at the level of the individual research project. At this level, there would seem to be no rational alternative but to go through a cost-benefit analysis; yet, in pure science, the benefits are, by definition, altogether unpredictable.

In popular lore or folk wisdom, there is a solution of sorts to this problem. It is sometimes called the "liberal" or the "pluralistic" approach and is, perhaps, best summed up in the adage that one ought not to put "all of one's eggs in one basket." The policy that the United States and many other societies have chosen to deal with the massive uncertainties inherent in pure research is to subsidize quite a number of _different_ research projects and approaches, even if that means the "best" or "most promising" project in a given area is less well funded than it would be if alternative approaches did not absorb some of the available resources. Even where "development" rather than research is at issue, the "parallel paths" approach is sometimes used, apparently on the grounds that even the "D" in "R&D" is sometimes subject to enormous uncertainty. There is surely something to be said for the familiar liberal or pluralistic approach to these problems.

Though the indivisibility of the ozone layer or of the climate system rules out a perfect analogy, the comparison with pure research nonetheless has heuristic value. Though, in some sense, we cannot experiment with the ozone layer or climate a piece at a time, we can avoid putting all our eggs in one basket. Our society can avoid placing full confidence in any _one_ group of scientists in a given specialty, even if they are more numerous or generally more distinguished than the scientists who disagree with them. We can also avoid forms of economic development that might also threaten the stratosphere or our economic system. If we proceed with only one technology for performing some indispensable social function, and that one technology should surprise us by proving to be ecologically very harmful, we are then in a fix. If, on the other hand, we have developed alternative methods, we may have tolerable options even if one of these methods proves to be harmful to the ecosystem. Much of the rest of this paper will endeavor to justify this general approach to the matter in a slightly less impressionistic way.

We are still left with the question of how cost-benefit analysis can best proceed when there is next to no knowledge of the production function. Obviously, there can be no ideal or complete solution, for decisions in situations that are dominated by uncertainty can be as good, *ex post*, as decisions in certain situations dominated only by lucky accident. Still, some ways of proceeding are presumably better than others. Which ways are these?

It might seem that the answer has already been provided in the existing literature about uncertainty and statistical decision theory. It might be said, for example, that cost-benefit calculations, as students of these matters explained long ago, can be done in the presence of uncertainty in the following way: Outcomes are expressed in monetary values; if the decision maker has a constant marginal utility of income, the expected money value (EMV = probability times money value of outcome) is an appropriate measure of the value of the project or lottery; if the decision maker does not think each dollar gained or lost has equal utility, then the expected utility (probability times utility of outcome) must be calculated, and this can be converted into money terms by eliciting what sum of money, if offered with a probability of one (the *certainty monetary equivalent* or CME), is determined by the decision maker to be equivalent to each uncertain outcome. Once the EMVs or CMEs are found for the uncertain variables, then the cost-benefit analysis can proceed in the usual way. This approach, though logically correct, and in some degree useful, is, for present purposes, rather like the counsel that a robin can be caught by putting salt on its tail.

The use of CMEs (or EMVs) is not, in itself, an answer for the cases that are of special concern here because these cases are what we might call "uncertainty dominated." A somewhat less restrictive definition of an "uncertainty-dominated decision" may ultimately be better, but, for now, it is perhaps sufficient to define an uncertainty-dominated decision as one such that the cost-benefit ratio, obtained by calculations with certain monetary equivalents, will be greater or less than one, depending on which of the two or more discrete and *equally* acceptable subjective estimates of the probability of a given outcome is used. For the class of environmental problems that are at issue here, there is no hope of getting objective probabilities based on experimental experience. We are in situations that are indisputably Bayesian, and in which the decision maker or cost-benefit analyst is so much in the dark that he regards two or more discrete subjective probability estimates (say, estimates from two equally competent scientists who disagree) as equally likely to be true. If the choice between or among these alternative probablity estimates determines whether the project gets a green or a red light, the situation is "uncertainty dominated" and the procedures in the existing cost-benefit manuals do not provide a sufficient basis for dealing with the problem.

What kind of model, then, should we try to develop to deal with uncertainty-dominated situations? If the earlier part of this paper was at all correct, the model must be one which <u>integrates research with actual policy choices</u>. There is a precedent of sorts for this in the literature of psychology and marketing. As Peter Doyle puts it, in a discussion of advertising, "The most valuable contribution of behavioral science [is that] learning is accepted as a central feature of buying behavior."[6] The behavioral scientist Donald Schon put this very well when he wrote:[7] "The learning agent must be willing and able to make the leaps required in existential knowledge. . .from the experience of one situation to its use as a projective model in the next instance. These are leaps because they cannot be justified except by what happens after they are made. <u>They are conditions, not consequences of knowledge</u>" (underlining added).

Since new developments or technologies that imperil the environment can also bring definite gains, and possibly also reduce the hazards from energy shortages or political upheavals due to interruptions in economic growth, they should not be ruled out whenever they cannot be proven to be perfectly safe. New technologies that <u>might</u> be environmentally disastrous could sometimes also prove to be environmentally preferable to those we are using now.

As Stephen Hanke[8] has noted, new information that is "purchased" by "buying" or trying some amount of a given good or technology, is itself a public good, and one which existing agencies of government (like the National Science Foundation), with the responsibility for the production of new information, are not normally expected to provide. Thus, it is probable that this type of information is not provided to society to an optimal extent. Hence, the rational environmentalist must turn away from the "don't do anything for the first time" philosophy, and turn toward one which says "it's better and even safer to spend some of our income to obtain information." The extra information may sometimes require some "leaps"--some danger of some environmental damage. But, at least for those who are risk avoiders, there should be, if possible, a succession of short leaps rather than one long one. There should be a premium on small-scale innovations and on those that could not bring disaster to the whole ecosystem.

These suggestions can best be elucidated in the language of the "decision trees" of decision theory. A decision tree, as is well known, consists of "decision forks," which mark the points at which choices may be made among alternatives, and "chance forks," which mark the points or stages of a process in which one among the different possible but imperfectly predictable (or unpredictable) outcomes emerge. Some works on decision theory, with their characteristic focus on balls drawn from urns, list "deciding not to play the game" among the options at the first decision fork, but there is no such alternative for a society concerned about public investments or environmental policy. Laissez-faire is, of course, an option for governments, as is the policy of "never allowing anything new." But these policies obviously mean that society plays the game in particular ways, and in ways which, in addition to their other shortcomings, probably also entail relatively substantial risks for the society: the former increases the chances of ecological disaster, the latter the danger that the society will not have the resources and technical capacity to deal with future environmental problems or be able to cope peacefully with demands for additional output.

An approach to decision making in conditions of ignorance which gives no place to the suggestions made above will entail a decision tree with a relatively small number of large branches, such as is depicted in extreme form in figure 10-1. The decision fork is conventionally depicted with a rectangular shape, whereas the chance forks are round. The alternatives presented in figure 10-1 are very stark, but the illustrative numbers are characteristic of many decisions of environmental concern, in that a highly likely gain of significant magnitude is promised on one fork, along with a tiny chance of catastrophe; whereas another fork, with the nonde-

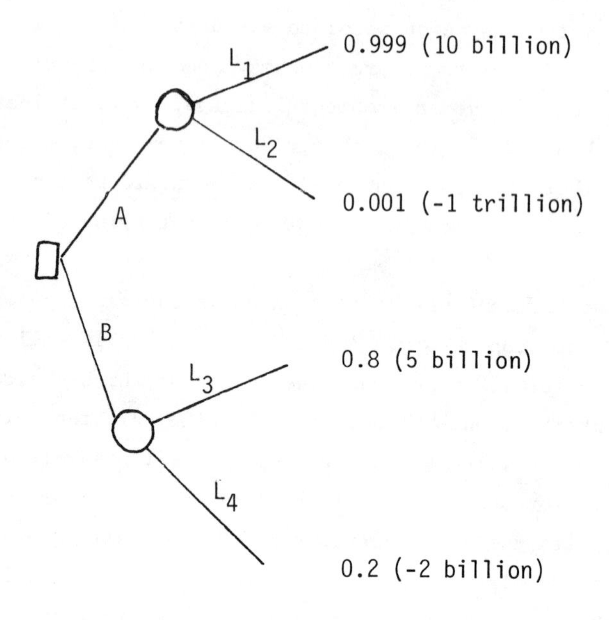

Figure 10-1. Hypothetical decision tree characteristic of many environmental decisions.

velopmental or less innovative alternative, promises little gain and no danger of such a disastrous outcome. Suppose, as the numbers on figure 10-1 indicate, that option A with lucky form L_1 appears to have a 0.999 probability of a gain of $10 billion, and that the unlucky or L_2 fork for option A appears to offer a one-in-a-thousand probability of a trillion-dollar loss. Option B appears to offer a 0.8 probability of a $5-billion gain and a 0.2 probability of a $2-billion loss.

Clearly, if, for the moment, we assume no risk aversion or preference, so that estimated monetary values can be used as a basis for decision, and accept that the probabilities attached to each possibility at the chance forks are correct, option A should be chosen. The trouble is that, if the foregoing arguments in this paper were even half-right, there is no objective basis for determining the probabilities at each chance fork: we are not able to count the balls of different colors in the urns with which decision theorists play, but must decide on the probabilities in a Bayesian fashion. For example, we must ask scientists what the odds are on the outcomes of experiments that have never been performed. Our particular scientist has put the odds on a disastrous outcome from option A at one in a thousand, but an optimistic scientist might say the chance of a catastrophe is one in a million, a pessimistic scientist may say it is one in a hundred, and an ecologically enthusiastic scientist might say it is one in ten. Obviously, if either the pessimist or the ecological enthusiast is correct, option B has a greater expected value than option A, so we are in an "uncertainty-dominated" situation where an arbitrary choice among equally distinguished experts determines which alternative should be chosen.

The foregoing argument suggests it would often be expedient to create a far different decision tree. The "learning by trying" (or "buying") proposal would entail a search for the full range of possible options, including various mixes of alternative technologies and policies and particularly small-scale experimentation with any possibility that has much to be said for it. Thus, there would be a much larger number of possibilities, and each of these possibilities, in turn, would lead not only to new chance forks, but also to new decision forks presenting still further possibilities that could be chosen later. The number of options is a function, among other things, of the amount of effort (and even formal research) that is put into finding options. And, when these options that involve a wide range of techniques are followed, they branch out into exceptionally wide fans of alternatives at the next decision fork, and the information that was "bought" by the fact that many things were tried promises to make possible better decisions on major matters in the next period. The options involving the possibility of major irreversibilities or other catastrophic losses are still there and may continue to have attraction, but an awareness of other options that buy information, which eventually open up better alternatives, will tend to reduce the likelihood that these all-eggs-in-one-basket alternatives should be chosen. The suggestions in the foregoing section, in short, have the effect of grafting additional branches onto the decision tree, albeit at some expense, including some which, through the information they purchase, yield such enticing options at the later decision forks that, despite discounting, they will probably often have a greater "path value" than the starker alternatives that are characteristically debated. When the decision tree beomes a luxuriant bush, it is much more likely that at least one branch will bear fruit that is safe to eat.

Some Bayesian decision theorists like to point out that a vague and insecure subjective hunch about probabilities should enter into a decision tree in exactly the same way, and with the same weight, as probabilities that are objectively and correctly known. It may be that the counter-intuitive, but demonstrable, character of this conclusion has led to an overemphasis of it, and a neglect of a simpler but practically very important point, namely, that extremely doubtful estimates of the probability of different outcomes (such as those obtained from scientists when there is no consensus among them) should, in fact, lead the practical decision maker to act in a different way than he would have done with correct and objective information on probabilities; it should induce him to spend more to obtain information.

Some may agree that a wider array of options would be good to have, but also argue that the decision trees that result from the suggestions in this chapter are likely to be such bushy thickets that it will not be feasible or, in any event, economic to find the best branch. How can the best branch to follow from the trunk be found when each branch forks out to new chance and decision forks, each of which branches out, in turn, to further forks, and so on. Clearly, the path with the highest estimated monetary value (or alternatively, certain monetary equivalent) ought to be chosen, but this cannot be determined without knowing the overall expected value of a long series of contingencies and decisions.

There is no difficulty at the formal level. As Baye's theorum makes clear, the decision tree can be "flipped," so that the expected value of each path, given _ex ante_ optimal choices at each decision fork, can be calculated from the payoffs and losses, the probabilities that each will occur, and the costs of the information purchased. All that needs to be done is to start at the outer tips of the branches with the possible payoffs, to determine the average expected value of the gamble at each chance fork, and to "fold back" as one proceeds toward the trunk, calculating the expected value on the assumption that the best ex ante choice is made at each decision fork. The expected value of each branch may be determined by this procedure and the branch with the highest expected value chosen.

Unfortunately, the problem may not be so simple in practice. In real-life public decision making, there is no "end of the game" with determinate payoffs at the top of the outermost twigs; the decision tree continues growing and generating new forks. The farther into the future the analyst tries to carry his decision tree, the less the likelihood he will know even what the options will be, much less be able to calculate the expected value of each. If new information not only improves estimates or probabilities, but has heuristic value as well, so that new possibilities that cannot be conceived of now are opened up, then there is no way of calculating the expected value of a given path out from the trunk; if we do not know, even in a probabilistic sense, what fruit might be on the twigs leading off from the branches, we cannot make an informed choice about what fork to take now. These difficulties need not deter us, but they do remind us that a good analyst must have good judgment as well as good techniques, so that he will have a sense of how far into the future it pays to carry the analysis.

It is hoped that these difficulties and the others we have encountered in this paper will also stimulate us to develop our formal techniques. If this paper is even close to correct, environmental problems, and, indeed, other problems that involve indivisibilities that obscure production functions, deserve fresh theoretical attention. They deserve this attention because of the fact that policymaking and knowledge-purchasing become one in the presence of public goods, and because the indivisibilities at issue entail losses that will be catastrophic or total. Some of the tasks are to elaborate our manuals for cost-benefit analysis in such a way that uncertainty receives more stress, to sensitize our paradigm to the necessity for policy options that are chosen because of the information they generate as well as their expected direct consequences, to include the cost and benefits of growing new branches on our decision trees, and to elaborate our tools for assessing the worth of new information.

10.4 Hysteresis, The Ozone Layer, and Climate

If the argument earlier in this paper is correct, the exceptional degree of ignorance and uncertainty that must characterize our knowledge of the effects of CFCs on both stratospheric ozone and climate, and the dangers that follow from this, are compounded by some long time lags. If the gas used in refrigerators and air conditioners leaks out only very slowly, and, in part, after the relevant appliances have been junked and slowly corroded, then it will presumably be a long time before we know what the effects of our current policy are. There is not only the problem that we have only one stratosphere and one global climate to experiment with or learn from, but also the problem that these samples of one sometimes yield observations about the cause-and-effect relationship only after a long delay. Nonetheless, some choice must be made now. Long time lags increase the ignorance and, in addition, the degree of risk associated with any given degree of ignorance. Though this economist is not competent to assess the length or extent of these lags, there is probably no one who would deny that some of them are measured in years and even decades. This means, of course, that the problem of "turning around" a policy that proves to be disastrous can be a serious one. We could conceivably discover after ozone depletion reaches level X or climate change reaches level Y, that unexpected and very serious consequences are emerging, and that substantial further change will lead to a catastrophe. But again, just conceivably, actions already taken will inevitably lead to, say, further depletion of ozone by 3X, or climate changes by 5Y, and all that can be done is to await the catastrophe. This situation involves a specific kind of irreversibility. In a sense, almost every choice is irreversible. However, if because of indivisibilities we cannot injure part of the stratosphere or global climate system without injuring it all, and long time lags mean that actions taken now will eventually lead to serious injury to this indivisible asset, then a special and very important form of irreversibility will occur.

There is no meaningful definition of "irreversibility" in the literature, yet there is an understandable intuitive concern about the process in many environmental decisions, and the foregoing is offered as explanation for such concern.

There is another more complex set of lags and that is perhaps best analyzed with the aid of a concept borrowed from the physical sciences, specifically "hysteresis," which involves a retardation of effect when the forces acting upon a body are changed. Edmund Phelps has used the concept in his seminal analysis of the microeconomic foundations of macroeconomics,[9] and Jordan Baruch, a distinguished engineer, has drawn an elaborate and intriguing analogy between hysteretic switches and certain decisions about technological development in large corporations. This account will borrow Baruch's more complete analogy,[10] but will apply it to a different class of problems, to which the analogy proves particularly apt.

Baruch begins by considering "hysteresis loops" in "bi-stable devices such as mechanical toggles, snap-action switches and flip-flop circuits." He provides a simple depiction of the hysteresis loops by postulating a device which converts an input, such as electrical energy, into some output, and assumes a nonlinear relationship between the input and the output. Baruch also postulates that, in addition, there is a "discontinuity" such that, as the input level is increased beyond a certain point, there is an increase in output so dramatic that there is more total output with less _total_ input, that is, the curve bends back and is bi-valued, as is shown in figure 10-2. By contrast, when input levels are decreased from sufficiently high levels, the ratio of inputs to output will be maintained over the abscissa interval mn, and only at point m does it drop to the same path it was on when inputs were being increased over time. Apparently, hysteresis loops of this kind are of interest in a number of engineering contexts, but the most pertinent engineering analogy for Baruch, as for us, is when phenomena of the sort described lead to a loss of control sensitivity, as in hysteric switches.

Baruch applies hysteresis loops to research and development projects he knows about in various corporations. Development expenditures, he hypothesizes, must reach some minimal level before they have any chance of even producing profitable output, or even engaging the interest of the firm's top officials. A development project, accordingly, will not begin or get anywhere unless it is expected to have a benefit-cost ratio considerably above unity to compensate for the uncertainty, and perhaps also an expected revenue that is absolutely large enough to overcome inertia

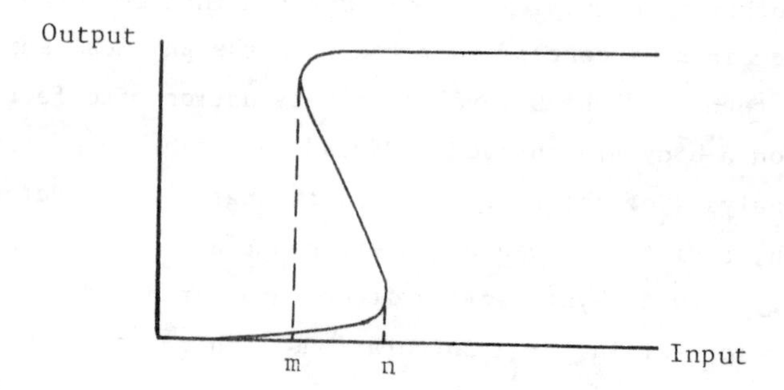

Figure 10-2. Determination of the transfer function.

and attract the attention of the firm's leadership. Thus, even in the large firm with considerable resources, it may be very difficult to get a major development project started or carried on at an adequate level.

By contrast, once there has been a major commitment of resources to the development project, the perceptions change, and the momentum is in favor of continuing the project, even if new information should suggest that the original cost-benefit calculation was much too optimistic. As in the hysteresis loop in some switches, the profitablity criteria that are used to decide whether to continue a major project are decidedly less stringent than those that must be met before a major project can get started. Baruch is not as specific as he might be about the mechanisms that might explain the pattern of behavior he observes or hypothesizes, but presumably a managerial psychology or evaluation system which makes it difficult or costly to admit that a mistake has been made plays a substantial role.

When we turn our attention to government and the political system, and most especially the role of pressure groups, the importance of hysteresis loops is surely far greater, and the mechanism that generates them far easier to describe. With rare exception, powerful pressure groups are a "by-product" of organizations created for other purposes. If there is a large cotton textile industry, there may be a trade association, and this trade association will tend to seek tariffs, tax loopholes, or other special interest legislation. But if there is suddenly a discovery of a profitable way to make widgets, the exploitation of that discovery is not likely to be subsidized by government programs attained by interest groups; the interest groups will not have formed until after the widget industry has been in operation for some time, perhaps even a long time. Since this writer has set out the theory and evidence behind these assertions in detail elsewhere, no more will be said of them here.[11] All that need be said now is that the "by-product theory of interest groups" ensures that there will be a dramatic hysteresis loop in political processes, not to mention government bureaucracies. Often what has been done for some time will be protected and subsidized, even when this involves massive social inefficiency, whereas proposed activities of great potential benefit to the society will only rarely get any public support.

The implications for stratospheric ozone depletion and climate change from CFC emissions are, by now, so obvious that they need only be mentioned. If an existing industry produces a product or uses an input that, in time, proves damaging to stratospheric ozone or climate, the society is _not_ likely to be prompt in imposing an optimal effluent fee, or probably any appropriate regulation or fee. There is likely, at best, to be a significant lag which will be _in addition_ to the lag between physical cause and physical effect described at the outset of this section. This could conceivably even lead to catastrophic results. Symmetrically, if the government decides to regulate some activity in the interest of protecting the ozone layer or preventing climate change, a configuration of implicit bureaucratic lobbies and symbiotic private lobbies (that is, associations of producers of substitutes for the banned substance, or makers of emission control equipment) will probably eventually emerge, and these interests will tend to maintain the pattern of intervention, even if later information should show that this is irrational for the society. This lag also is additional to the lags described earlier. Because there are no comparable indivisibilities, the possibility of global catastrophe from a particular type of regulation is not present, but the social waste could, nonetheless, reach enormous proportions. The conclusion is that no basic decisions about policy for CFC emissions control can be made without giving a lot of thought to the wide hysteresis loops that are involved.

10.5 Researchable Policy Implications

It is not a good idea to go directly from theory--even valid and pertinent theory--to policy. Many economists do this, but the prevalence of the practice cannot justify the practice. Any good theory, by definition, involves simplification, and thus, for each policy decision, there needs to be an investigation to determine whether any of the aspects of reality from which the theory has abstracted are of decisive significance. As political officials sometimes say, the decision has to be "staffed out," at the least. In academic language, there is "a need for further research," even after valid and pertinent theorizing or conceptualizing has been done, or, at the least, the theoretical work needs to be used in conjunction with the results of other, empirically oriented researches.

Accordingly, a set of "policy hints" are now set forth, to be used in conjunction with other research; in addition, some of those hints should themselves be the subject of further development and research. In order to underline the heuristic character of these policy hints, they will be stated briefly and with little or no elaboration.

Policy Hint 1 -- The readily estimable costs and benefits are not a sufficient basis for policy decisions on policy for CFC emissions control. If the argument about indivisibilities offered here is correct, this much is clear even without further research. As Daniel Golomb has pointed out, the scientific estimates on which many cost-benefit calculations are based are reliable, if at all, only for a limited range, and the greatest concern is about developments outside that range.[12]

Policy Hint 2 -- There is a need for better monitoring, or "social and environmental indicators," to reduce the extent of the lags and the seriousness of the hysteresis loops.

Policy Hint 3 -- There is a need for "quick-response" research capabilities, so that unexpected results of the monitoring or "early-warning" system can be analyzed promptly for policy makers and the public.

Policy Hint 4 -- Beware of "monism" in sources of scientific and policy advice; if the "indivisibilities argument" in this paper is correct, even the best scientists may be wrong, sometimes even about some matters about which they are nearly certain.

Policy Hint 5 -- Beware of monism in policies, since a single policy that is followed consistently over a large area has a greater chance of causing a catastrophe than varied policies.

Policy Hint 6 -- Beware of even the small chance of the really catastrophic outcome; this is perhaps the most important result of the indivisibilities argument offered in this paper.

Policy Hint 7 -- There is no such thing as a *safe* lunch either.

Because there are a variety of environmental indivisibilities, no really feasible combination of policies can avoid *all* risk of ecological disaster. On top of the inevitable environmental hazards, there are indivisibilities of a political and social type. No one can be sure that a reduction in national incomes arising because of very restrictive environmental regulation would not lead to some sort of upheaval, or disastrous levels of inflation, or the end of the military equilibrium between the United States and the Soviet Union. Thus, though the argument here emphasizes the danger of catastrophic ecological outcomes and the need for caution about them, it also implies that all-around security is impossible and that there must be trade-offs, even among the possibilities of different types of disaster.

Policy Hint 8 -- Two risky technologies may be better than one.

The reason is, first, that this can decrease the extent of any *one* ecological hazard and, second, make it easier to change course if one of the risky technologies begins to threaten to do very serious damage.

Policy Hint 9 -- "Irreversibility" needs to be emphasized in policymaking and in further research, but the irreversibility of particular concern involves the case where long time lags mean that actions taken now cause irreversible damage much later to an indivisible resource, where damage to part injures all.

The usage here should be distinguished from the use of the term in much prior writing, for irreversibility in that latter sense is true of practically all economic choices, the only exception of note being the purchase of goods for which there is a "money-back guarantee."

Policy Hint 10 -- The research process and the policy-making process cannot be altogether separated where the effects of CFC emissions are concerned.

As argued earlier, when there is only one indivisible member in the "scant" experimental set, we must "learn by doing." Thus, the information obtained by policy choices must be taken into account in choosing policies, and, consequently, research-oriented people should have a role at all levels of policymaking.

References

1. Institute for Defense Analysis, Economic and Social Measures of Biologic and Climatic Change, U.S. Department of Transportation, Climatic Impact Assessment Program, Monograph no. 1 through 6, 1975.
2. Jill Williams, "Carbon Dioxide, Climate and Society," in Options 1978 (Laxenberg, Austria, ILIASA, 1978) p. 1.
3. Zvi Griliches, "Hybrid Corn: An Exploration in the Economics of Technological Change," Econometrica, vol. 25 (1957) pp. 501-522; Zvi Griliches, "Research Costs and Social Returns: Hybrid Corn and Related Innovations," Journal of Political Economy, vol. 66 (1958) pp. 419-431.
4. Allen Kneese, "The Faustian Bargain," Resources no. 44 (September 1973).
5. Larry Sjaastad, personal communication.
6. Peter Doyle, "Advertising Expenditure and Consumer Demand," Oxford Economic Papers, vol. 20 (1968) pp. 394-416.
7. Donald Schon, Beyond the Stable State (New York, Random House, 1971), p. 235.
8. Stephen Hanke, personal communication.
9. Edmund S. Phelps and others, Microeconomic Foundations of Employment and Inflation Theory (New York, W.W. Norton, 1970); and Edward S. Phelps, Inflation Policy and Unemployment Theory (New York, W.W. Norton, 1972).
10. Jordan J. Baruch, "Technical and Management Note," IEEE Transactions on Engineering Management (August 1974) pp. 105-197.
11. Mancur Olson, The Logic of Collective Action (Cambridge, Mass., Harvard Economic Series) 1971; and Mancur Olson, The Rise and Decline of Nations (New Haven, Yale University Press, 1982).
12. Daniel Golomb, Discussion at Port Deposit Conference, Port Deposit, Maryland (1978).

Chapter 11

CONTROLLING REFRIGERANT USES OF CHLOROFLUOROCARBONS

Peter Bohm

11.1 Introduction

The global use of chlorofluorocarbons (CFCs) as a propellant has been considerably reduced by the banning of nonessential uses of aerosols in the United States and a few other industrial countries. In particular, since the detrimental effects (on consumption, employment, and other activities) of such a ban appear to be modest, other countries may well follow suit. If so, the remaining major use of CFCs will be as refrigerants (as indicated in table 11-1).

The refrigerant use of CFCs presents two major policy problems. First, existing substitutes for CFCs are expensive, flammable, toxic, or corrosive. Hence, there can be no simple adjustment of technology with commensurate small effects on consumers and producers, as was the case with aerosols. Second, the durability of existing cooling equipment means that the current stock of such equipment will remain an important source of CFC emissions for a long period. This implies the dual objective of reducing CFC releases from existing cooling equipment as well as from new equipment, differing again from the aerosol case, where product durability and releases from existing sprays were not a matter of major concern.

Table 11-2 lists available estimates of CFC emissions in 1973 by major forms of cooling equipment. (In that table, E denotes emissions with a first subscript indicating stage of use--M, manufacture; U, normal use; S, repair and service operations; and D, disposal. A second subscript distinguishes between new (N) versus existing (E) equipment. For

Table 11-1. Percentage Distribution of 1974 End Uses of Chlorofluorocarbons F-11 and F-12

Source	United States	Total OECD
Aerosols	61.8	65.9
Refrigerants	23.8	18.1
Plastics	8.7	12.1
Other	6.2	3.9
	100.0	100.0

Source: Organization for Economic Co-operation and Development (personal communication).

example, E_{SE} denotes emissions from servicing of existing equipment.) Although the source of those data notes questions on their reliability,[1] they are useful in indicating probable orders of magnitude. Thus, disposal emissions account for some 23 percent of total CFC releases, whereas the rest can be attributed to emissions from use, repair, and servicing operations.

A reduction of future CFC emissions can occur at any of the emission stages identified in our classification. Table 11-3 lists the ways in which such reductions can be made--in addition, of course, to limiting CFC emissions by reducing the use of cooling equipment.

Table 11-4 lists selected policy instruments that could be used to bring about the changes given in table 11-3. Ideally, such instruments would be employed to equate the marginal social costs of reducing CFC emissions with the marginal social benefits of such reductions. However, the great uncertainty of the effects of CFC emissions implies that a social benefit function cannot be established. In fact, we are not even in a position to determine the social costs of the different policy instruments. To make some progress in evaluating policy options we shall then have to rely on assumptions. However, none of the following five assumptions seems unreasonable or unduly strong:

First, both regulation and economic incentives may be used for each instance of emission reduction (see table 11-4) granted that they do not have exactly the same effects in all respects.

Table 11-2. Releases in Million Pounds of CFCs (Primarily F-11 and F-12) from Cooling Equipment in the United States in 1973

Type of cooling equipment	Emissions from: Use and service (E_{UE} and E_{SE})	Emissions from: Disposal (E_{DE})	Total
Automotive air conditioners (some F-22)	49.6 (42.8)[a]	21.3	70.9
Home refrigerators, freezers, ice-makers and dehumidifiers	1.7[b]	3.3[b]	5.0[b]
Commercial cooling equipment (some F-22 and F-502)	28.4 (20.4)[a]	5.4	33.8
Chillers (some F-22 and F-502)[c]	31.1 (23.4)[a]	2.8	33.9
Total	110.8 (86.6)	32.8	143.6

Source: G.C. Eads and coauthors, Non-Aerosol Chlorofluorocarbon Emissions: Evaluation of EPA-Supplied Data, A Draft Working Note prepared for the Environmental Protection Agency by the Rand Institute (December 1977).

[a]Maximum recoverable at repair and service stage. Figures reported by A.D. Little as potentially recoverable through relatively minor modifications to current equipment design or service procedures.

[b]Probably too low. $E_{UE} + E_{SE}$ may be high as 3 million lb according to Du Pont Freon® Products Division, Information Requested by EPA on Non-Aerosol Propellant Uses of Fully Halogenated Halocarbons (Wilmington, Del., March 15, 1978), p. III-29; and E_{DE}, for 1976, is 4.5, according to another estimate reported by G.C. Eads, Non-Aerosol Chlorofluorocarbon Emissions, p. B-26.

[c]Chillers are large air conditioning units for commercial use (see G.C. Eads, Non-Aerosol Chlorofluorocarbon Emissions, app. D).

Table 11-3. Possible Measures Reducing CFC Release, by Source

Existing equipment		New equipment	
		E_M:	Change of production technology Reduce CFCs per unit Substitute other refrigerant
E_{UE}:	Speed up service Replace old units	E_{UN}:	Design changes Speed up service (Replace old units)
E_{SE}:	Service procedure changes Scrap instead of service/ repair	E_{SN}:	Design changes Service procedure changes Scrap instead of service/repair
E_{DE}:	Recovery of CFCs Destruction of CFCs	E_{DN}:	Recovery of CFCs Destruction of CFCs
+	reduced use of cooling equipment		

Second, it is likely that economic incentives would not be very effective in reducing E_U and E_S. Given that the cost of recharging and servicing cooling equipment will primarily consist of labor costs, a tax on the refrigerant (with a present sale price on the order of \$1 to \$3 per pound) would have to be very high to speed up service or replacement and stimulate more careful refrigerant-saving servicing operations.[2] For example, servicing a home appliance and adding a pound of refrigerant with a tax of 100 percent of the present price would add only a couple of dollars to a service charge of at least \$20. Hence, reduction of E_U and E_S would probably have to be managed primarily by regulatory instruments.

Third, economic incentives are likely to be the best way to reduce E_D, given our experience with laws on littering or improper dumping. Such laws have had limited effect, probably due to difficulties of supervision.

Fourth, disposal of cooling equipment constitutes a problem even without taking CFC emission effects into consideration. Old refrigerators and air conditioners are often dumped or stored where they create hazards to children and damages to aesthetic values.[3] Moreover, if dumped into the municipal waste stream, they add significantly to waste treatment costs. If disposed of in other ways, they have a certain reuse

Table 11-4. Selected Instruments to Enforce Measures Reducing Release of CFCs (F-11 and F-12 Only)

Emissions	Regulation	Economic incentives
E_M	Ban on high CFC-releasing production technology	Tax on CFC use or subsidies on CFC savings
	Ban on high CFC contents per unit	
	Ban on CFC as a refrigerant	
E_U	Standards on design of new equipment to reduce leakage under normal use	Tax on recharge
	Inspection of units at regular intervals and service if required (or disposal)	
E_S	Standards on design for easy service without CFC release	Tax on recharge
	Standards for service and repair work	Subsidy on refrigerants collected
E_D	Ban on improper storage or disposal of CFC-containing equipment	Subsidy on proper disposal of CFC containing equipment

or scrap value, which might increase in the long run because of design changes once used units tend to be recovered on a large scale. All these factors taken together may significantly reduce the net social costs of managing the disposal of cooling equipment on account of the effects of CFC emissions.

Finally, if the disposal problem is handled by a subsidy related to CFC content, and if the subsidy is coupled with a tax on the use of CFCs--creating, in effect, a deposit-refund system--we will have a policy package which avoids or moderates effects on the government budget. The possibility of modest financial effects may be quite important from the viewpoint of political acceptability.

In what follows, we shall analyze this suggested tax/subsidy or deposit-refund system. In Section 11.2, we present some alternative versions of the system. Its overall effect on the economy is the subject matter of Section 11.3. Finally, in Section 11.4, we shall propose a particular version of the deposit-refund system as a possible "first-draft" solution to the CFC problem caused by the use of cooling equipment, to be combined with suggested regulatory instruments in the future.

11.2 Alternative Versions of an Incentive System

If used CFCs are to be retrieved instead of being released into the atmosphere, we can ask who is to collect them and what is to be done with them. We can also ask how the refunds on collected units are to be determined and how they are to be financed. Then, we will examine alternative institutional ways of determining the deposit (tax) on CFCs. Finally, we must consider the particular problems caused by emissions from existing (nondeposit) equipment. We shall deal with these issues, in turn.

11.2.1 Destruction or Reprocessing?

Technically speaking, it is possible to destroy CFCs collected from used equipment without leakage into the atmosphere, or to refine collected CFCs for reuse.[4] Once the net costs of these two alternatives have been determined, the government may select the least expensive alternative and require its implementation. As the two alternatives may be cost-ranked in different order for different types of CFCs, for different locations of the collected CFCs, the requirement may be structured on the basis of such factors. However, administrative advantages (simplified control procedures) may outweigh such considerations and call for a uniform type of disposal.

Over time, there may be changes in the costs of the different disposal alternatives and new ones may appear. This is likely to occur not only as a result of normal technical change, but also because of industry changes induced by the policy measures that are taken. A government might respond to such changes by adjusting its rules on how final disposal should be made. However, the uncertainty of such a response (given pressure from industry and unions) is likely to reduce industry's incentives to develop new alternatives and improve existing techniques. Instead of the government determining the disposal of used CFCs, it could determine certain required properties of disposal, in terms of units of atmospheric CFC releases, and let the market establish the reprocessing industry which minimizes costs of CFC disposal. This would allow for permanent competition among reprocessing alternatives to minimize costs of reduced CFC emissions. In particular, it would lead to optimal switching among alternatives over time and in accordance with cost changes due to technological change. Moreover, there would be incentives to invest in research and development of new reprocessing solutions and to redesign cooling equipment on the basis of expected reductions in disposal costs.

Regardless of whether the collected CFCs are destroyed, reused, or reprocessed in any other sense, the industry may reprocess the discarded _equipment_ which contained the CFCs. Although induced technical change may produce easily detachable CFC containers or coils as the only parts which the reprocessing industry necessarily handles, the rest of the equipment may be worth reprocessing. This already seems to be true for certain types of equipment. But it would be more universally true if scrapping or reuse of the remaining parts carried a (higher) positive market value once the CFC part is collected. Moreover, collecting the remaining parts of the equipment could eliminate certain negative external effects. If so, the government would have to be relied upon to transform these reductions of social costs into price signals. Since it is now the case that certain types of cooling equipment are dumped where they cause externalities, and since the equipment is expensive to treat when it appears in the municipal waste stream, a subsidy or refund equal to these social cost savings could make the reduction of CFC releases less costly in the context of an overall recovery program for used cooling equipment.

In summary, we have the alternatives of CFCs being reused or destroyed in a reprocessing stage and of reprocessing only the CFCs or the cooling equipment as a whole. Given economic incentives set equal to the amount of the negative externalities avoided, and given the rules for reducing or eliminating CFC emissions by reprocessing, the choice of optimal reprocessing can be left to the market.

11.2.2 The Transfer from Users to Reprocessors

At each point in time, there is a given market value V_j of the CFCs embodied in form or equipment type j and returned for reprocessing. In some instances, V_j could be positive although we would usually expect V_j to be negative. Now, the transfer from the point of cooling equipment use to the point of reprocessing may be organized in several alternative ways. The user may deliver it himself at an estimated cost of T_j^U and collect V_j as well as the refund. Or it may be collected by the reprocessor at $V_j - T_j^r$, where T_j^r is the pick-up cost. Or it may be picked up and transferred to the reprocessor by a third party, for example, the seller of new cooling equipment, at a price T_j^S. Given that the user is provided with adequate information about the various alternatives (for example, from the Yellow Pages of the local directory or from a local government information service), he may choose the alternative that minimizes his return costs (C_R^i), that is:

$$C_{Rj}^i = \min(T_j^U, T_j^r, T_j^S), \text{ for user i.} \qquad (1)$$

Thus, the collection or transfer activity may be left to the market. There is a caveat, however. If there are economies of scale in this activity, the market may not identify the least-cost alternative or the alternative established may use some form of monopolistic pricing. In such cases, the local government might give concessions to a collection firm at regulated prices (T_j) for limited periods, ensuring _ex ante_ competition among potential holders of the concessions.

11.2.3 The Refund Payments

We have assumed the reprocessor and the collector are paid for their services by V_j and T_j, which are market-determined or regulated as indicated above. Since the reprocessing industry can be expected to be more concentrated than the collectors, it would be practical (for control purposes) to have the refund payments introduced at the reprocessing stage. Thus, for each unit received by a reprocessing firm, it pays a refund R_j (for the CFCs) directly or indirectly to the user. This means that the user receives a total amount of $R_j + V_j - C_{Rj}^i$ when he returns a used piece of equipment (or the relevant part of it). To avoid fraud when there are several reprocessors who could sell units among each other for additional refunds, each unit of equipment would have to be identifiable, say, by a registration number introduced by the producer or importer of new cooling equipment, or by the agency which administers the refund payments, the latter being relevant especially for initially existing equipment. As an alternative, the agency could be made formally responsible for supervising the inflow of all equipment or CFCs at the reprocessing plants.

As these remarks suggest, the government would directly finance the refund payments. However, in certain versions of a deposit-refund system, the payment may be organized in a different fashion. If the most efficient reprocessing activity is conducted by the original producer of the CFCs or of the cooling equipment, and if the producer has to introduce a deposit (equal to a tax) on new CFCs (equipment) sold, he may use the deposits to finance the refunds and only deliver the surplus to the government (or cover the deficit of refunds over deposits from government sources). Or the government may even abstain from requiring that a certain deposit--in the sense of a tax--is to be paid on new equipment. For example, this is true for beverage containers in certain states where only a minimum refund is determined by government. Then, the deposit is simply part of the price for the beverage and thus accrues to the producer to be used for refund payments on containers returned to the producer.

As explained in more detail elsewhere,[5] the latter system has a disadvantage in that the refund rate can hardly be sustained at a level above what the return is worth to the producer. Hence, a government subsidy (refund) will be needed when the socially efficient volume of product returns calls for a refund rate above this value. This may often be true for discarded cooling equipment. More important, perhaps, is that the refrigeration industry does not have as rapid a turnover as does the beverage industry. With durability of cooling equipment on the order of ten to twenty years, the deposit and the refund payments would have to be separated even when the producer happens to be the optimum reprocessor. Therefore, we expect the optimum version of a deposit-refund system to be that where refunds are paid from government to reprocessor.

11.2.4 The Optimum Refund Rate

Assume, to begin with, that, at each point in time, there is a given number of units of cooling equipment being discarded (X_{Rj}). Without the return alternative, users will dispose of these units in other ways (stored, sold to be scrapped, dumped in the municipal waste stream, illegally dumped, etc.). The costs of the disposal alternative then chosen is C^i_{dj} for user i. Introducing the return alternative at costs $C^i_{Rj} - V_j$, these costs may fall short of C^i_{dj}. If this is true for all users and if the difference $C^i_{Rj} - C^i_{dj} = \Delta Cj$ is the same for all, users would appear to benefit from the introduction of the new alternative even without any refund. Thus, if $\Delta Cj < V_j$--and, in particular, if this remains or becomes true in the future after changes of product design or relative prices have taken place--the market could take care of the CFC disposal problem on its own. But if $\Delta C_j > V_j$, a refund is required for the return alternative to be chosen and sustained by the market. Provided the social willingness to pay (SWTP) for reducing CFC emissions per unit of equipment j, $SWTP_j$, is at least $\Delta C_j - V_j + \varepsilon$, where ε is the minimal amount necessary for users to perceive an advantage of the return alternative over the dumping alternative, a refund in this amount will be called for. (The social willingness to pay would be derived from the SWTP per pound of CFC not released into the atmosphere.) In practice, V_j is likely to be subject to cyclical variations as well as structural changes which will require adjustments in R_j over time. Given that V_j is market-determined, R_j may be formally tied to V_j so that changes in V_j immediately result in changes in R_j ($\leq SWTP_j$) to insure that $R_j = \Delta C_j - V_j + \varepsilon$. Changes in C_j may be more complex and, thus, may have to be countered on an *ad hoc* basis.

In general, we may expect $C^i_{Rj} - C^i_{dj}$ to differ among users due to differences in locations, transportation facilities, storage facilities, replacement decisions, etc. Ranking units to be discarded in the order of increasing $C^i_{Rj} - C^i_{dj}$, we may derive a disposal cost difference curve such as that labeled ΔC^i_j in figure 11-1. Unless maximum $C^i_{Rj} - C^i_{dj} = \Delta C^i_j$ is below $SWTP + V_j - \varepsilon$, not all discarded units will be returned. Thus, at $R_j = SWTP_j$, the optimum level of units returned will be X^o_{Rj} in figure 11-1.

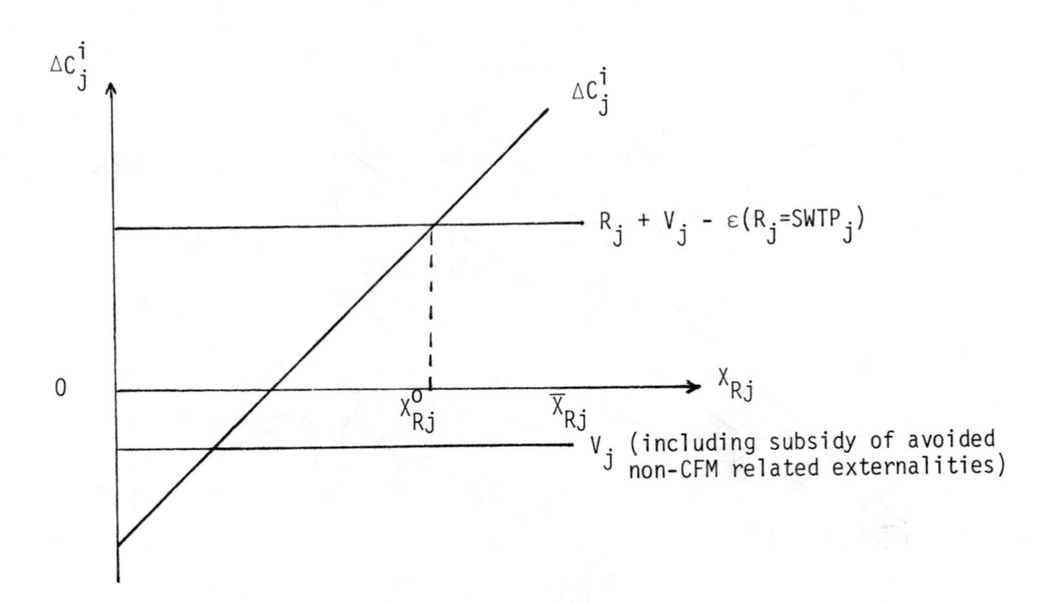

Figure 11-1. Disposal cost difference curve, showing optimal level of units returned (X_{Rj}^o).

The introduction of the return alternative and of a refund may be expected to speed up the rate at which existing equipment is taken out of service.[6] This effect can be expected to be a stronger the smaller the ΔC_j^i. Thus, the introduction of the return alternative alone will increase disposal at each level of ΔC_j^i as shown by ab in figure 11-2. Accumulating these increases from minimum ΔC_j^i to $\Delta C_j^i = V_j$ will give a total increase equal to ac at $R_j = 0$. Raising R_j from 0 to $SWTP_j$ will add further to X_{Rj}, giving an accumulated propensity to return as shown by PR_j in figure 11-2. For $\Delta C_j^i > R_j + V_j - \varepsilon$, the return alternative will not be used and, thus, disposal decisions will not be affected here, leaving unchanged the volume of discarded units not returned. In other words, $\overline{\overline{X}}_{Rj} - X_{Rj}^{opt}$ in figure 11-2 equals $\overline{X}_{Rj} - X_{Rj}^o$ in figure 11-1, where $\overline{\overline{X}}_{Rj}$ is the total disposal of cooling equipment _j_.

To conclude, the refund rate R_j will equal the maximum social willingness to pay ($SWTP_j$) unless all discarded units will be returned at a lower level of R_j, in which case optimum R_j equals the minimum value

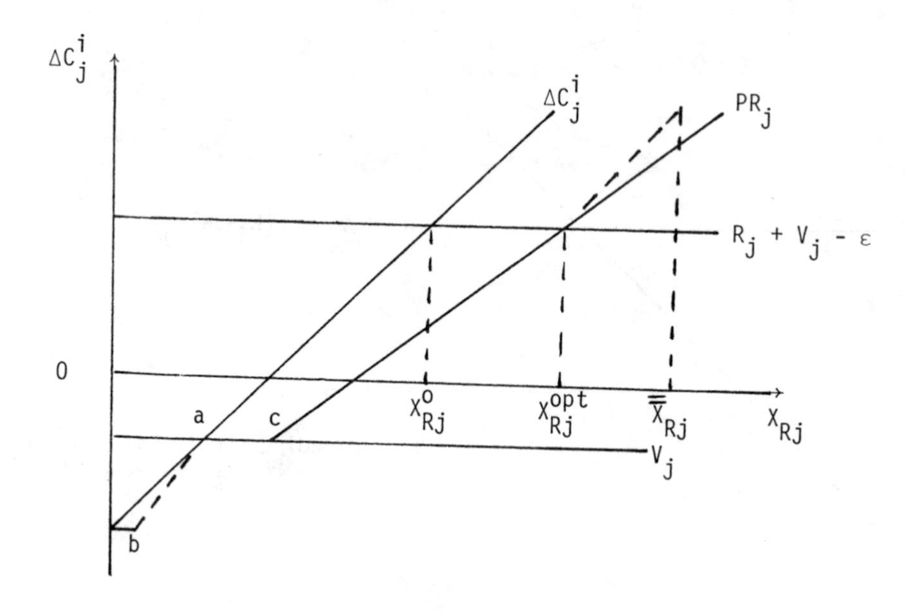

Figure 11-2. Effect of introduction of return and refund alternative.

required for a 100 percent return rate. This means that refunds per unit of CFC will be equal for different types of equipment to the extent that maximum ΔC_j^i exceeds $SWTP_j + V_j - \varepsilon$. When this is true for all j, the total payment received for units returned (R_j+V_j) will differ due to differences in CFM content and differences in the reprocessing value V_j, R_j being unaffected by changes in V_j and ΔC_j^i. However, when a 100 percent return rate is achieved at $R_j < SWTP_j$, optimum R_j, as well as optimum refunds per unit of CFC, will vary with changes in maximum ΔC_j^i (or ΔC_j) and V_j. The latter variable is assumed to be market-determined and easily observed; thus, R_j can be tied to V_j to keep $R_j + V_j$ constant over time until a change in maximum ΔC_j^i (or ΔC_j) requires a change in the sum $R_j + V_j$. Policy cost aspects may, of course, call for less variable refund rates over time and possibly for equalizing R_j over a subgroup of equipment which otherwise would have different optimum refund rates.

11.2.5 The Deposit Instrument

If the durability of cooling equipment were very small, the deposit D_j would simply equal the refund R_j and the return alternative would be open to all discarded units. With the substantial durability that actually exists, we have to deal with two separate issues: the disposal problem for items already in use (see the next section) and a deposit-refund system for items sold from now on. Here, R_j may deviate from D_j.

Let us assume, to begin with, that V_j on existing equipment or SWTP during the lifetime of existing equipment are so small that only the future disposal of equipment sold from now on may call for governmental intervention. Thus, one possible solution is to calculate an expected optimal R_j at the time of disposal and determine a deposit rate corresponding to this refund rate. With an expected real rate of return on deposits turned over to the government equal to $\underline{r}$, the deposit rate would be $D_j = R_j e^{-rtj}$ where $\underline{t_j}$ is the expected average lifetime of equipment type $\underline{j}$. In this version of the system, the buyer of a piece of cooling equipment would know the real value of the future refund. However, at time t_j, the expected R_j at time zero may not be the appropriate refund rate, given the actual ΔC_j^i and SWTP at $\underline{t_j}$. If the appropriate refund were lower than R_j, R_j would still have to be paid out under a traditional deposit-refund system. But, if the appropriate refund rate were higher than R_j, a decision would have to be made whether to add a (true) subsidy to R_j. With this version of the system, the expected government payments (that is, refunds plus optimal subsidies) will be higher than the value of the deposits at the disposal date in the 100 percent return case.

Another possible solution in the situation now discussed is to separate the actual refund payment completely from the deposit rate actually paid and let the refund be determined as discussed in the preceding section. The refund prospect would then turn from one of a fixed minimum refund rate in money (adjusted for inflation) to one of a refund rate "in kind;" that is, one that will bring about an economic stimulus to return instead of dumping cooling equipment and return it a rate sufficient to meet a given and well-specified policy objective. Thus, for any individual year, the refund payments may deviate from the value of the deposits once made for the units involved. But, if we assume that there would be no bias in current estimates of future optimal refund rates, the deposit rate could still be determined as suggested above, and, in the long run, deposits would tend to cover actual refund payments, though there would be increased uncertainty about real refund rates facing buyers of cooling equipment.

If this kind of uncertainty were considered a serious burden for buyers, an extreme corrective would be to have the government specify an $R_j + V_j$ in real terms and in advance. However, since ΔC^i_j at the disposal date would not be known in advance, and since the refund prospect is unlikely to be high enough to cause real concern among buyers, we shall not pursue this particular alternative further. So, given two alternative versions of a deposit-refund system with the same uniform principle for determining the deposit rate as $D_j = R_j e^{-rt_j}$, we may turn to the disposal problem for the existing stock of equipment for which no deposits will have been paid at the purchase date.

11.2.6 Disposal of Existing Equipment

Given that, in the near future, SWTP and V_j (for existing equipment) are high enough to spur some action concerning disposal problems prior to the disposal of yet unsold equipment, let us return to the analysis of optimal refund rates R_j. For R_j to remain a _refund_ rate instead of a pure subsidy rate, owners of existing equipment would have to pay a _deposit ex post_, unconventional as this may seem. As one extreme alternative, owners of cooling equipment could register their holdings of cooling equipment specified as to age and type at a date when they are already required to report to government; say, in conjunction with income tax returns for households, property tax statements for owners of housing and business property, registration of automobiles for owners of air-conditioned cars. They would later be billed and their statements could be checked on a sample basis. In contrast to cooling equipment on premises open to the public, refrigerators and freezers in private homes might be technically difficult or legally or morally objectionable to check. Households may therefore be subjected to an assessment on the basis of average holdings where currently almost all households have a refrigerator and about 50 percent have a freezer.[7] Or as another extreme, the deposit for all equipment may be determined by average holdings, turning the deposit into an earmarked tax and thus a pure financial instrument.

If an ex post deposit solution is not chosen due to high policy costs or for other reasons, the only way to finance the R_j, now a pure subsidy, is to use other sources of government revenue. Aside from the traditional options of raising taxes and/or reducing other government expenditure, there is the special option of adding a tax for current disposal costs to the deposit D_j collected for new equipment. As we have assumed the deposit rate to be designed, this tax is certain to be socially inefficient. But given that all other financial options open also create inefficiencies, a disposal tax could be an optimal choice given the decision to subsidize proper disposal of cooling equipment.

Finally, the subsidies could be financed by government borrowing which may have a primary effect of reducing the investment volume in the society. Given that actions taken to reduce CFC emissions, in fact, are an investment in future environmental quality, borrowing (or rather, reduction of other investments and, hence, reduction in future production) is a natural financing alternative. An obvious source of borrowing is the deposit funds from the sales of new cooling equipment. Thus, instead of investing these funds in projects giving a real rate of return of r, they would partly or wholly be used to finance the R_j subsidies at a capital cost of r. (Financial decisions regarding outstanding refunds, if and when deposits are withdrawn, would be left to future generations.) Sweden has used this system in disposing of junked cars since 1975. A fee (deposit) is paid on the sale of new cars and a subsidy (refund) of the same amount is paid on cars turned over to authorized junk dealers.

Given the special problems of dealing with the disposal of existing equipment, the short-term SWTP may fall systematically below the long-term SWTP relevant for returns of new cooling equipment. Thus, an optimal R_j on existing equipment will tend to be correspondingly lower than discussed earlier or even zero for some or all existing equipment. Furthermore, to avoid an initial overreaction to the return alternative at R_j equal to a short-term $SWTP_j$ through an excessive supply from stored up, used equipment, a phasing-in period may be required with gradually rising refund/subsidy rates over time allowing a steady growth of the reprocessing industry.

11.2.7 Emissions During Use and Service of Cooling Equipment

If emissions from servicing operations have the same impact as emissions from nonreturn disposal of cooling equipment, the same refund rates could be offered for CFCs collected by repairmen. However, in the short run, when these emissions may be particularly important, a refund rate higher than that for disposal may be called for. As the optimal deposit for recharges may be higher than for initial CFC charges due to a shorter time to final disposal, and as the volume of recharges is certain to exceed the volume of CFCs recovered in servicing operations, the total amount of service deposits should be expected to exceed the total amount of service refunds. This might not turn out to be true if other refrigerants with no or smaller expected emission effects and, hence, smaller deposits could replace CFCs with high refund rates. But otherwise, we should expect a surplus to arise which partly or wholly will eliminate the financial problems discussed above.

11.3 Effects of a Deposit-Refund System

We now examine major effects of a deposit-refund system in terms of impacts on disposal behavior, on service behavior, and on second-hand markets; effects on prices, profits, output, and product design of new equipment; and consequences for various economic groups and for certain macroeconomic variables.

11.3.1 Choice of Disposal Alternative

Given the decision to dispose of a piece of cooling equipment and given perfect information on disposal alternatives, the choice of alternative is simply assumed to be that which minimizes costs, C_d^i or $C_R^i - R - V$ (dropping the subscript for product type j). For the user, R + V is directly quoted in money, whereas C_R^i and C_d^i may not be. As indicated earlier (in Section 11.2.2), C_R^i is assumed to be the minimum cost of transferring the unit (or the relevant part containing CFCs) to the reprocessor. In case C_R^i is not a transportation cost paid to someone else it is an imputed cost of the transfer. The cost of the "dumping" alternative, C_d^i, is even more likely to be an imputed cost. However, the consumer is not interested in anything more precise than whether the imputed cost difference, $C^i_R - C^i_d$, is larger or smaller than the payment received, R + V. Given information from local sellers of new equipment and/or collectors of used equipment about transportation costs as well as about the present R + V, the consumer simply chooses one of the alternatives, thereby revealing a maximum or minimum imputed cost difference $C^i_R - C^i_d$. And the consumer is assumed to collect this information unless he has reason to believe (say, from the behavior of others) that he is indifferent between the two alternatives, which we shall interpret to imply that, in fact, $(C^i_R - C^i_d)$ approximately equals (R + V).

11.3.2 Effect on Disposal Dates

The analysis of when to dispose of used equipment is less straightforward than the choice of disposal alternative. In fact, the choice of an optimal disposal date presupposes an analysis of what disposal alternative to choose, an analysis which, in principle, will have to be done repeatedly over time. This is because the implications of optimal disposal have to be known before the decision is taken to discontinue using (or servicing) the existing unit. As long as the nonreturn alternative for disposal turns out to be optimal, only the effects of higher servicing prices on the choice of disposal date need be considered. But whenever the return alternative is optimal and implies a significant reduction of costs, disposal dates typically will be affected and occur at an earlier date.[8] The only general exception is when the unit breaks down to an obvious beyond-repair state. Otherwise, the introduction of the return alternative may reverse a decision to service the unit or to keep the unit in use. Servicing can imply additional risks of CFC emissions, and retaining the unit is likely to add to CFC emissions. The introduction of a return alternative which speeds up disposal should imply reduction in CFC emissions, at least as long as replacements are not sufficiently discouraged by price increases for new equipment due to the deposit. If not, R_j should be correspondingly higher for units with a high risk of servicing emissions or "old age" emissions.

11.3.3 Effects on Servicing Behavior

We have seen that the introduction of a refund on used equipment will tend to reduce demand for servicing. Additionally, if a deposit is introduced on recharges of CFCs in used equipment, service operations will tend to be more expensive. Although the importance of this effect is in doubt, it should contribute to shifting disposal to an earlier date, regardless of the disposal alternative chosen. Moreover, it will tend to speed up service on equipment suspected to have a CFC leakage. The direction of the short-term effects on the servicing volume is therefore not certain. Nor is it certain to be a positive effect on the recovery behavior by repairmen and the servicing industry. The fact that there is a refund on CFCs recovered is a very weak incentive when the refund is to be transferred to the owner.

11.3.4 Effects on Second-Hand Markets

The introduction of a refund will, in principle, raise the price of second-hand equipment which now also contains the right to a refund. This effect, however, is likely to be important only for old equipment with a not so distant disposal date, given fairly clear expectations of the size of the future R + V payment. The demand curve and the supply curve will tend to rise to the same extent, leaving the volume traded unchanged. The deposit should yield similar effects, increasing equilibrium price even more.

11.3.5 Effects on the Market for New Equipment

Assuming product design remains unchanged, both the refund and the deposit can affect product prices. However, given the present durability of cooling equipment, the effect on demand of a refund prospect as late as twenty years from now will be limited. This effect could well approach zero, meaning that the refund is discounted much more heavily by the owner than by the government interest rate (r) which determines the deposit rate, and implying that many buyers will simply overlook the fact that there is a refund to be collected some time far into the future.

It therefore seems reasonable to assume, as an approximation, that the refund prospect has no effect on demand but that the deposit has the same effect as an ordinary price increase in the same amount. And with a completely elastic, long-term supply, the result will be a price increase equal to the deposit. Given the available estimates of the price elasticity of demand (-0.4 for refrigerators[9]), the effect on output should not be very pronounced. For example, at r = 0.05, a $10 refund in real terms in ten to twenty years, for products now priced at $200, would call for a deposit (price increase) of $4 to $6, or at most 3 percent of the price, and, hence, lead to a long-term reduction in demand for refrigerators by, at most, 1.2 percent. In addition, there is the possible effect of refunds and higher servicing costs speeding up replacements and leading to an increase in demand. Hence, the net effect on output and on industry profits is likely to be small, particicularly if there is a long phase-in period for the deposit-refund system.

11.3.6 Effects on Product Design

Chlorofluorocarbon producers and manufacturers of refrigeration equipment are said to be constantly working on reducing CFC emissions and to have made some progress.[10] Although the basic goal is improved product performance, this fits the interpretation that current CFC prices have provided some incentive to reduce initial CFC charges and CFC recharges. Deposits on new CFCs would add to these incentives, probably to a significant degree. The reason is that refund rates per unit of CFC-containing equipment would have to be relatively high to have any effect at all and hence, that refund rates when measured per pound of CFCs recovered would be quite substantial. For example, a $5.00 refund on automotive air conditioners with an average remaining charge of two pounds would imply a refund of $2.50 per pound (recovered CFCs are assumed to be uncontaminated and to have a V = 0). With an average CFC capacity of four pounds per car, a ten-year average durability, and r = 0.05, the deposit would be about $1.50 per pound, or $6.00 per unit. This would imply that a deposit would add about 100 percent to the 1978 CFC cost of one to $2.00 per pound.[11]

If the order of magnitude of these figures were realistic, the consequent design and system changes to reduce CFC emissions could eventually be quite extensive. A number of system changes seem close to implementation, for example, suggesting that most CFCs could be mixed with leak detectors to observe leakage by visual inspection and facilitate prompt servicing; that hermetically sealed automotive air conditioners could replace current non-hermetically sealed units; that on-site repair kits could be used for large refrigeration systems; that F-22 and ammonia could replace F-11 and F-12 in some uses.[12] The durability of products could be changed so that disposal dates and emissions at disposal would be delayed, or so that non-CFC components would be less durable, requiring an early replacement of the unit before leakages appear and possibly making the CFC component reusable. And there would be incentives to adjust procedures of manufacturing, storage and transportation of CFC products as well as intensified research and development of new refrigerants.

11.3.7 Distribution Effects

The basis for a SWTP > 0 is that there is a nonnegligible risk of detrimental repercussions of CFC releases today on the health and environment of future generations. This would mean that there is non-zero probability that future generations inside, as well as outside, the policy-active country will benefit from the introduction of an incentive system, as suggested here.

Now, focusing on the present generation, we shall first note that littering of used equipment may be reduced, thus benefitting those who otherwise would have been exposed to hazards and eyesores of littered units. In addition, we should observe the effects on different end users of cooling services.

Those who now discontinue using a piece of cooling equipment that will not be replaced will benefit from the refund option now open to them. This may be true in particular for low-income households with a low C_R^i, due to low opportunity costs of time.

The outcome for those who discard an old piece of equipment and replace it with a new one may be positive or negative. There is certain to be a price increase on new equipment (given quality), whereas the future benefits of the return alternative are uncertain. Only those who have an $(R + V - C_R^i + C_d^i)$ on the now discarded unit that is larger than the deposit on a new unit (assuming the same deposit for all relevant alternatives) are sure to benefit from the system.

First-time buyers of new equipment would definitely lose by the price increases due to the deposit, assuming (as we have done here) that the buyers' discount rates exceed r in $D = Re^{-rt}$.

Buyers of second-hand equipment are likely to be confronted with an increase in market prices larger than the present value of the refund prospect, as we indicated above. Thus, this group, which probably contains a large fraction of low-income households, would tend to lose, in particular if they are first-time buyers.

Commodities produced using refrigeration will tend to become more expensive, causing losses to consumers. The expansion of industries using refrigeration as an input will be confronted with increased factor prices, and replacements will imply increases in costs. However, the use of refrigeration in manufacturing as well as in the service industry will often involve shorter replacement cycles than for most household uses of refrigeration. Thus, refunds will often be a net addition to the income of the firm and appear at an early date, making the present value of the refund nonnegligible.

To finance refunds (subsidies) on existing equipment, we have discussed solutions in which consumers are taxed. Let us consider effects on these taxpayers. If the tax is on all consumers, regardless of holdings of refrigeration equipment, there are certain to be losers, although tax rates will tend to be low. If, at the other extreme, the tax is levied in relation to such holdings, the gains of those who discontinue holding this kind of equipment will be reduced and, perhaps, even turn into losses. And there will be fewer net winners among those consumers who have old equipment replaced by new units. Where an excise tax is added to the deposit on new units, shifting the tax completely onto buyers, the frequency of losers or the size of the losses will increase among end-user groups B through E, as defined above. However, if borrowing is used to finance the refunds on existing equipment, future generations will take the place of present consumers suffering the loss noted here.

Let us now look at the main effects on producers or stockholders. As was the case for users, the phase-in pattern will be a major factor in impacts. However, given some expectation that government will impose further regulation (based on the aerosol ban), the market should have already registered such expected losses in its valuations. Hence, some of the costs to be noted, if expected by industry, have already affected market values.

The introduction of a deposit-refund (DR) system will reduce short-term profits for CFC producers. It should be noted, however, that the DR system will lead to an increase in the supply of used CFCs which may benefit this industry to the extent that it is also active in reprocessing. Effects on long-term profits are likely to be negative as well, but the industry has the option of reducing negative effects by adopting product changes. Manufacturers of cooling equipment will be in a similar position. It should be noted, however, that if policy speeds up replacement of old equipment, there will be a demand increase for the products of these manufacturers and possibly for the CFC industry as well.

The repair and servicing industry should be confronted with an uncertain effect on demand for its services, since some factors favor disposal instead of repair and others increase demand for servicing and/or shift demand to an earlier date.

We have already touched upon the effects on industries using refrigeration as an input, noting that short-term cost increases probably would occur due to factor price increases, but that in many instances these effects would be small or nonexistant.

Finally, the reprocessing industry, dealing with CFC destruction or reclamation, will essentially be a new industry, presumably providing benefits for those who engage in it. We have assumed competition to be established here, so long-run profits should tend to zero. To the extent that existing firms are involved in reprocessing, the effect on short-run profits would tend to be positive.

It is not easy to assess the general impact of all these effects, since many are offsetting; however, it seems plausible that total effects on both consumers and industry will be mildly negative. Government may also be interested in other impacts, including effects on employment, wages, the price level, and the balance of payments.

Traditionally, studies of deposit-refund systems (for example, for beverage containers) and similar policies have counted jobs gained and jobs lost. However, given a likely long phase-in period, the effect on employment, if any, should be one of increased mobility rather than changes in employment levels, and the resulting effect on wages even guesswork at this general level.

The effect on the price level (and the indirect distribution effects therefrom) runs a risk of being misconceived. Deposits (taxes) will certainly add to prices in the affected sector and, thus, to the general consumer price level. However, the price-cost increases are investments in future environmental quality, with an expected return equal to or greater than its cost.

The effects on international trade present special problems dealt with in more detail elsewhere (see chapter 3). However, if deposits (taxes) on CFCs exported from one country were replaced by non-taxed supplies from other nations, the case could be made for exports being exempted from taxes. This, however, should be discussed in the more general context of international ozone policy.

11.4 Outline of a Simple, Approximative, Deposit-Refund System

At present, there are no data available on which to base a specific and elaborate proposal for a deposit-refund (DR) system with respect to CFC use. The development of such data, and the time for government deliberation and approval, should take at least two or three years. The following simple numerical example has an illustrative purpose only; it is based on estimates made for 1978 and refers to a possible introduction of a DR system not earlier than 1982.

An initial distinction turns on whether the disposal cost difference, $C_R^i - C_d^i$, influences disposal behavior. Little influence is to be expected for the large units used by industry, large service centers, and in modern residential and business buildings. When these large units are discarded, they can hardly be dropped anywhere or left in the "backyard;" they have to be transported to a destination where some sort of waste processing takes place. Therefore, once there is a system for handling products containing CFCs, it would automatically yield the new destination for these units. An economic incentive system would merely reinforce this "automatic" tendency. Hence, the design of the system and the choice of deposit and refund rates would hinge on the factors determining disposal of the small units used by households, car owners, junk dealers, and small service firms. The kinds of equipment involved are refrigerators, freezers, water coolers, dehumidifiers, ice machines, and vending machines used in households, small retail stores, restaurants, etc., and air conditioners in cars. (Other small air conditioners do not use F-11 and F-12.) For these units, the imputed cost of the disposal alternative otherwise used, C_d, may well fall short of the minimum cost of "returning" the unit for reprocessing, $C_R - V$, where V is the "reuse" value including waste management costs and externalities avoided (aside from ozone effects). Thus, the choice of the refund rate would be crucial for disposal behavior regarding these units.

To determine the volume of F-11 and F-12 remaining in these small units may constitute a problem affecting the design of the return alternative. Measuring the CFM content of equipment may be time consuming and require special instruments, at least for existing types of refrigeration devices. If so, this would add to disposal costs. On the other hand, to avoid measuring the actual content and to use an approximate average figure would eliminate part of the incentive to keep emissions at a low level. For illustrative purposes, we assume that actual measurement is to be made for each unit.[13]

Next, we estimate the maximum value of the unit entering the reprocessing plant, assuming that the whole unit, and not only the CFC container or coil, is brought there. That estimate includes the cost of recovering the CFCs, the reprocessing cost, the resale value (if any) or the destruction cost, as well as cost of scrapping, dumping, or recycling of the rest of the unit. This would, of course, have to be made for different types of applicances. For illustration, assume that there is an average of two pounds of CFCs in the small units, with a resale value of one dollar per pound. (New CFCs today have a retail price of about $1.00 to $3.00; see DuPont.[14]) Assuming measurement, recovery, and reprocessing costs of around $2.00 per unit, and scrapping/waste management net costs for the remainder of the unit of another $2.00, the net internal value of the unit as a whole would be an average of minus $2.00. (The EPA lists a national average of $2.00 to $2.50 for present disposal costs for refrigerators, freezers, etc., including pickup costs in 1977.)

Now, the "external" cost of alternative disposal (still excluding the ozone effect) would have to be added to this internal value to get V per unit as defined earlier. For units dumped into the municipal waste stream, this would be at least the listed scrapping/waste management net cost. Then, there is the external cost of temporary littering with some of these applicances. The littering is taken to include not only illegal dumping, but also outdoor storage on the owner's lot affecting the view of adjacent lots. Assuming that temporary littering occurs for half of the units for at least some period of time and that the average social WTP to get rid of them is \$5.00, the average external costs per unit would be \$2.00 (average eventual waste management cost) plus \$2.50 (average external cost of littering), totaling around \$5.00.

Adding the internal value (minus two dollars) and the external costs avoided (\$5.00), we would end up with an average reuse value, V, equal to \$3 per unit. This approach would mean that a government subsidy in the amount of the external costs avoided, \$5, would be turned over to the reprocessing firms. If so, part of it would be financed by reduced waste management costs and the rest by general taxes or a product charge corresponding to some \$3. These external costs, however, could be financed as part of the DR system, in which case it would be appropriate to calculate the marginal external costs, say, \$2 for waste management avoided (average costs) and at least \$5 for littering; i.e., \$7 in all. In this case, local taxes could be reduced by \$2 per unit (waste management costs avoided), and \$7 would be added to the refund rate to be determined with respect to the ozone effects.

Now let us turn to the direct disposal costs of the owner of the unit; i.e., C_d and C_R. In many areas, there is a free pick-up service; taking this as the general case we have C_d equals zero. We may assume, for simplicity, that pick-up service for reprocessing would be arranged in the same fashion, i.e., financed by local taxes, so C_R equals zero, also. This simplifying assumption does not mean that no positive net payment to the owner is required in order to have all units returned. The lack of such a payment would make it less attractive to collect and transfer information about disposal alternatives, less urgent to get rid of a unit taken out of use, and less urgent to make up one's mind whether to keep the unit as a reserve appliance or for some other possible use. The existence of a free pick-up service now does not mean that all discarded units are actually picked up.

Given that $C_R - V = 2 > C_d = 0$ per unit for the owner, we arrive at the question of the total refund rate. With seven dollars as a minimum refund for avoiding other externalities than the ozone effect we can ask: what is the maximum amount the government should impose on the owner of a unit of cooling equipment to ensure that disposal will be carried out at a time and in a way which avoids CFC emissions from discarded appliances? And would this amount lead to undelayed return of all of these appliances given that $C_d^i = C_R^i$? Let us look at these questions from another angle and assume that an average net payment of $10 per unit to the owner would be high enough to keep him informed about how to dispose of a replaced or "beyond-repair" refrigeration unit and make him call for a pickup without much delay. Given the $7.00 minimum refund mentioned above and the V of -$2.00, this would require a $5.00 refund for the avoided ozone effect from two pounds of CFCs, or $2.50 per pound of CFC. (If the CFC container is empty, he would, of course, receive only a total of $5.00.) Assume, finally, that $2.50 per pound of CFC recovered in such a way that the CFC will not be released into the atmosphere is below what the government is willing to impose on behavior that eventually would lead to CFC emissions.

So far, we have been discussing net payments and refunds (in today's prices) as if we were in a long-run-equilibrium deposit-refund system. As suggested earlier, the introduction of a DR system may create certain _transition problems_. Thus, for example, an "immediate" (present plus three years) introduction of average net payments at the $10 level might cause a too abrupt expansion of the authorized reprocessing industry (authorized to receive money for refunds from the government), leading initially either to very low internal reuse values or to employment, capacity and storage problems. This would be true, in particular, if there were a backlog of equipment for disposal stored by households, etc., some of which were awaiting the introduction of a high refund. If such start-up problems were expected to arise, the refund could be introduced at a low rate, and then later be increased towards a moving, long-run equilibrium rate. In addition, such a gradual increase of refund rates may be useful in finding out more about the propensity to return equipment as a function of net payments or refund rates.

A possible version of an optimal transition period might look like this. Prior to the date of the formal introduction of the DR system, experimental research and development in reprocessing may be subsidized by government using equipment already being collected by municipalities but so far turned over to ordinary junk dealers, ordinary waste treatment plants or dumps. When the system is formally introduced, refunds may be set at a level that makes average net payments equal to $5 per unit for three to five years (in constant prices). With an internal value of minus $2, this means a refund rate of $7 per unit. After three to five years, the short-term rate may be changed, and to attain long-run equilibrium, the refund per unit of CFCs (F-11 and F-12), as well as the refund per unit of equipment, may have to grow approximately with the rate of productivity increase to provide a guarantee of unchanging real incentives.

Now, if the figures we have used here are at all correct or relevant, (for 1978), what would total refund payments from the government equal? We shall try to give an idea of an upper limit to these payments, using the Rand Study.[16] That study distinguishes among home appliances (excluding air conditioners which use F-22), automotive air conditioners, commercial refrigeration, and chillers. Here, we shall treat the first two groups as small appliances, and, for simplicity and lack of detailed data, the remaining two as large appliances (with no littering problem).

In 1973, some five million units of <u>home appliances</u> were scrapped with an average of two-thirds of a pound of CFCs remaining (see Eads[17]). Let us assume that, by the mid-eighties, this will be some 7.5 million units or, at most, 10 million units initially, with a refund stimulating disposal behavior. With a gradual increase in the percentage of more CFC-heavy appliances, freezers in particular, we may assume an average CFC content of at most one pound per unit. And, with an initial refund rate of $7.00 per unit ($2.50 for one pound of CFC + $4.50 for the unit), the total refund payments here would be some $70 million per year. If a refund rate of $10 ($2.50 + $7.50) were to be introduced some years later, at which time the initial disposal backlog may be assumed to have been eliminated, the total disposal figure would remain around ten million units (given the normal growth in disposals), and total refund payments would be some $100 million per year (see table 11-5).

For automobile air conditioners, the situation should be somewhat different. If hulks are abandoned, the air conditioner does not contribute to the external effects from littering. If hulks are scrapped, the air conditioner itself (that is, without CFCs) can hardly be assumed to affect either scrapping values or scrapping costs. Hence, no unit refund is relevant here. Still assuming a resale value of $1 per pound of CFCs and a recovery/reprocessing cost of the same order of magnitude, the internal reuse value would be around zero. According to Eads,[18] at most 50 percent of the CFC charge (3.5 to 4.0 pounds) remains in junked cars with air conditioners and, in the mid-seventies, the number of junked cars with this equipment leveled off at some five million units per year.[19] Assume that this latter figure holds for the mid-eighties (higher energy prices may contribute to auto air conditioning becoming less frequent, since it now accounts for some 5 percent of fuel consumption). Assume also that there will be an average of two pounds of CFCs in air conditioner-equipped junked cars (perhaps a reasonable figure in five to ten years given the on-going attempts to reduce leakages). Then, there would be a maximum of ten million pounds of CFCs recovered from this source. At an R of $2.50 per pound, the car owners would get a refund of $5 in addition to the payment for the hulk (now about $25 on average). Hence, total refund payments per year would be some $25 million.

Applying the same refund rate per pound of CFCs and no unit refund for large equipment, we arrive at the following figures. According to Eads,[20] there were some five million pounds of CFCs (the major portion of which was F-11 and F-12) in commercial refrigeration units scrapped in 1973. Given the growth in sales of these devices and improved servicing and design, there may be close to ten million pounds in the units scrapped in the mid-eighties. For chillers, some two million pounds of CFCs are estimated to have been found in discarded units in 1973. Say that, at most, four million pounds will be found twelve years later. Thus, we would end up with annual refund payments on the order of $25 million for commercial refrigeration and for chillers (see table 11-5).

Table 11-5. Base Period and Projected Mid-1980s Levels of CFC Related Consumption, and Projected Mid-1980s Costs of Deposit-Refund System

Consumption item	Base period values					Projected values, 1982-1988			
	Date of base period	Million units		Million units F-11 & F-12		Disposal		Annual refunds in million dollars[a]	Deposits on recharges per year[a]
		New	Disposal	New	Disposal	Million units	million lbs F-11 & F-12		
Small appliances									
Home applicance	1973-1976	10	5.3-5.9	8	3.3-5	10	10	70-100[b]	--
Automobile air conditioners	1973-1975	5	5	17.5-20	5-21.5	5	10	25	--
Large appliances									
Commercial refrigeration	1973	--	--	--	5.4	--	10	25	--
Chillers	1973	--	--	--	2.8	--	4	10	--
Recharges	1973	--	--	≥150	--	--	150	35[c]	375
Sum	--	--	--	--	--	--	--	165-195	375

Note: Dashes indicate not applicable or data not available.

Sources: The figures for 1982-1988 are based on the author's projections. For home appliances, the figures for 1973-1976 are based on G. C. Eads and coauthors, Non-Aerosol Chlorofluorocarbon Emissions: Evaluation of EPA-Supplied Data, A Draft Working Note Prepared for EPA by the Rand Corporation (December 1977), pp. B-18 and B-26; for automobile air-conditioners the data for 1973-1975 are based on Eads and coauthors, p. C-12; for chillers, the data are based on Eads and coauthors, p. D-8. For recharges, the data for 1973 are based on DuPont Freon[R] Products Division, Information Requested by EPA on Non-Aerosol Propellant Uses of Fully Halogenated Halocarbons (Wilmington, Del., March 15, 1978) p. III-29.

[a]Refunds paid are calculated on the basis of a price of $2.50 per lb of CFC plus additional refunds per unit in the case of home appliances. Values are in 1978 dollars.

[b]Initially, refunds are calculated as $2.50 per lb of CFC plus a $4.50 refund per unit with a CFC charge of 2 pounds, yielding the $70 million figure. After 3-5 years, refunds will be $2.50 per lb of CFC plus a $7.50 refund per unit with a CFC charge of 2 pounds, yielding the $100 million figure.

[c]Accounts for CFCs recovered in servicing operation.

The sum of refund payments for the disposal of equipment is $130 million. To this, we must add the refunds paid for CFCs recovered in servicing operations, a completely unknown but probably small amount. Let us assume that it will be around 10 percent of CFC recharges.

DuPont estimates the total recharge volume of CFCs in 1976 to be around 220 million pounds.[21] Of this amount, at least some 150 million pounds appear to refer to F-11 and F-12. Moreover, the total recharge volume grew by almost 2 percent per year from 1973 to 1976.[22] To estimate the volume in the mid-eighties, however, we would have to take into account the effects of the DR system on servicing practices, system design, and choice of disposal dates. To make another heroic assumption, let us say that these effects outweigh the 2 percent growth and cause the recharge volume to remain at some 150 million pounds of F-11 and F-12. With a deposit on recharges much higher than for initial charges (because of expected differences in emission patterns), say, equal to the refund rate at the same date ($2.50 per pound), these revenues would amount to $375 million. It may be added here that, for automotive air conditioners, the most recharge-intensive type of cooling equipment,[23] present prices for recharge operations are $10 to $15 plus $3 per pound of CFC.

The revenue from recharge deposits ($375 million) turns out to exceed the total amount of refund payments, $165 million, including $35 million in refund payments in servicing operations (that is, 10 percent of the recharge volume). If this surplus would remain after more careful calculations have been made, the deposits on new units of equipment could be reserved for financing the future refunds on these units in line with the operation of an ongoing, completely self-financed DR system. So, if we were to determine the volume of deposits on new equipment, we would not have to deal with it as a financial source for current refund payments, nor would we need to discuss any addition to deposits in terms of disposal tax to finance current refund payments. Our main interest in the deposit rates for new equipment would, instead, refer to the effects on product prices and demand. To estimate these effects even in the short run, we would need to know the order of magnitude of the effects on future equipment design and future F-11 and F-12 use. As this does not seem feasible, we will have to settle for an estimate of an upper bound to the deposit rates for equipment, based on current technology and design.

Now, if the real value of the future refund rate would have to rise with the growth in real income per capita to provide a constant incentive to check CFC emissions (and if that real value would remain below the politically-determined willingness to pay for avoiding certain risks of CFC emissions), the real refund rate would have to grow at the rate of productivity growth which we take to be approximately equal to the real social discount rate. Then, deposit rates per CFC would tend to equal future refund rates; i.e., some $2.50 per pound of F-11 and F-12 in the mid-eighties (in 1978 dollars). If, on the other hand, refund rates would not be required to grow in real terms to maintain incentives, or if they could not grow due to binding limits to the willingness to pay for CFC emission protection, the deposit rates would be lower than the future refund rates. For a constant real refund rate of $2.50, a discount rate of 5 percent, and an average durability of cooling equipment of fifteen years, the deposit rate would be on the order of $1.25 per pound of F-11 and F-12.

A deposit rate of $1.25 to $2.50 would imply an average price increase for cooling equipment of, at most, 5 percent. For refrigerators priced from $200 and up, with an initial charge of about 0.6 pound CFC,[24] and a unit deposit of, at most, $9 ($7.50 excluding the CFC deposit), the price increase would be, at most, 4.5 percent. For automotive air conditioners with a four-pound initial charge, and with a price of about $400, the price increase would be, at most, 2.5 percent. Subsequent changes in system design are likely to make actual price increases in the mid-eighties much lower than these maximum figures indicate.

The main purpose of this numerical exercise has been to provide a baseline case for a better understanding of the DR system in operation, rather than to evaluate the actual effects of such a system. However, this exercise lends support to the provisional evaluation that such a system would be effective; hence, further investigation of the design and probable effects of a deposit-refund system is strongly recommended.

11.5 Concluding Remarks

Because economic incentive systems do not seem to be the first choice in actual policy, we may conclude by comparing the deposit-refund system discussed here with the policy of a purely regulatory approach to emission control of refrigerants. The regulatory alternative may look more natural or more attractive to politicians and administrators, at least at first glance. The basic requirement of the regulatory approach is that the producer or seller of a product having problems connected with disposal, bears responsibility for disposal and takes back the product when the user wants to get rid of it. In a long-term market economy perspective this means that prices will go up and, thus, that the user will pay for the liability of legal waste disposal being transferred to the producer/seller. Here, the rules for proper disposal could be exactly the same as under a deposit-refund system. The main differences in comparison with the incentive system would be: (1) that returns have to be made in a specific way (that is, to or *via* the producer/seller), and (2) that no refunds are offered.

There are certain problems connected with this solution, however. First, return costs may be higher due to the constraints on the form of the transfer of the discarded product. Second, as refunds are ruled out, return rates may be kept at too low a level, assuming a significant willingness to make sacrifices for reducing the risks of CFC emissions. This might be true even if the producer/seller were required to pick up the used equipment and this could be done efficiently. Third, and probably most important, the product durability of cooling equipment would create problems under this kind of regulation. Neither the seller nor the producer may be around when the user eventually wants to dispose of the product; or the user may have moved to a new area where prohibitive transportation problems may arise. These two problems would contribute further to low return rates and would probably call for additional measures. Moreover, new producers/sellers will have an advantage over established ones, once the system has been operating for some time. In addition, there will be an incentive to stay in the business for a limited time only (until returns start appearing). Thus, there will be detrimental effects on efficiency as well as distributive effects both among producer/sellers and among users of different brands, etc.

Part of these problems could be solved by making sellers liable to pick up old units regardless of their origin. However, given that costs of transportation and handling often would outweigh reuse or resale values, other problems would arise. For example, hidden or overt rebates would be offered on new equipment when there is no "trade-in," This means that we are back at the original problem where proper disposal is more costly for the user than improper disposal.

In conclusion, a regulatory return alternative with no economic incentives would limit return rates, possibly to very low levels and possibly much below what a social willingness to pay calls for; it will introduce inefficiencies in the markets for new equipment as well as for collecting used equipment; and it would, of course, have no direct effects on CFC emissions up to the point of disposal. In other words, regulating disposal without the assistance of economic incentives may fail to bring about a satisfactory solution to the emission problems of cooling equipment. In contrast, an economic incentive system of the type discussed in this paper would:

1. Have at least some impact on emissions at all points
2. Allow return rates to be determined by the revealed social willingness to pay for reductions in the risks of ozone depletion
3. Not interfere with efficiency in the markets for new equipment or in the collecting and reprocessing markets
4. Allow costs for refunds/subsidies (in addition to administrative costs) to be covered by deposits/taxes, hence removing the financial problems which might otherwise be treated as a rationale for regulation
5. Not seem to have any substantial and politically unattractive distribution effects--especially not against low-income households for which the payments offered by the return alternative net of the imputed time costs will provide a relatively strong incentive.

This system, which includes a tax (deposit) on CFCs used for refrigeration, can be combined with a tax on all CFC uses, if this is part of a preferred policy with respect to foam blowing, solvent, and remaining aerosol uses. The tax rate could be the same for all uses (which may require an adjustment of the non-CFC part of the deposit for cooling equipment), or it may differ among uses. The latter approach, though perhaps more cumbersome from an administrative point of view, could be motivated by the fact that different uses have different emission patterns. For example, aerosol use involves a more or less immediate release of all of the CFCs contained in the product, whereas refrigerant uses imply a possible CFC release as late as twenty to thirty years after production. Thus, a deposit-refund system for refrigerant uses of CFCs can be combined with any kind of regulation of such uses as well as with any kind of regulatory or tax scheme for other CFM uses.

References

1. G. C. Eads, et al. Non-Aerosol Chlorofluorocarbon Emissions: Evaluation of EPA-Supplied Data, A Draft Working Note prepared for EPA by the Rand Corporation (December 1977).
2. Ibid., pp. A-1, C-15.
3. Resource Conservation Committee, Staff Background Paper No. 4 (Washington, D.C., U.S. Congress, October 1977) p. 7.
4. Dupont, Freon® Products Division, Information Requested by EPA on Non-Aerosol Propellant Uses of Fully Halogenated Halocarbons (Wilmington, Del., DuPont, March 15, 1978) p. III-46-49.
5. Peter Bohm, Deposit-Refund Systems: Theory and Applications to Environmental, Conservation and Consumer Policy (Baltimore, Md., Johns Hopkins University Press for Resources for the Future, 1981) pp. 51-52 in particular. This book contains a general discussion of deposit-refund systems and theoretical properties as well as actual and potential applications.
6. Ibid.
7. Eads and coauthors, Non-Aerosol Chlorofluorocarbon Emissions, app. B.
8. Bohm, "Deposit-Refund Systems," p. 22-25.
9. Institute for Defense Analysis, Economic and Social Measures of Biologic and Climatic Change (Washington, D.C., U.S. Department of Transportation, Climatic Impact Assessment Program, 1975).
10. Dupont, Information Requested by EPA, p. III-6.
11. Eads and coauthors, Non-Aerosol Chlorofluorocarbon Emissions, app. A.
12. Dupont, Information Requested by EPA, ch. III.
13. Eads and coauthors, Non-Aerosol Chlorofluorocarbon Emissions. See app. B for measurement techniques available now.

14. DuPont, Information Requested by EPA, p. III-46

15. U.S. Environmental Protection Agency, Analysis of Environmental and Economic Impacts of Waste Reduction Procedures and Policies, Task Order #6, Subtask #2 (Washington, D.C., EPA, December 1977).

16. Eads and coauthors, Non-Aerosol Chlorofluorocarbon Emissions.

17. Ibid., pp. B-18 and B-26.

18. Ibid., p. A-8.

19. Ibid., p. A-3.

20. Ibid., p. C-12.

21. DuPont, Information Requested by EPA, p. III-29.

22. Ibid.

23. Ibid.

24. Eads and coauthors, Non-Aerosol Chlorofluorocarbon Emissions, p. B-24.

Chapter 12

CFC EMISSIONS CONTROL IN AN INTERNATIONAL PERSPECTIVE

Peter Bohm

12.1 The Problem

Everybody needs the protection of the earth's ozone layer. But some, such as light-skinned people who are more prone to skin cancer, need it more than others. Similarly, climate change involves adjustment costs, even if it brings long-run benefits to some; and climate change bringing benefits to some generally imposes costs on others. Further, given the delayed effects of today's CFC emission activities, people (and governments) have different tradeoffs between short-term and long-term costs and benefits. So, even if all individuals and countries would gain from keeping the ozone layer intact and climate unchanged, their willingness to make sacrifices for achieving those goals may differ a great deal.

We know there is a considerable risk that certain activities, such as the use of CFCs, reduce the ozone layer and increase temperatures. We also know that any such effect of a local activity of this type will be global. Thus, the ozone layer and climate are common property resources and public (capital) goods in the pure textbook sense. But in contrast to national public goods, those goods are dependent on what all governments around the globe decide to do about their protection. In principle, therefore, we regard managing the ozone layer and climate as problems where all nations (governments) are the consumption units and where the optimum level of protection should be defined from an evaluation of services of the public good by all of the units. In effect, this would apply the paradigm from standard public good theory, substituting a global public good for a national/local public good and national governments for individual consumers.

Now, this application of public good theory can pose problems. Ideally, we would like to know each country's willingness to make sacrifices for reducing risks of damage to public goods in order to create a basis for international policy. The well-known problems of revealing preferences for national or local public goods exist, of course, for global public goods as well, and may even be more difficult or prohibitive in the latter case. And there is the more fundamental question of whether the analogy between international and national public goods really holds.

In the national (or local) setting, estimates of individual willingness to pay (WTP) could be used as a basis for decision making. For example, with a policy of income redistribution, there could be fairly general agreement for using and aggregating such estimates without adjusting them. Or there could be fairly general agreement within government on how to weight or otherwise transform information about individual WTP. This is because a basic institutional agreement exists whereby a government can be taken to represent - and be respected by - the people within the jurisdiction. Thus, taking the institutional rules for national and local decision making as given, it is conceptually possible to put WTP estimates to some use for resolving public good problems. If we could imagine a nation without any constitution or government, it would definitely be difficult to interpret the meaning of WTP estimates (an academic question, to be sure). In particular, since the existing income distributions were not in any way the indirect result of a constitution, we would not be able to speculate on "improving" the welfare of this anarchy by making decisions about a public good on the basis of such estimates.

If it is difficult to conceive of a national community without a constitution and a government, this is not the case for the global set of nations taken as a community. The lack of global constitutional government means that it would be difficult for anyone to know what to do with or how to evaluate a set of national WTPs for ozone protection. No individual or government is bound by any agreement or constitution to accept the expressed values of other nations as a basis for determining how much worldwide CFC emissions should be eliminated or how much any nation should contribute to reducing the emissions. Anyone is free to reject the relevance of the income distribution, the social rate of discount, and the power structure influencing the true (by assumption) response by another country. It therefore seems that not even the principles of traditional public good theory are a relevant starting point for global CFC management.

What seems to remain, once a global vision of global CFC management is taken to be institutionally nonexistent or inconceivable, is a set of national visions of global CFC management. Equipped with such a vision, a government could join the others at the bargaining table for attempts to reach an international agreement. The preferred outcome for each government would presumably be to have all others eliminate their contribution to ozone depletion and climate change and make one's own contribution only to the extent that it is net-beneficial for one's own country. As a bargaining strategy, this "free rider" policy would hardly be successful. The latter part of the strategy--reducing national CFC emissions to the extent that marginal benefits exceed marginal costs within the nation--provides a minimum solution, but one for which international bargaining is hardly required. We might expect individual nations to be willing to make further contributions only as a payment for the commitment of further contributions by others. However, governments may disagree on the terms-of-trade involved, perhaps to the extent that there will be a stalemate in these negotiations.

The issue to be discussed here is whether the picture of international policy negotiations now given--presumably relevant for many other areas of international bargaining--is the appropriate one for the CFC policy field and whether it provides the appropriate background for U.S. strategy. Or is it more appropriate to analyze the problem of U.S. policy on CFC emissions in the international context as one where U.S. action should be evaluated by (1) the internal U.S. costs and benefits (as before) and (2) the overall effect that U.S. actions may have on other governments' decisions (instead of the narrow *quid pro quo* approach mentioned above)?

12.2 Dimensions of Policy Interdependence

U.S. policy on CFC emissions may influence the extent of control measures taken by other countries in at least two major dimensions. First, by offering to undertake certain measures that will reduce CFC emissions from what they otherwise would have been, the U.S. government may influence the *willingness* of other governments to take actions with respect to their own CFC emissions. Second, by using specific control measures, the U.S. government may increase the technical know-how of other governments in the relevant policy field and, hence, influence their *ability* to take actions. We shall deal with these two main avenues of policy influence in turn.

The willingness to reduce CFC emissions in other countries may be influenced by the United States (or any other major contributor to emissions) along two alternative routes. Let us call them the push and the pull alternatives, respectively, and discuss them in detail, in turn.

In the *push* case, the results would be that of the *quid pro quo* approach mentioned above. The U.S. government would promise to accomplish a certain reduction of its CFC emissions (beyond what it would do purely in its own self-interest). Then other governments--either members of an established international organization like the OECD or some selected group of governments--would know that if they fulfilled their part of the agreement, made precise after negotiations and mutual concessions, they would benefit from the reductions made by the United States and other participating nations. As a major contributor to the CFC problem, the United States may have offers to make that will appear important to other nations and thus may strongly influence their willingness to do something in exchange for the U.S. commitments.

A number of factors will determine the extent to which other governments will agree to take action in exchange for a commitment by the United States. We have already noted some of them. Some governments may know or feel that ozone protection and climate change are not important to them. Or they may discount these effects heavily, giving priority to more immediate policy goals. Or they may prefer not to complicate present policy efforts by adding another field of intervention, for purely technical or for political reasons. Or straightforward economic costs of intervening in the production and consumption activities that affect the ozone layer and climate may simply be considered too high.

All the aforementioned factors may be regarded as involving different aspects of social efficiency pertinent to the individual country. Another factor of possibly great importance is international distribution or equity. Here, nations may reveal even wider differences in disputing what the ozone policy "terms-of-trade" should be.

At one extreme, the United States may argue that it has already done so much in drastically reducing its aerosol consumption (by some 90 percent of aerosol consumption, or some 55 percent of total CFC consumption) that it should not embrace new and more complicated areas of CFC control policy without other countries doing anything. It may be argued that the United States would otherwise impose losses on its population and industry while primarily providing benefits to the rest of the world, without getting anything in return.

At the other extreme, evidence such as that of table 12-1 (and similar evidence in Chapter 3) can be used to argue that the United States has such a large responsibility for the present CFC problem that it would have to do much more before other countries had any reason to make sacrifices in reducing the problem. Consequently, it could be felt that U.S. per capita "consumption" of the earth's ozone layer and climate, both global common property resources, should be cut down to some global or OECD average per capita level before the U.S. can rightfully claim that measures should be taken by others. In fact, given that CFCs emitted in the past are expected to have impacts for a long time to come and that these effects may be irreversible, accumulated emissions per capita can be viewed as the relevant measure. Here, table 12-2 shows the U.S. as even more unique, because it has for some time--albeit unintentionally--allowed its citizens to benefit from low-cost consumer products, particularly refrigeration, at the possible expense of the well being of future generations in the world as a whole. In this perspective, the United States might be required to do much more before others feel at all morally committed to make sacrifices in reducing their CFC emissions.

Table 12-1. Relative Shares of the Production and Consumption of F-11 and F-12 in OECD Countries, 1974

Country	Production	Consumption
U.S.	49.8	51.5
Germany	11.7	8.7
U.K.	9.5	6.8
France	9.5	6.4
Italy	5.0	6.4
Japan	4.5	4.6
Netherlands	3.8	2.0
Canada	3.1	3.2
Spain	1.2	9.4
Others	----	9.4
	100.0	100.0

Source: OECD, Economic Impact of Restrictions on the Use of Fluorocarbons (Paris, OECD/ENV/Chem/77.2, Jan. 24, 1978).

Table 12-2. Comparisons of U.S. and World Production of F-11 and F-12

	U.S.	World	U.S. Percentage of World
	Metric Tons		
Production of F-11			
1975	121[b]	357.3	34
1958-75 cumulative[a]	1541	3265.0	47
Production of F-12			
1975	178[b]	416.3	43
1958-75 cumulative[a]	2468	4520.0	55

Source: National Research Council, Halocarbons: Environmental Effects of CFM Release (Washington, D.C., NAS, 1976) p. 39.

[a]Quantities prior to 1958 were small and, if anything, would increase the U.S. share of accumulated production.

[b]U.S. production decreased for the first time in 1975.

Given that such equity considerations may make other countries less prone to being influenced by "push" offers from the United States, let us turn to the "pull" alternative. In this case, the U.S. government could influence other countries' willingness to undertake CFC control measures by taking unilateral actions which impose noticeable costs on production and consumption activities in the United States. Given that other countries follow suit by taking voluntary actions (as have Canada, Norway and Sweden), the power of the moral example set by the United States might well have more effect than more "businesslike" negotiations or actions which some countries may view as "blackmail." More specifically, if the U.S. government continues to provide the international community with information about ongoing U.S. research activities and policy plans, there may be some feedback on what other countries would do if and when the United States implements further policy actions in the field. This information may be firm enough to be added to the (soft) estimates of the internal benefits of these actions for the United States, and hence, increase total benefits enough to warrant a stronger or faster U.S. policy effort than the internal benefits alone call for.

Closely related to this pull, or moral suasion, approach to influencing international willingness to take action for the control of CFC emissions, is the second dimension of policy influence, that of increasing the _ability_ of other governments to intervene in those activities in their countries which cause CFC emissions. The choice of policy instruments in the United States may affect this ability in various ways. Here, we shall distinguish two ways in which this influence may work, that of providing _technical_ solutions in the relevant fields of production and consumption and that of providing _policy_ solutions to control the CFC emission activities. In the first case, policy action may produce new technical alternatives (1) in manufacturing with respect to chemicals used as well as processing techniques, (2) in methods of using and servicing the products concerned, and (3) at the disposal stage, involving new destruction or reprocessing techniques. The fact that these technical alternatives are made available and put into use, may lead to imitation by other countries. In the second case, the introduction and practical experience of new policy instruments will broaden the policy know-how of other governments and possibly contribute to speeding up concrete policy actions on their part. For example, it may be demonstrated how regulatory as well as incentive schemes can be designed and what their consequences would be.

As policy instruments are likely to differ in impact on the policy choice of other countries, it may be worthwhile to analyze the potentials of different measures in this regard. For example, it may be worthwhile to look into the prospects of political acceptability of different policy solutions within other countries and possibly let such findings have some influence on the choice made in the United States. Or a case may be made for applying less commonly used policy instruments, thus broadening the policy menu and reducing policy preparation time in other countries.

12.3 Effects of U.S. Policy on the International Control of CFM Use

We shall comment here on the possible influence of the U.S. government's choice among alternative policy instruments on measures taken by other countries with respect to CFC use. We will organize our discussion in terms of the end uses, and their distribution, shown in table 12-3.

The U.S. ban on nonessential uses of _aerosols_ has been emulated by Canada, Norway and Sweden, so there are now some examples of direct imitation of policy development in the United States. Aside from giving rise to pure imitation, which may be taken to be the most flagrant example of policy interdependence, the U.S. aerosol ban is likely to have other effects on ozone policy abroad. The most important, perhaps, will reflect the development of new propellants and consequently, the practical experience gained through consumer acceptance of those substitute propellants. In addition, administrative experience of the ban, as well as _ex post_ estimates of the impact on industry and employment, will reduce uncertainty for other countries considering similar measures.

So far, no actions have been taken with respect to the use of CFCs as blowing agents for _plastic foams_, whose uses include insulation, furniture manufactures and packaging. For most of these products, recovery of the CFC blowing agents seems difficult or uneconomical.[1] And in most cases, no suitable non-CFC substitutes seem to exist (see chapter 3).[2]

Table 12-3. Percentage Distribution of End Uses of F-11 and F-12 in the OECD, 1974

End Use	U.S. (%)	OECD (incl. U.S.) (%)
Aerosols	61.8	65.9
Refrigeration and Air Conditioning	23.2	18.1
Plastics	8.7	12.1
Other (Solvents and other uses)	6.2	3.9
	100.0	100.0

For CFCs used as solvents, the emission problem is essentially specific to the United States; only 0.7 percent of F-11 and F-12 use in OECD countries, excluding the United States, falls under this heading (see table 12-3). Thus, it has little relevance in the present context. However, though CFC use for plastic foams and for other uses (solvents and other uses) are minor parts of total CFC use, they resemble aerosol use in that emissions occur mainly in manufacturing or in early stages of the use of the end-product. Thus, in comparison with CFCs used as refrigerants, where CFC releases appear over much longer periods of time, these CFC uses have a more important role than is revealed by the data presented in table 12-3, given some rate of discount larger than zero.

On this basis, the important product categories facing additional CFC control internationally are plastic foams and refrigeration. Unlike most aerosol uses, technical substitutes without ozone layer or climate effect but leaving prices and properties of the end-product essentially unaltered do not exist for these product categories. Because of its relative importance, we concentrate here on the problem of controlling CFC use in cooling equipment.

Given that an across-the-board ban, as in the aerosol case (with only a few specific exceptions), does not seem to be a realistic policy alternative for the refrigerant case at the present time, the regulatory policy alternative would have to be fairly detailed and perhaps quite intricate. It might require differentiation among types of equipment, standards for servicing (possibly different for existing and future equipment), standards on design, and bans on certain forms of storage and disposal. A regulatory package proposed for, and adopted by, the United States would probably facilitate the adoption of similar measures elsewhere. In addition, the direct and indirect effects of this specific regulation on product design in the United States would further enhance chances of policy implementation abroad; for example, the development of a new refrigerant without detrimental side effects or the design of leak-proof equipment may enable other countries to proceed directly to a ban on certain product designs existing initially. If regulation continues to be the preferred policy instrument in environmental policy around the world, the implementation of a regulatory package in the United States would probably have important spillover effects on international CFC emission policy. In addition, a policy of this kind may help to promote and speed up the adoption of international standards in the manufacturing of such equipment, compared to the case where different countries adopt widely different policy instruments.

The other main group of potential policy instruments are subsumed under the heading of economic incentive systems. A deposit-refund or tax-subsidy system for CFCs is one possible example of such instruments (see chapter 11). The principal effect on international CFC policy of adopting any such incentive system in the United States would be to expand the set of policy alternatives. If, furthermore, experience accumulates indicating a realization of the presumed efficiency properties of such systems, it may make CFC emission policy look more attractive. Thus, these types of policy instruments may have an important influence on the outside world, although of another form than that of the regulatory alternative. A joint research effort by economists, lawyers, and political scientists may clarify the probable magnitudes of the merits of the two alternatives and make possible a more complete evaluation of the policy choices open to the U.S. government.

Now it should be added that optimal control of the use of CFCs as a refrigerant may include both regulatory and economic incentive systems. The reason is that regulation is not likely to be an effective remedy for the problem of improper disposal--as a ban on improper disposal may be either inoperative or very expensive to enforce--whereas economic incentives could be designed to cope with this problem (see chapter 11). On the other hand, even very high relative subsidies on CFC recovery or taxes on CFC releases may be ineffective for controlling emissions during the use of the cooling equipment or in servicing operations (labor cost being a dominant cost item here); hence, product standards and regulation of servicing operations may be required (especially for automobile air conditioners). If optimal national policy (considering only national benefits and costs) turns out to include a limited use of both types of policy instruments, it would presumably bring about the advantages in terms of external policy influences of both of these instruments, as well. Thus, the addition of international considerations can support the choice of a mix of regulatory and economic incentive instruments. A possible conclusion for national policy planning in a new policy area such as CFC emissions management is that, unless a policy mix is favored on the grounds of national efficiency alone, it should be worthwhile to estimate the policy influence on other nations of a policy mix as well as of a pure regulatory and a pure incentive approach.

12.4 Concluding Remarks

We have argued here that CFC management policy is unlikely to benefit much from the principles of traditional public good theory. Nor is it likely to be successfully carried out along the lines of traditional international bargaining with interdependent concessions as a requirement for actions to be taken. We may also note that the typical approach of conducting negotiations on an industry-by-industry basis--in this case, separating negotiations concerning aerosols from those concerning refrigerants, etc--may be less likely to produce significant results in terms of overall protection than if different nations were "allowed" to choose areas for policy intervention in the order of their individual preferences.

Our main point has been that the United States, a major producer of possible effects on the ozone layer and climate, and at the same time a pathfinder in international CFC emissions policy, may contribute to maximizing global efforts in this area by choosing its own policy with regard to the stimulating effects on the policy of other countries. As we have indicated, the U.S. choice of policy may influence both the willingness and the ability of other governments to take actions to control the CFCs. The effect on the ability to take protective actions may occur by way of furnishing information about policy options and about the practical experience of policy instruments chosen; such may be best achieved by choosing a policy mix of regulatory and economic incentive instruments.

This kind of policy would essentially be one of pulling other countries along in a fight against ozone depletion and climate change, instead of focusing on an attempt to push them into multilateral agreements at the bargaining table. The former policy seems, in fact, to have been chosen by the United States up to this point, but there may be a growing internal resistance to this approach in the future, given the rising social costs of additional CFC emissions control measures.

References

1. duPont, Freon® Products Division, Information Requested by EPA on Non-Aerosol Propellant Uses of Fully Halogenated Halocarbons, March 15, 1978, p. IV-26.
2. Ibid., p. 120; and the National Research Council, Halocarbons: Environmental Effects of CFM Release, Washington, D.C. (NAS, 1976).

Chapter 13

ENVIRONMENTAL REGULATION, FACTOR MOBILITY AND INTERNATIONAL TRADE

Martin McGuire

In the interest of reducing the emission of ozone-depleting chemicals into the atmosphere, the U.S. government has taken various preliminary regulatory steps to reduce the production and sales of offensive products in the United States. These steps have been unilateral, involving no binding commitments nor informal understanding with other governments that they also place complementary restrictions on such chemical emissions. The consequences of such a unilateral U.S. initiative is investigated in this chapter by combining elements of general equilibrium international trade models with those of pure public goods models to examine the economic structure of worldwide pollution emissions and control. At a theoretical level we examine the economics of production, and trade of a good which is produced and consumed in more than one country, which is traded among countries, and which generates pollution in the production/consumption process. At the regulatory or policy level, the purpose of this synthesis is to judge the likely effects of one form of regulation or another, or of varying degrees of regulation, to anticipate how manufacturers, counties, or consumers will react to alternative methods of control. The investigation is new in that it examines the effects of unilateral U.S. action on the allocation of factors of production within and among interdependent countries in the world economy. Both short-run and long-run effects are examined.

Author's Note: The later stages of this research were supported by the National Science Foundation, SES-80-14238.

The major policy-relevant conclusion to emerge concerns the effects of unilateral U.S. regulation on the long-run viability of the regulated U.S. industry. It is shown that under certain economic assumptions, including the assumption that one factor of production is mobile across national boundaries, unilateral regulation will tend to bring about the extinction of the regulated industry in the regulating country. U.S. policy should anticipate these harmful consequences of unilateral action. In addition to seeking cooperative action, U.S. policy should consider compensatory instruments to offset the harmful effect of unilateral regulation. Such instruments would include commercial and tariff policies, and other taxes or subsidies to factors of production in the domestic industry.

In contrast to the economics of multicountry production of pollution-related products, the economics of multicountry consumption are rather well understood. For instance, it is well established that the level of ozone in the upper atmosphere is a global collective consumption good. That is, the quantity of ozone benefits the entire human race. This is not to say that ozone benefits everyone equally but rather that every individual benefits from atmospheric ozone independent of any other's benefit. The effect of unilateralism on the consumption side has been thoroughly analyzed. When unilateral U.S. policy succeeds in reducing ozone depletion at a cost to the United States, everyone throughout the world benefits. Even though the benefits to U.S. citizens may exceed their cost, efficiency and equity considerations dictate that multilateral cooperative regulation is preferable. On efficiency grounds, acting alone the United States would not push ozone protection to its efficient frontier; unilateral regulation provides only a second-best degree of protection. On equity grounds, every unilateral U.S. effort amounts to an unrequited grant-in-kind of special value to the rich, populous, fair-skinned countries of the Northern Hemisphere.

These consumption or collective-good considerations favor multilateral cooperation, but do not decisively disfavor unilateralism. The outcome of our examination of the production side of the international economy, by way of contrast, is to alert policymakers to the positive futility (possibly to the point of counter-productivity) of unilateral U.S. regulation of ozone-depleting materials.

The main points of the analysis can be outlined as follows. Section 13.1 develops the proposition that in the production of CFC-emitting products, the use of the environmental sink can be regarded as a factor of production no less than other inputs such as labor and capital. This is a commonly recognized approach to pollution, and it allows application of tools of production and trade theory to the problem of regulation.

In the absence of any regulation, different countries will produce and export or import diverse goods in accordance with their perceived comparative advantage. The most complete, systematic, and precise explanation of comparative advantage in trade derives from the Heckshер-Ohlin model.[1] This model is based upon a number of assumptions about production within and between countries, the most important of which are as follows:

1. Perfect competition and factor-of-production mobility holds within countries, but not between them.

2. Different countries have the same technologies for producing any particular good, and these technologies exhibit constant returns to scale.

3. Transportation costs are ignored.

4. Free trade establishes a single world price for each commodity.

Using these assumptions, a model of world trade can be constructed. The model implies that each country has a comparative advantage in producing those goods which use its relatively abundant resources most intensively; within this model the effects of technical change, resource expansion or depletion, and changes in world demands as well as tariffs and other commercial policies can be unambiguously predicted. The analysis to follow uses the Heckshеr-Ohlin model to examine the expected effects of U.S. unilateral CFC regulations on international production patterns of CFC-related products.

By treating the environment as "just" another factor of production we place at our disposal a powerful set of analytic tools. However, we require a preregulation point of departure from which to apply these tools to analyze the effects of controls. To establish this point of departure in this paper it is assumed that polluting emissions are generated in fixed proportions with the utilization of another factor of production. This assumption yields a rational explanation for the amount of pollution which obtains in the absence of regulation. The specific example used is that pollution is generated by capital in the offending industry--that each unit of capital employed in the offending industry emits a certain fixed amount of pollutant. The analysis is completely general so that symmetric results would obtain on the assumption that pollution is generated by labor, land, or any other factor in the offending industry. The dual production and consumption character of industrial pollutants is addressed in more detail in section 13.1. Then, using the fixed proportion assumption as to the technology of pollution in the production process, sections 13.2 to 13.4 examine the effect of pollution control on factor allocations, factor incomes, specialization, and trade under alternative assumptions as to factor mobility and consumer demand.

The first-order effect of pollution control is to substitute nonemitting factors for the pollution-intensive factor in the offending industry. On the assumption that pollution is proportional to usage of one factor or another, pollution control is equivalent to a tax on a factor of production. As argued in section 13.2, this equivalence means that the powerful body of technique following on Harberger's original analysis of the general equilibrium incidence of the corporation tax is applicable to the effects of regulation when a factor of production emits the pollution in fixed proportions.[2]

With an equivalence between pollution control and tax on a factor of production (capital) established, section 13.3 considers the consequences within the boundaries of a country if _any_ factor on production is free to cross its national boundaries. This depends on three considerations: whether pollution control is unilateral or coordinated among countries, whether it is the pollution-related factors of production or others which are mobile, and on the relative importance of pollution related goods in the consumption preferences of the mobile factors. Section 13.4 extends the analysis to a system of countries, in an international trade context. Here, the viability of unilateral regulation is examined, the conditions under which coordination among countries is necessary or not. Last, section 13.5 presents a summary, conclusions, and policy recommendations.

13.1 The Dual Production-Consumption Character of Industrial Pollutants

13.1.1 General Features

In building a general normative model the crucial feature of CFCs to be captured is that they are simultaneously a consumable commodity (a negative one) and a factor of production in manufacturing--a public bad in consumption and a private good in production.

First, the release of CFCs creates damaging worldwide atmospheric change, which is "consumed" willy-nilly by everyone on earth. In this respect CFCs are as pure a case of negative public good (that is, a public bad) as is known to exist. There is, of course, a subsidiary technical relationship between atmospheric change and CFC release, and further relationships exist between atmospheric change and harm to people and property. Since these technical relationships use the simple sum of all CFCs released (or a variable monotonically related to this sum) there is a one-to-one correspondence between ΣCFC and damaging impacts. Therefore, it is natural and valid to call CFCs the public bad. All individuals must jointly "consume" the consequences of ozone depletion and climatic change. Although the disutility suffered may vary among subpopulations because of differences in locale, in genetic makeup and other factors, all suffer inescapably from every ton of CFC released to the atmosphere, and the proper social measure of CFC disutility is the sum of all these individual losses. Here we will ignore the options open to individuals to protect themselves through individual action against the undesirable consequences of ozone depletion, although individual actions (for example, less sunbathing, etc.) could be incorporated at the cost of further complexity.

The second crucial feature of CFCs, as of any industrial pollutant, is their role as a factor of production. In the case of CFCs, exploitation of the atmospheric sink is a "factor of production" in the production/consumption of certain aerosol and refrigerant commodities. This is to say that implicitly some technical relationship can be written relating aerosol-refrigerant output to labor, capital, land, and CFC discharges as inputs.

Production/consumption processes in the manufacture of aerosols and refrigerants most generally can be characterized by some functional relationship among the variables T (for CFCs), K and L (for factors of production such as capital and labor), and X (for product): $\Phi(X, T, K, L) = 0$. In one sense CFCs are an output of the production process, and we may speak of the factors of production L and K producing the joint products X and T, with CFCs being considered as intermediate products that ultimately cause atmospheric damage and generate disutility. This idea is captured by writing $f(L, K) = g(X, T)$. It is equally valid to consider the atmosphere a common resource, and CFC an intermediate input. Viewed from this perspective we should write $X = h(T, L, K)$ as the production function for product X, and label the input axes L, K, and T. The advantage of this production function approach is to allow us to draw on many aspects of the theory of international trade, comparative advantage, and the efficient allocation of factors to different industries across nations.

13.1.2 The Technology of Pollution: The Case of Fixed Proportions Between Emissions and Capital

To implement this inquiry therefore, we will adopt the standard Heckscher-Ohlin (H-O) model from trade theory but modify it in two ways. The model to be used is identical to H-O in all respects but for the introduction of environment. Thus, we will assume fixed factor supplies of L^A, L^B, K^A, K^B in countries A and B. Identical linear homogeneous technologies produce two goods X and Y in the two countries. Perfect competition and factor mobility within countries entails equal unit-factor rewards across industries. Transportation costs are ignored. Free trade and competition among countries generates common worldwide commodity prices.

We now wish to make the minimum alteration in the H-O framework to accommodate the environmental factor. The environment is a factor of production, which is "used up" in industrial and agricultural processes. From this perspective, the environmental factor can be incorporated in H-O by adding the environment T as one factor in one industry. We choose X^i as the polluting industry in country i, and Y^i as the nonpolluting industry. Accordingly, to represent production we can write the two homogeneous production functions

$$X^i = f(L^i_x, K^i_k, T^i) \qquad (1)$$

$$i = A, B$$

$$Y^i = \Phi(L^i_y, K^i_y) \qquad (2)$$

L^i_j denotes i's employment of labor in industry j. T^i, which indicates i's "usage" or depletion of the environment, is measured in tons (or some physical quantity of effluent output). In either country the factor endowments ($L^i_x + L^i_y = \overline{L}^i$) and ($K^i_x + K^i_y = \overline{K}^i$) are fixed. However, there is no constraining physical limit on the amount of effluent which can be discharged. In principle, T^i could exceed all bounds. But it is only realistic to assume even in the absence of regulation that pollution reaches a finite equilibrium level because of the technology of the polluting industry. In order to limit preregulation pollution it will be assumed that T^i and K^i_x are combined in fixed proportions. (The alternative assumption of smooth substitution possibilities between T^i, and L^i_x or K^i_x is considered in another phase of the author's research.[3])

13.2 The Effects of a Tax on Emissions in the Fixed Proportions Case When Productive Factors Are Not Mobile Between Countries

In the absence of effluent fees, the world economy represented by country A's and country B's production functions X^A, B^B, etc., will reach a trade, production, and consumption equilibrium identical to the standard H-O outcome in all respects. Now suppose a tax, or equivalent mandated maximum rate of effluent per unit of output is imposed on the polluting X-industry. Under our assumptions, such a tax is in all respects equivalent to a tax on capital in the offending industry. Accordingly, we can refer to a "tax on pollution" as a "tax on capital" or another factor throughout the remainder of this chapter. To analyze the effects of such a tax one may turn to a vast literature on the incidence of partial factor taxes which has appeared since Harberger's original article, extending his analysis to growing economies, to economies with immobile factors, and to trading economies.[4] Similarly, extensive and highly complex analyses exist of taxes and tariffs, factor market distortions, and factor mobility in the Hecksher-Ohlin tradition.[5] In general terms these studies predict that an effluent tax under conditions of fixed proportions and zero factor migration such as we have assumed will cause production of the pollution intensive good to be <u>curtailed but not eliminated</u> in all countries which impose such a tax. As production of the pollution-intensive good is curtailed and L^j_x and K^j_x are accordingly reduced, industry Y^i will expand to absorb the factor of production so released. This process in turn will alter the distribution of income between labor and capital (and other factors of production). Any new, postregulation short-run equilibrium between supply and demand will in all probability bring about a change in the relative rewards (wages, rent, real interest payments) to all other factors of production throughout the entire economy. This effect may be small if the industry regulated is relatively small; or the effect may be great if the regulated industry is significantly large. In any case the qualitative nature of this disturbance is definite and unambiguous. Whether environmental regulation helps labor versus capital or vice versa depends on the particular circumstances in the regulated industry which is a subject deserving further analysis. Such analysis would require a data base as to the relative capital, labor, and land intensity of regulated industries in comparison with unregulated ones.

13.3 The Effect of a Tax on Emissions in the Fixed Proportions Case When Productive Factors can Migrate Between Countries

For the long run, the assumption that capital cannot migrate across national boundaries is, of course, unsatisfactory, especially in the day of the multinational corporation. Consequently, we now propose to relax the assumption completely and to provide a simple treatment of the consequences of a tax or regulation on one factor in one sector of the economy when that factor or some other one is free to migrate into or out of that economy. We concentrate on the case of an infinite supply elasticity for the mobile factor of production, since the analysis is easily extended to less extreme cases.[6]

The analysis here is essentially an extension of Mieszkowski.[7] Although the results of this analysis may be implicit in more complex studies, they have not been hitherto explicitly recognized. It turns out that the effect of a factor tax under these conditions does not depend at all on whether the taxed industry is relatively capital-or labor-intensive (this being a crucial parameter when all factor supplies are fixed). Nor do the elasticities of substitution in either industry affect the result. Rather, the crucial determinants of tax incidence are shown to be first, whether the "mobile" factor or the "immobile" factor is taxed, and second, whether the product of the taxed or the untaxed industry counts predominantly in the mobile factor's consumption package.

Where labor (or any factor) has perfectly elastic supply, its price becomes a constant in the economy. But prices measured in terms of which good? If we take good X as the "wage good" or "numeraire," then P_L/P_X becomes constant; if Y is the "wage good" P_L/P_Y becomes constant (P_L being the nominal price of labor). As we shall see, it matters which good we choose as wage good or numeraire. Moreover, a partial factor tax under the assumed conditions has a systematic effect on the production possibility curve as in the Harberger and Krauss-Johnson analysis,[8] sometimes contracting but also sometimes expanding production possibilities, depending again on which factor is taxed (mobile or immobile) and on which good is the primary consumption good of that factor.

The assumption of significant factor supply elasticity suggests that the model applies to rather long run consequences, where capital may easily be relocated across national boundaries, or labor migrations may cumulate to a significant amount.

13.3.1 The Two-Sector Model with Perfect Factor Elasticity

As in the Harberger model we assume two sectors each producing homogeneous goods X and Y according to a linear homogeneous production technology: $X = f(L^X, K^X)$, $Y = \phi(L^Y, K^Y)$ where L^X, and L^Y (K^X and K^Y) are the allocations of labor (capital or the composite factor capital plus the environment) to industries X and Y. In the absence of any taxation, competition ensures that value marginal product equal unit factor cost

$$P_L = P_X\, MP_L^X = P_Y\, MP_L^Y$$

$$P_K = P_X\, MP_K^X = P_Y\, MP_K^Y \qquad (3)$$

where P_L indicates wage rate, P_K the capital rental rate, P_X and P_Y commodity prices, and MP_L^X (MP_K^Y etc.) indicates marginal product of labor (capital) in industry X (Y).

Before examining the effect of a tax it is useful to establish the shape of the production possibility curve when one factor is perfectly elastically supplied. Suppose for illustration that the supply of labor is fixed, while capital is available at a fixed price either in terms of good Y or of good X. Production possibilities under these conditions are conveniently analyzed by expressing the equilibrium conditions in terms of the numeraire, say Y. Before the tax

$$\frac{P_L}{P_Y} = \frac{P_X}{P_Y} MP^X_L = MP^Y_L = \text{constant} \tag{4}$$

$$\frac{P_K}{P_Y} = \frac{P_X}{P_Y} MP^X_K = MP^Y_K = \text{constant} \tag{5}$$

A perfectly elastic supply of capital in terms of good Y means that P_K/P_Y is constant, fixed outside the system by world capital rent for instance. It follows that MP^Y_K also is constant for all possible equilibrium outcomes. Drawing on the assumed linear homogeneity in production, we infer that the capital labor ratio in Y (K^Y/L^Y) is constant since MP^Y_K is constant. Therefore, MP^Y_L is also constant. Figure 13-1 then illustrates the possible equilibrium factor allocations between X and Y in the absence of a tax. The capital/labor ratio in Y is fixed according to the above argument. Therefore all equilibrium capital labor ratios in X are fixed also. A condition of equilibrium is that

$$\frac{MP^X_L}{MP^X_K} = \frac{MP^Y_L}{MP^Y_K} \tag{6}$$

The second ratio is constant; therefore so is the first which, in turn, implies a single K^X/L^X ratio (since X is produced with a homogeneous technology). These equilibrium factor proportions are shown as ρ^*_y and ρ^*_x in figure 13-1. Since the supply of labor in this economy is invariant, the horizontal dimension of the box diagram is fixed. Capital, however, can freely enter or exit the system at its set rental rate; as capital flows into this economy capital stock expands from K^a to K^b to K^c. The allocation of productive factors moves from a to b to c and output varies from a" to b" to c". The production possibility frontier is a straight line, since factor proportions in the two linear homogeneous industries do not change with changing factor allocations; geometrically, the lengths ga, gb, gc are proportional to aa', bb', cc':

$$\frac{ga}{aa'} = \frac{gb}{bb'} = \frac{gc}{cc'} \tag{7}$$

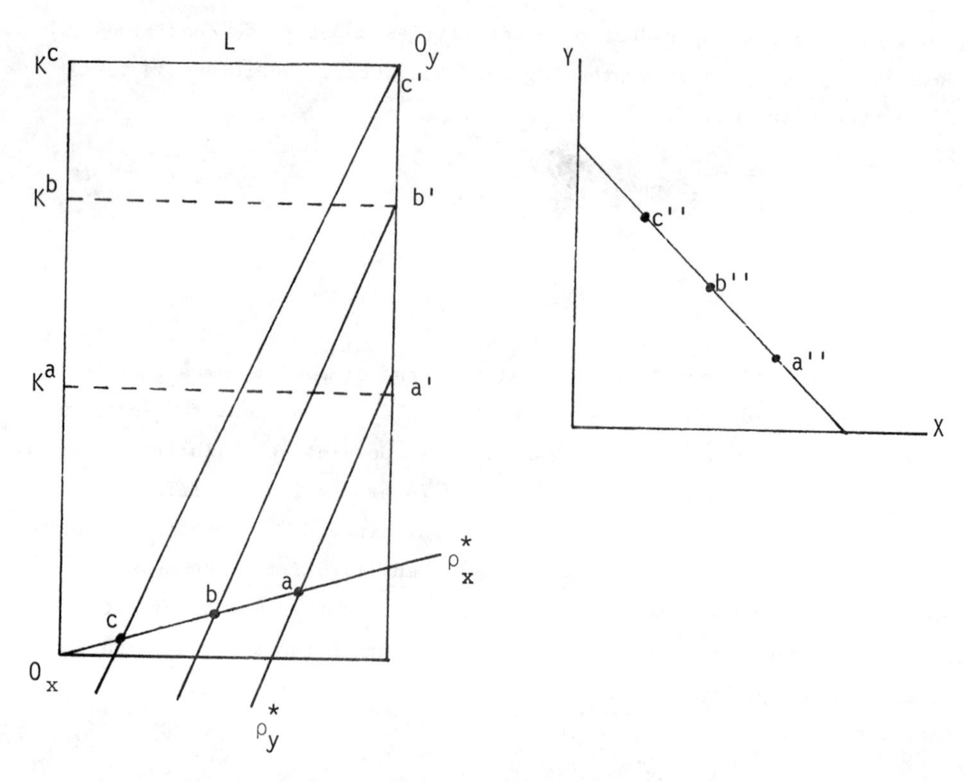

Figure 13-1. Geometric derivation of the transformation curve where the supply of a factor of production is perfectly elastic. Left, production box; right, production-possibility curve.

The equilibrium "size" of this economy, the production of X and Y, the quantity of capital in the system and the allocation of factor between X and Y will be completely determined by demand conditions.[9]

Note that the qualitative nature of this before-tax general equilibrium description would not change had we chosen good X as numeraire with P_L/P_X, therefore, a constant throughout the economy. Nevertheless, as the next sections will demonstrate, the effect of an effluent-charge factor tax does vary rather dramatically depending on which good is taken as the "wage good" or numeraire.

13.3.2 The Effect of a Factor Tax

We now will examine the effect of a tax on one factor of production in one industry. Since the relative capital/labor intensities in the taxed versus untaxed sectors will not change the qualitative outcome we can arbitrarily select the industry and factor to be taxed. Suppose, then, that capital or the environmental effluent in X is taxed, for simplicity, at an effective rate of 50 percent. To carry the analysis forward, we must now specify (1) which factor is perfectly mobile (has an infinite supply elasticity), that is, whether it is the taxed factor K or the untaxed factor L, and (2) which good X or Y to use to represent a constant real price for that mobile factor.

The numeraire may be either the good which employs a taxed factor of production or the other good neither of whose productive factors are taxed. To handle this problem we will simply do a case by case analysis of the four possibilities. By looking at both extremes we will bracket the sensitivity of the effect of a factor tax depending on which good predominates in the mobile factor's consumption.

13.3.3 Case 1: Fixed Total Supply of Labor, Perfectly Elastic Supply of Capital, Untaxed Good (Y) is Numeraire

To examine the effect of a 50 percent tax on capital in industry X, we can go back to Equations (2) and (3). A tax on capital which takes 50 percent of capital's return changes Equation (3) to

$$\left(\tfrac{1}{2} P_X MP_K^X\right)_{\text{with tax}} = P_Y MP_K^Y \tag{8}$$

since with a tax, capital will retain one-half its value marginal product, and factor mobility across industries equalizes after tax unit rents. Accordingly,

$$\left[\tfrac{1}{2} \frac{P_X}{P_Y} MP_K^X\right]_{\text{with tax}} = \left[\frac{P_X}{P_Y} MP_K^X\right]_{\substack{\text{without}\\ \text{tax}}} = MP_K^Y = \text{constant} \tag{9}$$

The two bracketed expressions are equal to the same constant by virtue of the supply elasticity and numeraire assumptions. From Equation (9) we conclude that

$$\frac{P_X}{P_Y} MP_K^X \tag{10}$$

must increase as a result of the tax. From Equation (4),

$$\frac{P_X}{P_Y} MP_L^X = MP_L^Y = \text{constant}. \tag{11}$$

These two conditions imply that as a result of the tax $\frac{P_X}{P_Y}$ increases, MP_K^X increases, MP_L^X declines. The reader can verify that a decline in MP_K^X and increase in MP_L^X are inconsistent with Equations (4) and (9). Of course, MP_K^Y and MP_L^Y are not changed by the tax. As a result, labor bears the entire burden of the tax through diminished opportunity to consume good X. Note that in making this argument we have made no references to the relative capital intensity in X or Y. The argument is valid whichever industry is capital-intensive. Similarly, the qualitative nature of the conclusion depends in no way on the substitution elasticities in either the taxed or the untaxed industries.

Figure 13-2 illustrates the effect of such a tax on the assumption X is relatively labor-intensive. Before tax the constant factor proportions are shown as before by ρ_X^* and ρ_Y^*. A tax on capital in X causes factor proportions in that industry to shift to say, ρ_X^{**}, to the point that

$$2\frac{MP_L^X}{MP_K^X} = \frac{MP_L^Y}{MP_K^Y} \tag{12}$$

The factor proportion ρ_y^* is unchanged due to the supply elasticity of capital being infinite measured in terms of good Y as numeraire. After the imposition of the tax, the variable supply of capital implies a flexibility in the vertical dimension of the edgeworth production box. The new production possibility curve is constructed by varying K at the factor proportion rays ρ_y^*, ρ_X^{**}. The new production possibility curve is a straight line, lying inside the old as shown. While the tax on capital in X has no effect on the maximum production of good Y, it reduces maximum production of good X from Og to Oh. We have illustrated this result for the case in which the taxed industry is labor-intensive. As the reader may verify, the same result carries through if X is capital-intensive.

13.3.4 Case II: Fixed Total Supply of Labor, Perfectly Elastic Supply of Capital, Taxed Good (X) is Numeraire

The sensitivity of the foregoing analysis to the assumption that the untaxed good is numeraire will now be tested. We replace Equations

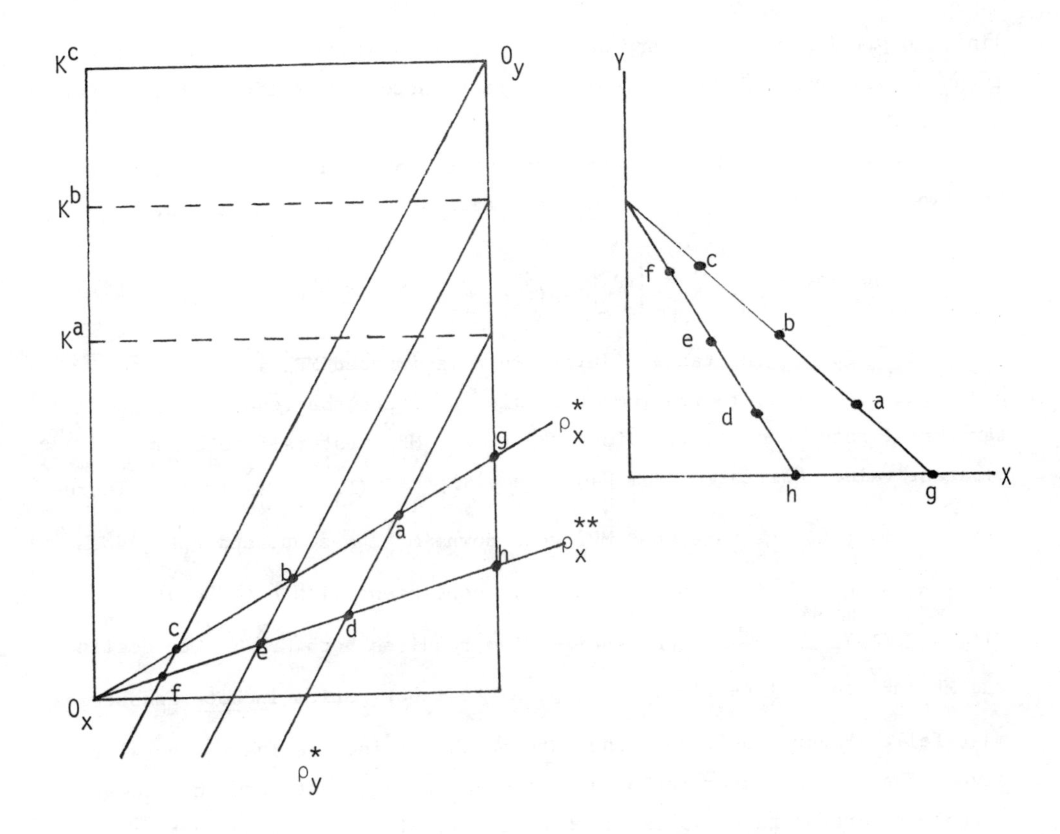

Figure 13-2. Effect of a pollution tax on a mobile factor of production when the nonpolluting industry produces the wage good. Left, production box; right, production-possibility curve.

(4) and (5) with Equations (13) and (14).

$$\frac{P_K}{P_X} = MP_K^X = \frac{P_Y}{P_X} MP_K^Y = \text{constant} \tag{13}$$

$$\frac{P_L}{P_X} = MP_L^X = \frac{P_Y}{P_X} MP_L^Y = \text{constant} \tag{14}$$

Before taxes the infinite supply elasticity of capital in terms of good X determines a constant MP_K^X; this, in turn, implies constant factor proportions (for a homogeneous production function) again yielding a constant MP_L^X. As before, therefore, the production possibility curve is a straight line. Capital mobility (entry and exit) plus equalization of factor returns across industries will imply constant factor proportions in industry Y as well.

To examine the effect of a 50 percent tax on capital in industry X, Equation (15) compares before and after tax equilibrium conditions,

$$\frac{P_K}{P_X} = \text{constant} = \frac{P_Y}{P_X} MP_K^Y = \left[MP_K^X\right] \text{ without tax } = \left[\tfrac{1}{2} MP_K^X\right] \text{with tax} \qquad (15)$$

Once a 50 percent tax on capital in X is imposed MP_K^X must rise to a new higher constant, in fact, must double to return the required after tax rental rate to capital. Since MP_K^X rises, MP_L^X must fall to a lower constant value. We infer from Equation (14) that $(P_Y/P_X)MP_L^Y$ falls. Therefore, it must be the case that MP_L^Y goes down, MP_K^Y goes up, and P_Y/P_X goes down; no other combination of changes is consistent with both Equations (14) and (15). Since the new equilibrium requires both MP_L^X/MP_K^X to decline and MP_L^Y/MP_K^Y to decline also, the capital to labor ratios in <u>both</u> industries must fall. Figure 13-3 shows the outcome when X is relatively labor-intensive. The capital labor ratio falls from ρ_y^* to ρ_y^{**}. At these new lower factor proportion rays capital in the economy will flow in and out in response to demands for final outputs X and Y. It follows from figure 13-3 that as a result of the tax the maximum outputs of <u>both</u> X and Y decline. The overall effect of such a tax, therefore, is to shift the entire production frontier inward. Similarly, if industry X is capital-intensive, the effect of a tax on capital in X is to shift the production frontier inward. Again, labor bears the entire tax burden. Wages measured in terms of good Y and good X decline.

13.3.5 Case III: Fixed Total Supply of Capital, Perfectly Elastic Supply of Labor, Untaxed Good (Y) is Numeraire

We now will turn to the case in which the untaxed factor is in perfectly elastic supply. Since we are assuming that the nominal tax falls on capital, this means that the horizontal labor-dimension of our edge-

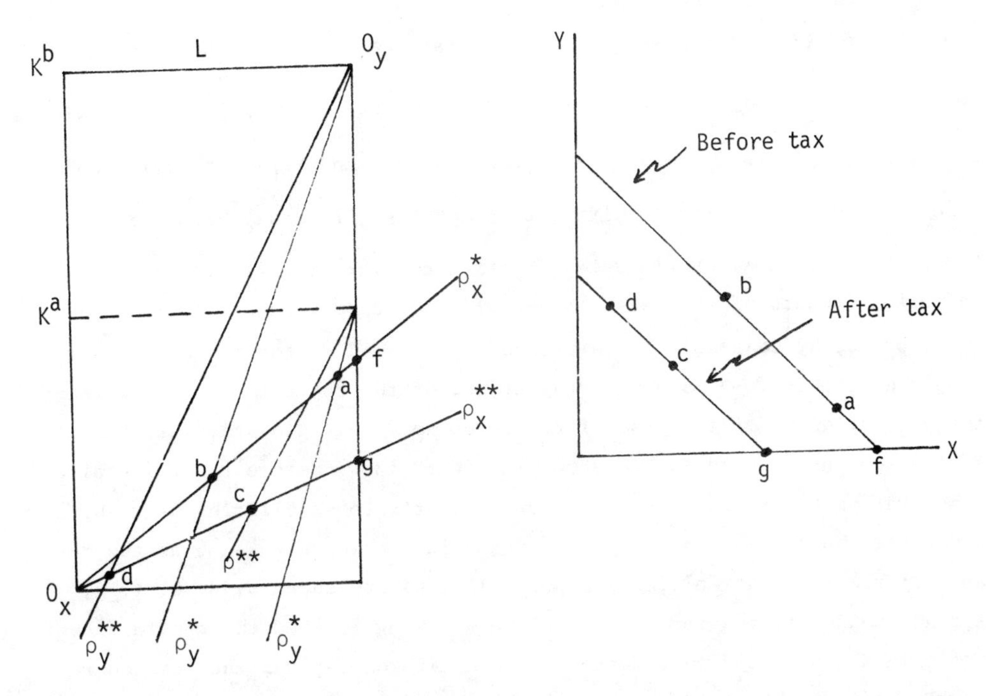

Figure 13-3. Effect of a pollution tax on a mobile factor of production when the polluting industry produces the wage good. Left, production box; right, production-possibility curve.

worth box is flexible. As in Case I, the equilibrium equations

$$\frac{P_L}{P_Y} = MP_L^Y = \frac{P_X}{P_Y} MP_L^X = \text{constant} \tag{16}$$

$$\frac{P_K}{P_Y} = MP_K^Y = \frac{P_X}{P_Y} MP_K^X = \text{constant} \tag{17}$$

Now MP_L^Y is constant because of the perfect elasticity of supply of labor, MP_K^Y is accordingly constant due to the linear homogeneity of production in the Y industry. As in Case I again, a tax on capital in X causes the capital labor ratio in X to fall. This follows from the constancy of MP_K^Y and MP_L^Y. Since

$$\left[\tfrac{1}{2} \frac{P_X}{P_Y} MP_K^X\right]_{\text{with tax}} = \left[\frac{P_X}{P_Y} MP_K^X\right]_{\text{without tax}} = MP_K^Y = \text{constant} \tag{18}$$

the value of $(P_X/P_Y)\ MP_K^X$ must increase after the tax. From Equation (16) $(P_X/P_Y)\ MP_L^X$ is unaltered by the tax. Since MP_L^X and MP_K^X change in opposite directions, MP_K^X must rise as a result of the tax and MP_L^X must fall. No other outcome is consistent with both Equations (17) and (18). Similarly, the price of good X must increase relative to Y; P_X/P_Y rises. In Case I unlimited amounts of the taxed factor of production were available at a fixed price; by contrast, in this case the untaxed factor has an infinite supply elasticity. As a result the effect of the capital tax is to expand the production possibility set for this economy enlarging the maximum potential output of good X. As shown in figure 13-4, before the tax capital/labor ratios ρ_X^* and ρ_y^* obtain; factor allocative possibilities of a, b, c, and g are shown. The capital labor ratio in X changes to ρ_X^{**}, as the result of the tax. Now allocation possibilities are shown by d, e, f, j, and h. Evidently the tax on capital in X having lowered the capital/labor ratio in X has forced that sector to economize on capital the limited resource and to substitute the potentially abundant resource labor. The point h is clearly on a higher isoquant than is point g.

13.3.6 Case IV: Fixed Total Supply of Capital, Perfectly Elastic Supply of Labor, Taxed Good (X) is Numeraire

This is the final case to consider. It parallels Case II except that now the untaxed factor is elastically supplied. Under these assumptions the equations of equilibrium are

$$\frac{P_L}{P_X} = \text{constant} = MP_L^X = \frac{P_Y}{P_X} MP_L^Y \tag{19}$$

$$\frac{P_K}{P_X} = \text{constant} = MP_K^X = \frac{P_Y}{P_X} MP_K^Y \tag{20}$$

As in Cases I through III, the abundance of one factor at constant price gives a straight-line production frontier. This time, however, the imposition of a tax on capital in X cannot alter MP_K^X. Instead, the

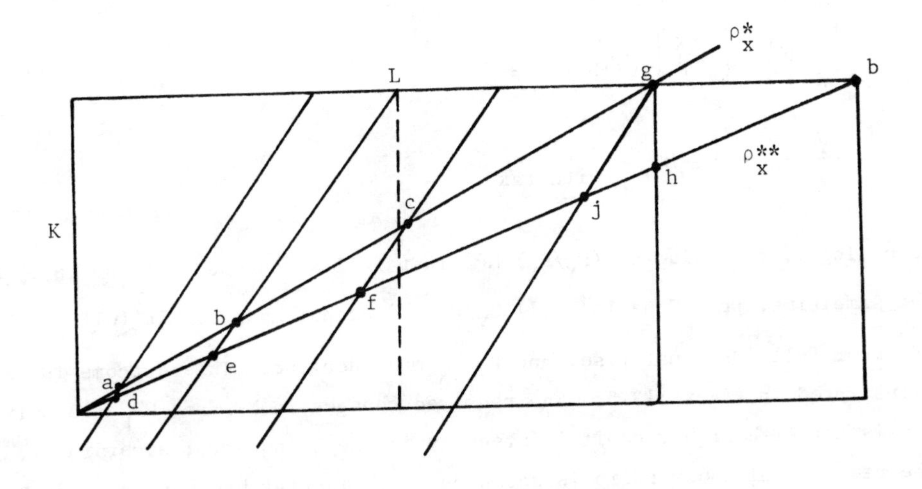

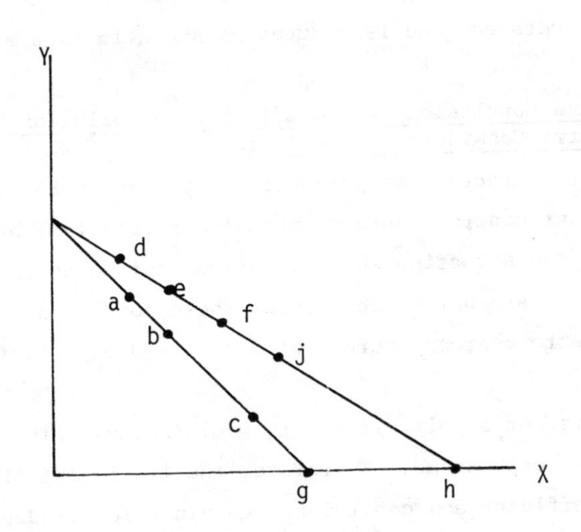

Figure 13-4. Effect of a pollution tax on an immobile factor of production when the nonpolluting industry produces the wage good. Top, production box; bottom, production-possibility curve.

entire effect must show up in the price of good X. This is seen from

$$\frac{P_Y}{P_X} MP_K^Y = \left[MP_K^X \right] \text{ without tax} \tag{21}$$

$$\frac{P_Y}{P_X} MP_K^Y = \left[\tfrac{1}{2} MP_K^X \right] \text{ with tax} \tag{22}$$

Accordingly, the value of $(P_Y/P_X)\ MP_K^Y$ must fall as a result of the tax. At the same time, from Equation (11) $P_Y/P_X\ MP_L^Y$ cannot change. It follows that MP_K^Y must fall, MP_L^Y must rise, and P_Y/P_X must decline. This outcome is illustrated in figure 13-5. The required changes in MP_L^Y and MP_K^Y are only consistent with higher capital intensity in the Y industry; accordingly, the new capital/labor ratio is shown as ρ_y^{**}. The tax has caused producers of Y to increase their use of the fixed resource of capital relative to the abundant, elastically supplied resource. Therefore, the maximum post-tax output of the untaxed good is reduced as shown in figure 13-5B.

13.3.7 Summary and Conclusions: The Effect of a Pollution Tax on a Country Considered in Isolation

In the long run, factors of production may respond to taxes, or to other tax equivalent control (such as mandated regulation) by migrating. We have looked at the situation when migration is free and other locational opportunities are unlimited. Under these conditions a mobile factor cannot be made to bear any burden of a tax, and may, in fact, benefit from a tax.

In reality, factor supply elasticity would not usually depend only on good X or Y but on real income. Therefore, the likely effect of an effluent tax when the effluent and capital are combined in fixed proportions and capital is mobile is somewhere between Case I and II, and when labor is mobile is somewhere between Case III and IV. Thus, a tax on the effluent-capital composite, if capital is mobile, will drive capital out of the economy contracting both industries, the taxed industry more than the untaxed industry. In contrast, a tax on the effluent-capital composite when labor is mobile has mixed effects, inducing labor to enter the economy for the production of the taxed good and to leave the economy in production of the untaxed good.

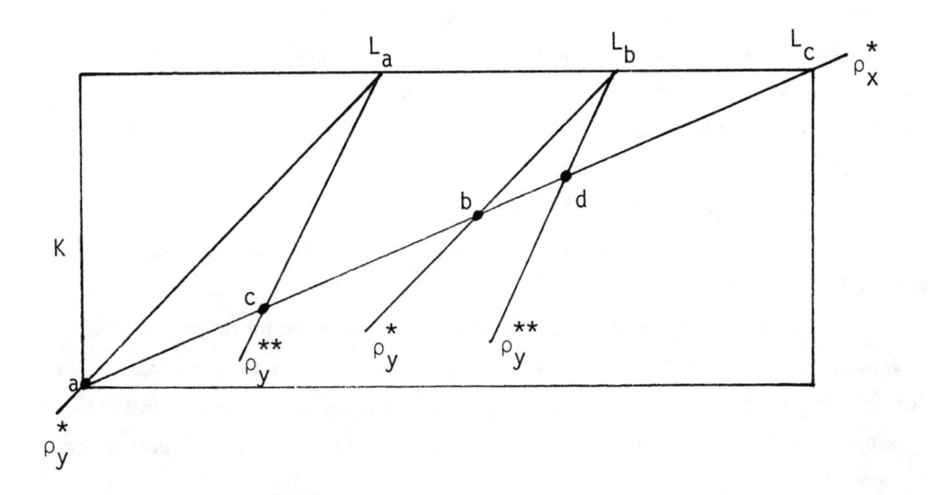

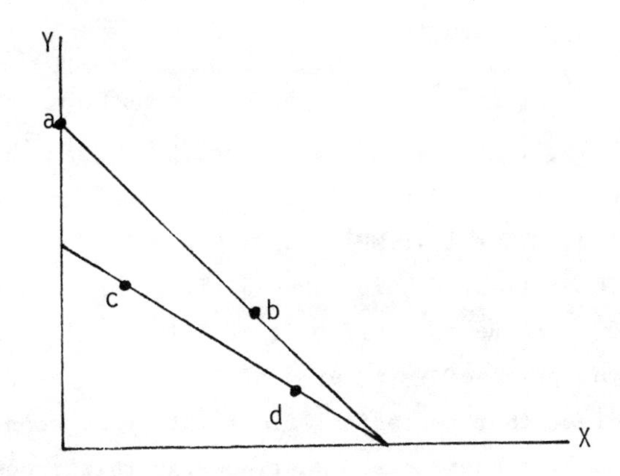

Figure 13-5. Effect of a pollution tax on an immobile factor of production when the polluting industry produces the wage good. Top, production box; bottom, production-possibility curve.

One of the more interesting features of these examples is the potential effect of a pollution tax (or equivalent regulation) on the size of the economy imposing the tax. Depending on which cooperating factors of production are mobile and on the significance of local products in their consumption the tax may attract or release such other factors, enhancing or depleting the local economic base.

Similarly, a larger economy may have a rather elastic supply of unemployed labor at a close to constant price because of minimum wage laws, an army of underemployed workers in the secondary economy or underemployed household members in the domestic sector. In these cases as well, the factor tax structure has been shown to interact in an intriguing way with worker mobility to encourage or discourage participation in the market economy.

Consider the following example to illustrate the point. Using two goods, X and Y, with market prices P_X and P_Y, each produced by an X and Y Cobb-Douglas technology in each industry. Let

$$X = L_x^{1/2} K_x^{1/2}; \; Y = L_y^{1/3} K_y^{2/3}$$

Assume that K is mobile, Y is the numeraire wage good, and that K is available at a real price in terms of Y of 1/3. It follows from these assumptions that: $MP_K^Y = 1/3$; $MP_K^X = 1/4$; $MP_L^Y = 4/3$; $MP_L^X = 1$; $P_X/P_Y = 4/3$; $K_X/L_X = 4$; and $K_Y/L_Y = 8$. The real production possibilities in this economy, as shown in figure 13-6, require one-half unit of X to be given up to produce one of Y, although the price of Y in terms of X is 3/4; the differential is explained by the fact that as capital flows into this economy, production of Y expands and of X contracts. Just how far this proceeds depends on consumption preferences of labor as between goods X and Y.

Suppose the total labor supply in a county is 128. If labor consumes only X and capital only Y, the equilibrium stock of capital is 768. When allocated $K_X = 256$, $K_Y = 512$, $L_X = L_Y = 64$, this produces \$512 worth of X and \$768 worth of Y, with all the X going to L and all the Y to K. Now suppose on the other hand that labor most prefers to divide its consumption equally between X and Y, while as before, capital consumes only its numeraire wage good Y. Now, an inflow of capital will raise its equilibrium level of 896. With allocations, $K_X = 128$, $K_Y = 768$, $L_X = 32$, $L_Y = 96$ the value of

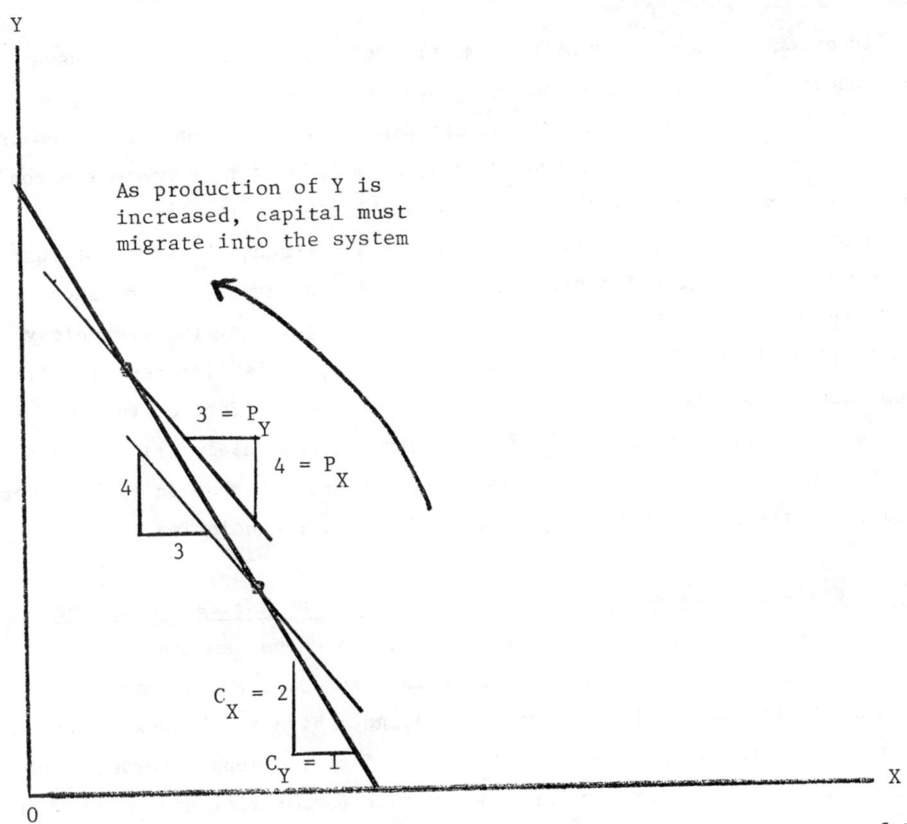

Figure 13-6. Comparison of relative opportunity cost of Y in terms of X when one factor of production is available at a fixed price

X produced is \$256 and of Y is \$1152. The income of L is \$512 spent equally on X and Y while the income of capital is \$896, all consumed as good Y.

13.4 Effects on Trade and Specialization

So far we have examined the impact of an effluent charge within the boundaries of a single economy. A crucial assumption has been that each unit of capital (or other factor) employed by the polluting industry emits a certain given amount of effluent. We have considered the alternative extremes of (1) no factor migration and (2) completely free factor mobility into and out of the economy in question. Now we will turn to an analysis of the effects of such pollution taxes (or equivalent regulations) on trade and specialization among countries.

In extending the analysis to a multicountry context, a major issue will concern whether, on the one hand, unilateral regulation is efficacious and desirable or, on the other hand, coordinated action among countries is essential or preferable. In other words, we will want to compare the consequences of unilateral with cooperative regulation to assess the viability or futility of going it alone. As we shall see, it turns out that the answer to this question and the efficacy of unilaterialism depends crucially on whether the productive technology (including the technology of pollution) is the same in different countries, or is different in different countries. Consequently, we will examine the effect of the pollution tax or control under two distinct alternatives, namely (1) that different countries employ different technologies and (2) that in each sector, X and Y, different countries employ the identical technologies.

13.4.1 Unilateralism Versus Coordination When Technologies Are Shared

In a multicountry system, if technologies are the same in all countries and each country functions as described in 13.3, factor migration will substitute completely for trade. Imagine a high total demand for the good which counts as wage good in the mobile factor's supply curve; then a large derived demand for immigration of that factor will accompany high production of the good. If total demand for that good is low, less will be produced with a lesser amount of the mobile factor.[10]

It also follows from the analysis of section 13.3 that if any factor is mobile across national boundaries, unilateral effluent taxes will cause the regulated industry in the regulating country to shut down in the long run. By contrast, if effluent taxes are coordinated all polluters and countries share in the reduction in output of pollution-related products. This result is most readily established by references to figures 13-2 to 13-5. In each case unilateral regulation raises the price of X, the pollution-intensive product, and in each case perfect factor mobility implies that no readjustment of factor allocations within the regulating country can effect a lower postregulation price of X; in the long run, all production of good X will relocate to the rest of the world. On the other hand, with identical technology and coordinated regulation across countries, the price of X will increase by the same amount in all countries; no country will derive a relative cost advantage in the production of X; and coordinated regulation will yield a balanced reduction in the production of the pollution-intensive good among all countries.

13.4.2 Unilateralism Versus Coordination When Technologies Are Different

By admitting technical differences between countries, we necessarily step outside the strict bounds of the Hecksher-Ohlin neoclassical tradition. Yet to sustain trade in goods simultaneously with freedom of factor migration requires some violation of this tradition as Chipman and also Jones have shown.[10] In any case, if we make this change in our assumptions, the conclusion emerges that unilateral regulation may be sustainable and desirable. When combined with external diseconomies such as environmental pollution, technical differences among countries can give rise to call false comparative advantage. This can be most easily illustrated for a simple two-factor Richardian model of trade. Figure 13-7 shows the situation in countries A and B which use one factor, capital to produce good Y and two factors, capital and environment, T, in fixed proportions to produce good X. We assume constant returns to scale in both production functions, but A uses more of factor T and less of K to produce a unit of good X than B uses. Figure 13-8 shows the preregulation and postregulation production possibility curves of these two countries. Without regulation, country A will tend to specialize in the production of good Y and country B in good X. In the absence of regulation, A perceives a comparative advantage in good Y, and B perceives a comparative advantage in good X. But these are false perceptions. In the presence of proper regulation, A should specialize in X and B in Y because of the relative cost of labor and environmental damage shown by the slope of lines C_AC_A and C_BC_B in figure 13-7. It follows that regulation by B is necessary and sufficient to change comparative costs. Regulation by A is unnecessary and superflous. (Thus, in imposing controls it may be important to examine the technologies not now in use, or the production opportunities in other countries not now economic, to forecast the effect of such controls.)

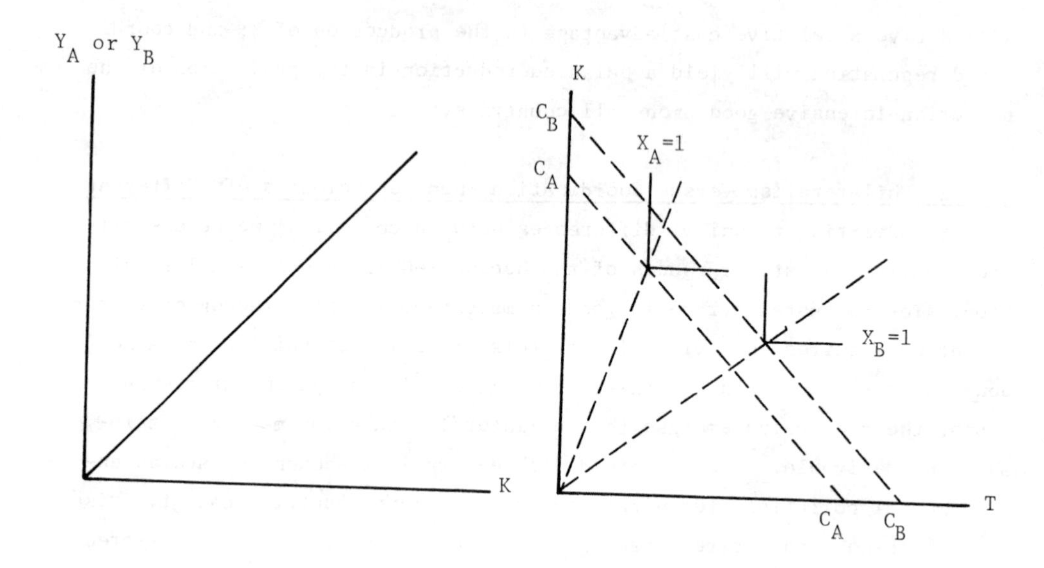

Figure 13-7. Example of technical difference between countries A and B in their exploitation of the environment in production. Left, production of Y; right, production of X.

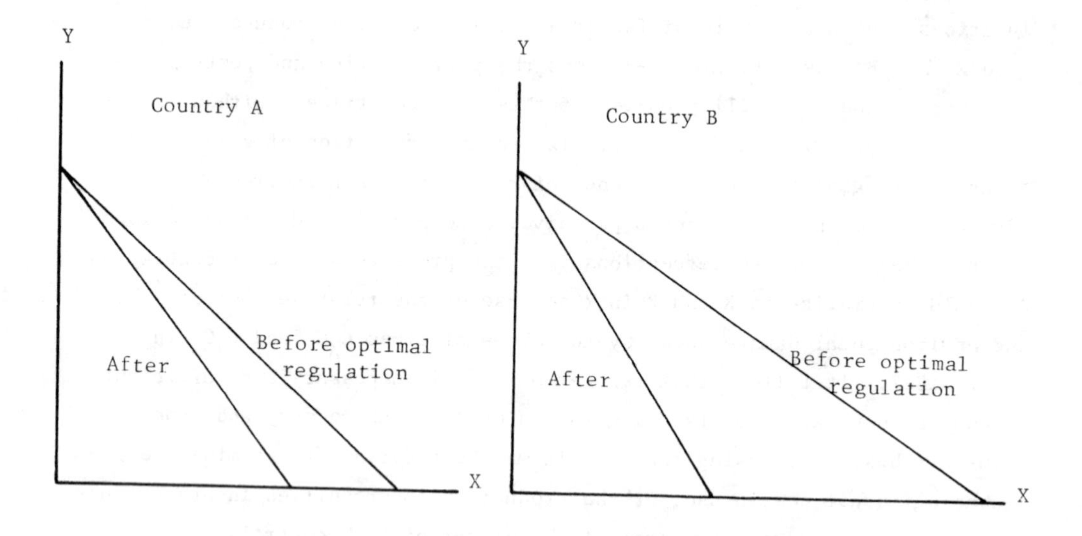

Figure 13-8. An example of false comparative advantage when technologies differ.

13.5 Summary, Conclusions, and Policy Recommendations

This chapter has examined the economic consequences of environmental regulation within an international general equilibrium framework. The study has been motivated by recent preliminary unilateral regulatory steps of the United States government to reduce the production and sale of ozone depleting chemicals.

Our strategy in this chapter has been to apply the two sector Hecksher-Ohlin model under a variety of assumptions as to factor mobility and trade. First, however, we have had to modify that model to incorporate the environment as a factor of production, and to relate the level of offensive emission to the level of output in the offending industry. We made the assumption therefore that polluting emissions and some other factor--most expectably capital--are combined in fixed proportions.

Pursuit of this line of thinking has led us to several analytical insights and conclusions. These have both technical and policy interest. First we have shown that environment-conserving regulation has the same effect on the industry producing the offensive product as a partial factor tax--on capital in the "offensive" sector--when capital and emissions are used in fixed proportions.

Second, when factors are not mobile as in the short run a _de facto_ partial factor tax (and increased costs of producing an offensive product), changes the short-run equilibrium between supply and demand such that in all probability the relative rewards (wages, rent, real interest payments) to all other factors of production change throughout the entire economy.

Third, in the long run when factors are mobile, U.S. unilateral regulation of environmentally damaging products is expected to be self-defeating and completely ineffective unless it is offset by precisely tailored trade and commercial policy. The analysis indicates wherever capital or any other factor becomes internationally mobile, then the country which regulates unilaterally will drive the regulated industry from its borders _completely_. Unilateral regulation, or even multilateral regulation if not properly coordinated among countries, in the long run is totally ineffective within the Hecksher-Ohlin framework.

The assumptions behind this model, of course, depart from reality (as does any model), so that the predictions made by the model (for example, as to the consequences of unilateral regulation) must be tempered by any countervailing evidence. Accordingly, until the predictions from this analysis can be tested through the collection and analysis of data, the inferences derived from our theoretical analysis should be regarded primarily as a warning that unilateral regulation may be totally ineffective. This consideration leads to a fourth conclusion or recommendation of this study. That is to verify these theoretical findings (e.g., to estimate by how much the real economic consequences of unilateralism diverge from our predictions), a detailed statistical specification of the process we describe should be undertaken, the appropriate data collected, and the model tested against the data. Appropriate multicountry data on production, consumption, and international trade in a variety of regulated products over a period of years should be collected. This data should discriminate between (1) products which have been regulated by only one country or in an uncoordinated fashion among countries, and (2) products which have been regulated in a coordinated (tacit or explicit) manner. The trade model should be specified in a form which would allow the locational and trade consequences of differential regulation to be estimated statistically. This would allow the analyst to filter out or correct for other factors influencing trade and the location of production and of consumption--such factors as world population growth, per-capita income change, exchange rate fluctuations, and so on. With empirical estimates of the trade consequences in hand, policymakers then could assess accurately the risks of unilateral U.S. regulation, and discriminate between those cases in which international cooperation is merely useful and those in which it is crucial. An adequate integration of environmental and trade policy requires statistical verification of the relocative effects of regulation.

References

1. A good introductory discussion of the Hecksher-Ohlin model appears in Richard E. Caves and Ronald W. Jones, World Trade and Payments: An Introduction (2 ed., Boston, Little Brown, 1977).

2. A.C. Harberger, "The Incidence of the Corporation Income Tax," Journal of Political Economy vol. 70 (1962) pp. 215-240.

3. Other related materials include M. McGuire, "Regulation, Factor Rewards, and International Trade," Journal of Public Economics (1982); and "Factor Taxes, Factor Migration, and Welfare," Canadian Journal of Economics (1982).

4. Harberger, "The Incidence;" and for growing economies, see Peter A. Diamond, "Incidence of an Interest Income Tax," Journal of Economic Theory vol. 2, no. 3 (September 1970) pp. 211-224; Marian Krzyzaniak, "The Long-Run Burden of A General Tax on Profits in a Neoclassical World," Public Finance vol. 22, no. 4 (1967) pp. 472-491, and "The Burden of a Differential Tax on Profits in a Neoclassical World," Public Finance vol. 23, no. 4 (1968) pp. 447-473, and "Factor Substitution and the General Tax on Profits," Public Finance vol. 25, no. 4 (1970) pp. 489-514; Kazuo Sato, "Taxation and Neo-Classical Growth," Public Finance vol. 22, no. 3 (1967) pp. 346-370; Martin S. Feldstein, Tax Incidence in a Growing Economy with Variable Factor Supply," Quarterly Journal of Economics vol. 88, no. 4 (November 1974), pp. 551-573; J. Gregory Ballentine, "The Incidence of a Corporation Income Tax in a Growing Economy," Journal of Political Economy vol. 86, no. 5 (October 1978) pp. 863-875; J. Gregory Ballentine and Ibrahim Eris, "On the General Equilibrium Analysis of Tax Incidence," Journal of Political Economy vol. 83, no. 3 (June 1975) pp. 633-644. For economies with immobile factors, see C.E. McLure, Jr., "The Theory of Tax Incidence with Imperfect Factor Mobility," Finanzarchiv vol. 30 (1971) pp. 27-48; and Melvyn B. Krauss, "Taxes on Capital in a Specific Factor Model with International Capital Mobility," Journal of Public Economies vol. 11, no. 3 (June 1979) pp. 383-394. For trading economies, see P.M. Mieszkowski, "The Comparative Efficiency of Tariffs and Other Tax-subsidy Schemes as a Means of Obtaining Revenue or Protecting Domestic Production," Journal of Political Economy vol. 74 (1966) pp. 587-599, and "On the Theory of Tax Incidence," Journal of Political Economy vol. 75 (1967) pp. 250-262.

5. See P.M. Mieszkowski, "On the Theory of Tax Incidence;" S.P. Magee, "Factor Market Distortions, Production, Distribution and the Pure Theory of International Trade," Quarterly Journal of Economics vol. 85 (1971) pp 623-643, and "Factor Market Distortions, Production and Trade: A Survey," Exford Economic Papers vol. 25, pp. 1-43; R.W. Jones, "International Capital Movements and the Theory of Tariffs and Trade," Quarterly Journal of Economics vol. 81 (February 1967) pp. 1-38; and R.W. Jones, "Distortions in Factor Markets and the General Equilibrium Model of Production," Journal of Political Economy vol. 79 (1971) pp. 437-459.

6. The effects of such taxes in the size of an economy and the welfare of its citizens is explored in greater detail in McGuire, "Factor Taxes, Factor Migration and Welfare."

7. See P.M. Mieszkowski, "On the Theory of Tax Incidence."

8. A.C. Harberber, "The Incidence of the Corporation Income Tax;" and M.B. Krauss and H.G. Johnson, "The Theory of Tax Incidence: A Diagrammatic Analysis," Economica vol. 39 (1972) pp. 357-382.

9. The analysis here is reminiscent of J. Chipman's of the conditions under which free commodity trade and free migration of one factor of production are compatible with nonspecialization of production in both countries. Our analysis, however, is not identical. Chipman assumes a fixed total amount of capital worldwide, to be allocated among countries. Our scenario assumes a variable quantity of capital worldwide available at a fixed price. See J Chipman, "International Trade with Capital Mobility: A Substitution Theorem," in J. Bhagwati, R. Jones, R. Mundell, and J. Vanek, eds., Trade, Balance of Payments, and Growth (Amsterdam, North-Holland, 1971) pp. 201-237.

10. R.W. Jones and R. Ruffins, "Trade Patterns with Capital Mobility," in M. Parkin and A.R. Nobay, eds., Current Economic Problems (New York: Cambridge University Press, 1975).

Chapter 14

THE EFFECTIVENESS OF EFFLUENT CHARGES WHERE "INDUSTRY" STRUCTURE VARIES

Wallace E. Oates
and
Diana L. Strassmann

14.1 Introduction

Economists have long argued for the use of pricing incentives for the control of pollution. The economic case is a compelling one: large savings in costs of abatement and a continuing incentive for the development of new techniques to reduce pollution.[1] By making pollution costly, a set of pricing incentives can mobilize the forces of the market so that abatement activities become directly profitable for individual firms and consumers.

Environmental policy, however, has relied almost exclusively on direct controls, and this may well remain the preferred option for the regulation of chlorofluorocarbon (CFC) emissions. The EPA strategy generally has been one of distinguishing "essential" from "nonessential" uses of CFCs, and simply banning the latter. This approach to curtailing the use of CFCs raises some difficult issues. Martin Bailey's estimates suggest (see chapter 6) that a complete cessation of CFC emissions appears unjustified by a reasonable benefit-cost calculation. This leaves the environmental authority with the task of achieving a partial cutback in emissions. Current policy seeks to attain this partial cutback by a set of direct controls specifying allowable uses for CFCs.

Author's Note: The authors wish to thank Peter Altroggen, William Baumol, Robert Dorfman, Albert McGartland, Eugene Seskin, and Jeffrey Smisek for their helpful comments on earlier drafts of this paper.

An alternative approach would rely on pricing incentives to realize the desired reduction in emissions. Under this approach, the market would effectively make the distinction between essential and nonessential uses, relieving the EPA of some subtle and often arbitrary decisions concerning permissible and nonpermissible uses of CFCs. The market approach could take any of several forms: a tax or "effluent fee" on the use of CFCs, a system of marketable emission permits (under which the EPA issues a limited stock of emission permits which can be bought and sold), or a deposit scheme of the sort suggested by Peter Bohm in chapter 11. Although the analysis in this chapter will run mainly in terms of an effluent fee system, we stress that the results are more generally applicable to other forms of pricing incentives.[2] We also note that although EPA's first instinct is toward direct controls, there has been some real interest in a system of economic incentives; a proposal, for example, appeared in the Federal Register in 1980 for a system of marketable permits with the recognition that a tax or fee on the use of CFCs represents another policy option.[3]

14.2 Market Incentives and "Industry" Structure

The focus of the analysis is the efficacy of pricing incentives in an economy where widely varying types of institutions are engaged in polluting activities. We mentioned earlier the desirable properties of a system of fees for the regulation of environmental damage, but we emphasize that the analysis of programs of effluent charges (and of alternative pricing schemes including marketable pollution permits) typically runs in terms of profit-maximizing firms in competitive markets. The establishment of the optimality properties of effluent fees usually depends on this framework and on a number of other fairly restrictive assumptions.

A cursory inspection of the real world reveals that the major sources of pollution encompass a wide variety of institutional structure. Large private firms, private but publicly regulated firms and even public-sector agencies are all massive contributors to environmental degradation. The application of the competitive model with its myriad of small firms acting as price-takers is therefore suspect for many classes of polluters.

These observations are clearly relevant to CFC emissions. Both the production and use of CFCs (as Walter and his coauthors show in chapter 3) involve a wide variety of enterprises that hardly fit the standard description of the "competitive firm." In 1976, for example, three large companies (Dupont, Allied Chemical, and Union Carbide) accounted for 85 percent of the production of CFCs in the United States.[4] On the use side, CFCs find their way into air-conditioners in motor vehicles, refrigeration units, aerosols, solvents, and various foaming agents. The users of CFCs thus engage in a broad range of activities that encompass a variety of enterprises in both the private and public sectors.

The economic analysis of market incentives for the control of CFCs and other pollutants must, therefore, push beyond the simple competitive model. We must ask how polluters with widely varying sets of objectives are likely to respond to these incentives. In this paper, we explore what some standard models of organizational behavior tell us about how decision-makers in different forms of institutions would respond to a set of effluent charges for protection of the environment. We then use these results to evaluate the implications of these responses for efficient resource allocation, taking explicit account of the distortions that market imperfections, bureaucratic behavior, and public regulation of private firms themselves introduce.

14.3 The Analytical Framework

For purposes of the analysis, we shall identify four different classes of polluters:

1. Competitive (profit-maximizing) firms
2. Imperfectly competitive firms
3. Public bureaus or agencies
4. Regulated firms

We shall use the first class, competitive firms, as a benchmark for purposes of comparison. Our second group of polluters, imperfectly competitive firms, make decisions which have some effect on the price they receive in output markets. We shall examine both imperfectly competitive firms that maximize profits and also those firms which pursue some form of "satisficing" behavior (like that embodied in Williamson's hypothesis that firms maximize an objective function that contains a number of variables in addition to profits).[5] To analyze the third class of polluters, public bureaus or agencies, we shall examine initially a variant of the Niskanen model of bureaucratic behavior.[6] This model postulates that public officials maximize an objective function containing, as arguments, their bureau's level of output and the level of "perquisites." We shall then explore the implications of Wilson's somewhat broader and richer view of bureaucratic behavior that relates bureaucratic decisions to the context in which the bureau is formed, to the economic and political environment in which the bureau operates, and to its internal hierarchy of decision-making.[7] Finally, we shall adopt the standard Averch-Johnson model of the regulated firm to analyze our fourth class of polluters.[8]

In addition to predicting the general form of response of individual polluters, it is important to establish the properties of the pattern of pollution-abatement activities across the many and varied polluting agents in the economy. This requires some point of reference. We shall employ two criteria in this study. The first is the global property of Pareto optimality. In a competitive world with perfect knowledge, Pareto optimality requires _both_ that the marginal social damage (MSD) from pollution be brought into equality with the marginal cost of abatement activities and that abatement quotas be allocated among polluters in the least-cost manner. Subject to a number of important qualifications, it can be shown that, in a competitive world, these conditions can be achieved through the use of a Pigouvian tax, or effluent fee, on polluting activities.[9] By internalizing the external cost, a unit tax on emissions equal to MSD can sustain the Pareto optimal level _and_ pattern of waste discharges by equating marginal abatement costs across all polluters.

In practice, however, knowledge is far from perfect, and this creates some formidable obstacles to the implementation of the Pigouvian scheme. First, the measurement of social damages is very difficult. The most serious of the harmful effects of pollution often takes the form of adverse health effects--the increase of mortality rates and the incidence of various diseases. In the case of ozone depletion, for example, the most frightening of the possible consequences involves the increased incidence of a variety of skin cancers caused by higher levels of ultraviolet radiation at the earth's surface. Polluters can also impose aesthetic costs by disfiguring landscapes. Quantification of these damages involves both uncertainty and the use of questionable value judgments. There are impressive attempts at such valuations,[10] but a comprehensive set of estimates of the MSD of a wide range of polluting activities is unlikely to be available in the near future.

Second, the optimal Pigouvian tax does not reflect the MSD at existing levels of pollution; rather, it equals the MSD at the optimal level of pollution. The determination of this optimal level is a perplexing issue. For example, it is not even clear that convergence to an optimal solution would occur from a process of iteration in which the effluent fee is set equal to MSD at each step.[11]

And third, the optimality of the Pigouvian solution requires that no distortions exist elsewhere in the economy; otherwise we enter the murky waters of the second-best. The problem of distortions assumes special importance in cases of externalities, since the presence of external effects raises the likelihood of violations of second-order conditions. In particular, we can say unequivocally that if the externalities are of "sufficient magnitude," the second-order conditions necessary for Pareto-optimality definitely will not hold.[12] In this case, the Pigouvian prescription may make the malady worse (for example, it might induce inefficient location patterns of polluting and nonpolluting activities).

As a practical alternative to the Pigouvian prescription, Baumol and Oates,[13] have advocated another, admittedly second-best, technique for using pricing incentives: the charges and standards approach. Here, the environmental authority determines a set of standards for environmental quality and then achieves these standards through the imposition of a system of effluent fees. The advantages of this approach are straightforward. It circumvents the need to measure explicitly the social damages of pollution. The environmental authority arrives independently at a set of environmental targets, presumably taking into account available evidence on the extent of damages. This procedure corresponds roughly to the way in which environmental policy has proceeded in most countries: authorities establish standards for ambient air and water quality which serve as objectives for the determination of levels of abatement activities.

Subject to some important qualifications, the charges and standards approach assures the achievement of the designated quality standards at the minimum aggregate abatement cost. A profit-maximizing firm will extend pollution abatement to the point where the marginal abatement cost equals the effluent fee; since all firms face the same fee, profit-maximizing behavior implies an equality of marginal abatement cost across all polluters. Assuming increasing marginal costs, it follows that abatement quotas cannot be reallocated among polluting firms so as to reduce the total costs of abatement. The environmental authority simply adjusts the fee to the level that reduces aggregate discharges to an amount consistent with the prescribed standard for environmental quality, and individual responses to the fee automatically generate the least-cost pattern of abatement among polluters.

The obvious deficiency of the standards and charges approach is that it ignores any distortions associated either with the level of pollution or the composition of final outputs: it focuses solely on minimizing the resource costs of pollution abatement in attaining a predetermined standard. Furthermore, uncertainty about the final magnitude of the effluent fee might inhibit firms from adjusting input combinations to current prices and from engaging in the appropriate levels of pollution-abatement research and development. Such responses on the part of firms could, in turn, make it difficult for the environmental authority to know how and when to adjust fees. Such "adjustment costs" admittedly detract from the appeal of the charges and standards technique; the magnitude of these costs may, however, be modest when compared with the control and abatement costs under other approaches. In particular, a market-incentives approach avoids the arbitrary nature of direct controls and the inevitably costly legal battles over individual decisions. In addition, direct regulations (especially those which require firms to install certain equipment) give firms few incentives to engage in long-run research and development of pollution-abatement technology. Effluent fees may not induce the optimal levels of such R&D either, but they do provide at least some incentives. Although the standards and charges approach is a second-best policy measure, it does point to a workable environmental policy that maintains some of the important properties of market incentives.

14.4 Effluent Fees Case-by-Case

14.4.1 Competitive Firms: A Further Qualification

In the previous section, we noted that in a competitive world with perfect knowledge, a Pigouvian tax can sustain a Pareto-optimal allocation of resources. Likewise, from the more limited perspective of the standards and charges approach, the establishment of a fee at a level sufficient to reduce aggregate emissions by the requisite amount will generate the least-cost pattern of abatement quotas among firms. Teitenberg and Ackerman have pointed out an important qualification to these results.[14] The optimality properties of both Pigouvian taxes and the standards and charges approach depend on the condition that a unit of waste discharge degrades the environment by the same amount, irrespective of the location or time of its emission. Obviously, this assumption need not, and often will not, hold. A firm which emits its wastes upstream may do more harm to water quality than one which discharges the identical wastes further downstream. Likewise, a factory spewing a given quantity of nitrous oxides into the air will have a more deleterious effect on air quality if it is located to the windward side of an airshed rather than leeward. Least-cost achievement of a given environmental quality standard requires that the unit fee be applied to the effect of the emission on environmental quality, not just on the _quantity_ of the emission. A schedule of fees, for example, might differentiate among zones of emission. Where such finer distinctions seem necessary, Pigouvian taxes and the standards and charges approach lose some of their simplicity.

In the case of the ozone layer and of climate this qualification is of little moment, for the two involve ideal examples of pure public goods. The ozone shield provides its protection on a global scale; the extent of damage to the shield and of climate change from CFC emissions appears independent of the location of the emissions. The atmospheric mixing processes seem sufficient to ensure that CFC emissions will ultimately exert their destructive impact on the global ozone shield and their likely adverse impacts on climate, regardless of their geographic source. The policy implication is that, in a competitive system, the fiscal penalty for CFC emissions should be everywhere the same.

14.4.2 Imperfectly Competitive Firms

The introduction of imperfectly competitive firms generates complications. Both classical profit-maximizing and satisficing models of firm behavior indicate that imperfectly competitive firms are unlikely to produce socially optimal levels of output. As Buchanan has pointed out,[15] the monopolist's suboptimal level of output is the source of a basic dilemma for the formulation of policy to regulate externalities. An effluent fee provides an incentive for needed pollution abatement, but, at the same time, raises the firm's marginal cost and thereby induces a reduction in output. The result is some gain in efficient resource allocation from reduced waste emissions, but some loss in efficiency from the contraction in output; the *net* effect on social welfare is uncertain. In short, an effluent fee (Pigouvian or otherwise) may represent too much of a good thing.

Moreover, a number of models of oligopolistic behavior indicate that firms in such industries may not produce at minimum cost, either intentionally because of managerial desires for such things as perquisites, or unintentionally because limits to rational decision-making cause firm managers to follow "rules of thumb" and "fire-fighting" approaches to the making of decisions. Other models suggest that excess capacity (one form of "technical" or "X-inefficiency") may stem from long-run, profit-maximizing behavior. By serving as a barrier to potential entrants, excess capacity may help firms in concentrated industries maintain above-normal profits. Also, if profits are high, it may pay for firms to keep extra capacity as a contingency against possible increases in demand.

In this section, we shall explore some of the many models of imperfectly competitive behavior and the implications of these models for the efficacy of effluent charges in reducing pollution from individual firms.

14.4.2.1 Profit-Maximizing Models. So long as firms maximize profits (and, hence, minimize costs), imperfectly competitive elements in output markets introduce no problems under the standards and charges scheme. Since all polluters set marginal abatement cost equal to the same fee, it follows that the predetermined level of environmental quality will be achieved at minimum aggregate abatement cost.

From the perspective, however, of full Pareto optimality, we must take into explicit account any distortions in the composition of final outputs, and this brings us to Buchanan's tradeoff between pollution abatement and monopolistic output restriction. Following Baumol and Oates,[16] we depict the nature of this tradeoff in figure 14-1. Let DD represent the industry demand curve confronting a monopolist, with DMR being the corresponding marginal-revenue curve. We assume that the monopoly can produce at constant cost (PMC = private marginal cost), but that its production activities impose costs on others. In particular, in the absence of any fees, the monopolist's (private) cost-minimizing technique of production generates pollution costs per unit equal to AB so that the SMC (social marginal cost) curve indicates the true cost to society of each unit of output. To maximize profits, the monopolist would produce OQ_m.

Suppose next that we subject the monopolist to a pollution tax, a fee per unit of waste emissions. This will provide an incentive to alter the production process in a way that yields lower emissions per unit of output. In figure 14-1, this would have two effects: it would raise the PMC curve and, over some range, would tend to lower SMC. This second effect results from the choice of what, from society's standpoint, is a lower-cost method of production (taking into account the costs of pollution). The minimum social cost of production will be reached when the pollution costs are wholly internalized so that $PMC_t = SMC_t$ (where the subscript t refers to costs in the presence of a Pigouvian tax). At this point, the firm's selection of a production process will be based upon a set of input prices (including a price of waste emissions) that reflects true social opportunity costs.

In figure 14-1, we see that the optimal output is OQ_o, which is produced at the least social cost. To achieve this optimum, we would require two policy actions: a Pigouvian tax on waste emissions in order to reduce SMC to SMC_t and a subsidy per unit of output equal to GF (the difference between marginal cost and marginal revenue at the optimal level of output). Because we have two types of distortion, full correction generally requires two policy instruments.

An environmental protection agency, however, will typically have neither the authority nor the inclination to offer subsidies to monopo-

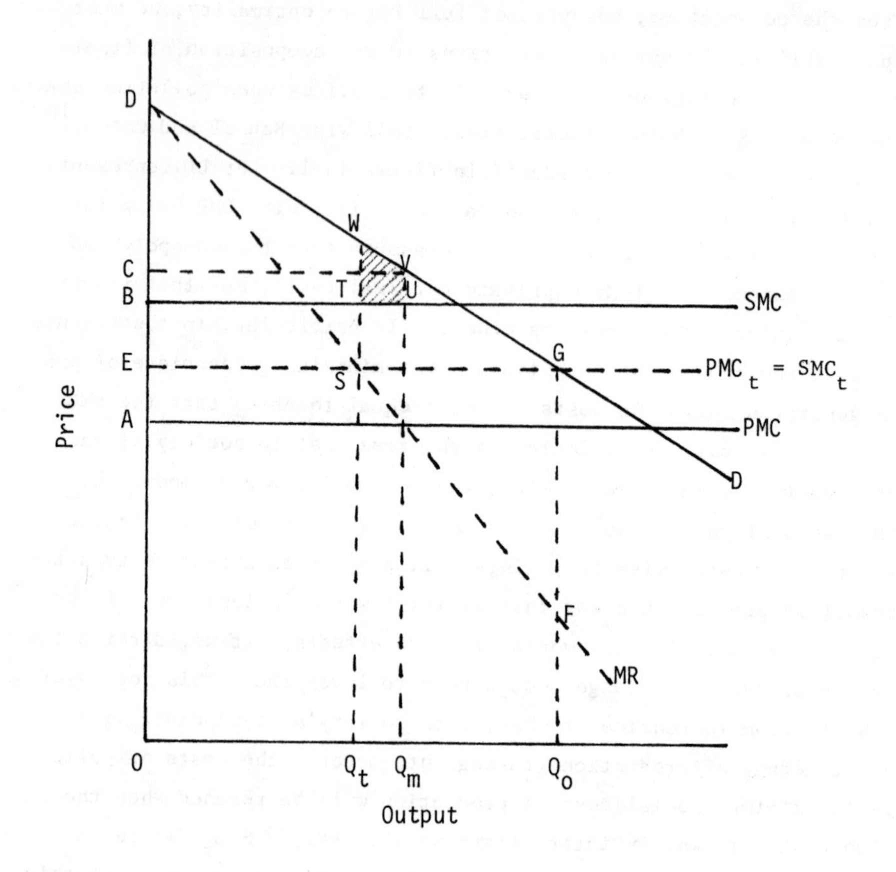

Figure 14-1. Monopolistic distortions.

lists. Suppose, more realistically, that it is empowered only to tax waste emissions. What would be the effect on social welfare of a standard Pigouvian tax on pollution? We see in figure 14-1 that this would result in a reduction in output from OQ_m to OQ_t. At this output, the tax would generate a cost saving to society indicated by the rectangle EBTS. On the other hand, it would be accompanied by a welfare loss represented by the trapezoid UVWT; this is the loss in consumer surplus resulting from the contraction in output to OQ_t. The net effect on social welfare obviously depends on the relative size of these two areas, leaving the environmental authority with the difficult problem of estimation of surpluses for its policy alternatives.

This raises as an empirical matter the issue of the relative magnitudes of the likely gains and losses from these two effects of a system of effluent charges. A definitive answer would require a general-equilibrium analysis that explicitly examined the effects of fees on all industries, incorporating estimates of both cost parameters and price elasticities of demand. Such a study extends beyond the scope of our efforts here. But we would like to offer some admittedly quite crude, partial-equilibrium calculations that we find suggestive; they point to losses in consumers surplus associated with any added restrictions in outputs that are quite small relative to the gains from a system of effluent fees in terms of reductions in the social costs of pollution. If the rough orders of magnitude are correct, the calculations would imply that monopolistic restrictions of output do little real damage to the case for pricing incentives for pollution control.

Returning to figure 14-1, note that the imposition of a Pigouvian tax generates a welfare gain to society equal to the area of the rectangle (EBTS); the tax also causes an additional reduction in output with a welfare loss equal to the area of the trapezoid (WTUV). The sizes of these two areas obviously depend on a number of key variables, including the magnitude of the cost increase from the fee, the level of marginal social damages, and the price elasticity of demand. More systematically, we can approximate the two areas in figure 14-1,

$$\Delta W_g = Q[SMC - SMC_t] \qquad (1)$$

$$\Delta W_\ell = \Delta Q[P - SMC], \qquad (2)$$

where ΔW_g is the welfare gain from pollution abatement, and ΔW_ℓ equals the welfare loss from reduced output. To determine the relative sizes of the two effects (ΔW_r), we divide Equation (2) by Equation (1), and then, to facilitate comparisons, divide both numerator and denominator by the price of output, P, yielding:

$$\Delta W_r = \left(\frac{\Delta Q}{Q}\right) \frac{[P-SMC]/P}{[SMC-SMC_t]/P} \qquad (3)$$

We now make some educated guesses on orders of magnitude of the terms in Equation (3). We note, first, that the percentage change in quantity in Equation (3) will depend upon the values of two other variables: the percentage change in price, and the price elasticity of demand (e_d). There are available a number of estimates of the increase in costs and prices for various polluting industries associated with the abatement necessary to achieve U.S. environmental standards. These range from 1 percent to a high of about 10 percent.[17] The price elasticity of demand will also vary among industries, but for the major industries of concern, including power generation, chemicals, pulp and paper, and so forth, a typical value of unity is probably a reasonable assumption. This gives us an estimated range for $\Delta Q/Q$ of:

$$\frac{\Delta Q}{Q} = |e_d| \frac{\Delta PMC}{PMC} = 0.01 \text{ to } 0.1 \qquad (4)$$

Turning to the numerator of the second term in Equation (3), we note that [P-SMC] is equal to that portion of the difference between price and private marginal cost that is not offset by the marginal social damage of waste emissions; that is,

$$\frac{[P-SMC]}{P} = \frac{[P-(PMC+MSC)]}{P} \qquad (5)$$

where MSC is the marginal external cost equal to AB in figure 14-1. We can make a very rough guess on appropriate magnitudes by noting that existing estimates of the benefits from air pollution control in the United States are about twice the level of abatement costs; Lave and Seskin,[18] for example, offer a conservative estimate of health benefits alone from meeting standards for ambient air quality of about $16 billion, as compared to EPA's estimate of $9.5 billion for abatement costs. Recalling that estimated abatement costs and the associated rise in prices in polluting industries range from 1 to 10 percent of price, we can estimate that MSD can range up to 20 percent of the price of industry output, depending on whether the industry is a heavy polluter. In the case where the industry generates a substantial quantity of damaging waste emissions, it is hard to believe that (P-SMC) is very large. If, for example, the industry's markup over private marginal cost is even as high as 20 percent, most or all of this markup is likely to be offset by MSD. Suppose we take for (P-SMC)/P what would seem a fairly generous typical value of 0.1.

The denominator of Equation (3) will, likewise, vary with the extent of pollution damages generated by the industry. If we assume, again, that benefits from abatement are roughly double their costs, then, for a heavy polluting industry, the reduction in social costs from abatement can run from 10 to 20 percent of price. If we take the lower figure as a typical value, we find that the numerator and denominator of the second term of Equation (3) are of roughly the same order of magnitude.

Taking the second term on the right of Equation (3) to be approximately unity implies that the size of the welfare loss from restricted output relative to the gain from abatement is of the same general magnitude as $(\Delta Q/Q)$. This suggests that the gains from abatement are likely to dwarf the losses from monopolistic distortions. Our estimate is that the latter losses are likely to range from only 1 to 10 percent of the welfare gains.

We must acknowledge the crudeness both of our partial-equilibrium measurement procedure and the underlying estimates of "representative" costs and benefits of abatement; we are trying to develop (both in conceptual and empirical terms) some better measures of these quantities. However, taking our results as a tentative judgment on the relevant orders of magnitude and placing this information in the context of the policy options available to the environmental authority, it seems to us to suggest that we should simply ignore the issue of monopolistic output distortions associated with a system of fees. The authority will surely be unable to vary the fee by market structure; the only real possibility is probably that of a discriminatory system of direct controls in which the environmental agency tries to tailor individual permits for polluters to take into account market structure. However, the potential gains from discriminatory permits to account for monopolistic distortions are surely too small to warrant such a cumbersome process with all its attendant inefficiencies.

14.4.2.2 Managerial Models of Maximizing Firms. Although distortions in outputs may not be a major issue, another line of work focuses on the technical inefficiencies inherent in the decisions of the firm.[19] Since simple profit-maximizing models provide no reasonable explanation for the existence of technical inefficiencies, we turn now to an exploration of models of the firm that are based on more complex objective functions.

The Marris model of the growth-maximizing firm has a number of theoretical justifications.[20] The favorable tax treatment of capital gains implies that at least some growth is necessary for the maximization of stock value. Moreover, Marris has postulated that managers of firms are likely to prefer an even higher level of growth than that which would maximize the value of company stock, because of the prestige and other perquisites associated with large companies. Managers are constrained, however, by the threat that the company will be taken over if the difference between the actual growth rate and that which maximizes stock value becomes too high. But since a takeover requires a premium to be paid on some of the stock purchased, managers are able to let the firm grow somewhat faster than the optimal rate from the stockholders' perspective.

Williamson has developed further the idea of managerial preferences by postulating an explicit managerial utility function.[21] The utility function formalizes what Williamson describes as managerial "expense preference." These preferences comprise expenditures on staff (S), managerial emoluments (M) (extra salary and perquisites), and "discretionary profits" (the difference between actual profits and minimum profits demanded). Discretionary profits enter directly into the managerial utility function: higher profits are a measure of executive achievement, and discretionary profits represent a source of funds which may be used both to increase perquisites and reported profits.

Although the above models vary in structure, an analysis of the Williamson model gives the flavor of the results that might be expected from firms whose managers maximize an objective function containing variable(s) other than short-run profits. Williamson postulates that firms maximize:

$$U = U[S, M, \pi_R - \pi_0 - T] \qquad (6)$$

subject

$$\pi_R \geq \pi_0 + T$$

or

$$U = U\{S, M, (1-t)\ [R(X) - C(X) - S - M] - \pi_0]\} \qquad (7)$$

where R = revenue = $p.X$; $\partial^2 R/\partial X \partial S \geq 0$

p = price = $p(X, S; \varepsilon)$; $\partial P/\partial X < 0$; $\partial P/\partial S \geq 0$; $\partial P/\partial \varepsilon > 0$

X = output

S = staff (in money terms) or (approximately) general administrative and selling expense

ε = a demand shift parameter

C = production cost = C(X)

M = managerial emoluments

π = actual profits = R-C-S

π_R = reported profits = π-M

π_0 = minimum (after-tax) profits demanded

T = taxes where t = tax rate

$\pi_R-\pi_0-T$ = discretionary profits

U = utility function

[Note that the substitution involved in moving from Equation (6) to Equation (7) assumes second-order conditions satisfied and disallows corner solutions. Williamson points out that an inequality-constrained maximization could be handled by making use of the Kuhn-Tucker theorem.]

To treat waste emissions and the effluent fee explicitly, we amend the Williamson model to distinguish between these emissions (E) and all other inputs (V). Equation (7) thus becomes:

$$U = U\{S,\ M,\ (1-t)[R(g(V,\ E)) - p_V V - fE - S - M] - \pi_0\} \qquad (8)$$

where p_v is the price of other inputs and f denotes the effluent fee. Maximization of this utility function yields as one of the first-order conditions the familiar result:

$$\frac{p_v}{f} = \frac{\partial X/\partial V}{\partial X/\partial E} \qquad (9)$$

This is the usual condition for the cost-minimizing combination of factor inputs: marginal products proportional to factor prices. It may seem surprising at first glance to find that firms that are technically inefficient (that is, do not produce at minimum cost overall) are effectively cost-minimizers with regard to pollution abatement. However, the rationale is quite straightforward: since abatement activities contribute nothing to staff or emoluments and reduce discretionary profits, the firm's managers have an incentive to minimize the expenditure on abatement (consisting of effluent fees plus pollution-control costs). We can thus extend our cost-minimization theorem under the charges and standards approach to encompass certain managerial models of maximizing firms: a world of such firms subject to an effluent fee will achieve the predetermined standard of environmental quality at the minimum aggregate abatement cost. This outcome will not, of course, be Pareto optimal because of technical inefficiencies and a distorted pattern of output.

14.4.2.3 Managerial Models of Nonmaximizing Firms. Nonmaximizing models of firm behavior (generally associated with Herbert Simon's group at Carnegie Mellon University) emphasize the constraints that information costs, time, concentration, and bounded rationality impose on managerial decisionmaking. The result is a "groping in the dark" approach to the making of decisions. Using nonmaximizing models of firm behavior, Cohen and Cyert distinguish between routine and nonroutine decisions.[22] The former follow established, satisfactory rules of thumb. In contrast, nonroutine decisions occur when some unexpected problem arises, and no method of dealing with the problem has even been agreed upon. The internal organizational characteristics of the firm dictate how such nonroutine decisions are made. Major decisions are the outcome of a bargaining process involving a coalition of the different interests served by the firm. The goals and policies of the coalition may shift over time with changes in the composition of employees and the amount of "managerial slack." Interest groups adjust their demands according to the amount of available slack; as a result, economic profits may disappear into a number of pockets.

What are the implications of such a theory for the efficacy of a system of effluent fees? The case for effluent fees rests on the presumption that an increase in the price of effluents will induce firms to use less effluents relative to other inputs; over the longer haul, fees will induce firms to engage in R&D that will allow them to develop even cheaper production technologies. However, although effluents themselves are unlikely to afford utility to any interest groups within a coalition, the incentives to change production policies quickly in response to relative price changes may be quite weak in a nonmaximizing context. Changes in production methods (particularly to less pollution-intensive methods) may involve major changes in equipment. If an important perquisite is "ease of management," the firm might conceivably employ some of the fat in its budget to avoid the effort and possible complications associated with the adoption and development of new abatement techniques. On the basis of the Carnegie Mellon model, there exists no clear presumption that firms cushioned by some managerial slack respond as quickly and as completely as would a competitive firm to the imposition of effluent charges. Although a complete discussion of the empirical literature on this issue is beyond the scope of this paper, some investigations of the diffusion of knowledge and new technologies have indicated that firms in concentrated industries do not respond as quickly to price changes and the availability of new innovations as do firms in more competitive industries.[23] The evidence on this issue, however, is not conclusive; yet the lack of consensus certainly provides some justification for skepticism about the belief that effluent fees will work as well in highly concentrated industries as they might in more competitive cases.

A theory of managerial behavior associated with the Harvard Business School also recognizes the importance of the internal dynamics and structure of the firm.[24] Rather than supposing that this structure determines the firm's strategy, the Harvard Business School model claims that the role of the chief executive and directors is to determine the strategy of the firm. Whereas the Carnegie Mellon theory posits that a firm's structure determines the firm's strategy, the Harvard Business School approach posits a reverse causality: it views the executives and directors as the agents who set major policies and determine the reward structure within the firm.

In a final analysis, the efficacy of effluent fees in reducing pollution from managerial firms depends on how much loss of control exists between top management and those who make the decisions about how "effluent-intensive" current and future production techniques are and will be. Insofar as the Harvard Business School model suggests that the problem of control loss will be less widespread than is predicted by the Carnegie Mellon model, the former model gives greater cause for optimism about the efficacy of effluent fees. Surely some loss of control will occur regardless of which model is more "realistic;" yet the real issue is the magnitude of these inefficiencies relative to the benefits of an approach incorporating market incentives. We shall return to this question again in our discussion of bureaucratic leanings toward "ease of management." What we can say is that all of the models of imperfectly competitive firm behavior that we have discussed establish some presumption that effluent fees will induce firms to pollute less; how closely the outcomes approach a cost-minimizing solution is less clear.

14.4.3 Public Bureaus

Since bureaucrats are neither profit-maximizers nor cost-minimizers, the response of public agencies to pricing incentives is problematic.[25] To explore the impact of effluent fees on public decisionmakers, we first examine a variant of the Niskanen model of bureaucratic behavior.[26] We then discuss the implications of Wilson's richer study of public agencies.[27]

In our variant of the Niskanen model, we postulate that the bureau's decisionmakers seek to maximize an objective function that contains as arguments the bureau's output (Q) and its level of perquisities (P):

$$U = U(Q, P) \tag{10}$$

Bureaucrats desire an increased output (or "size"), for this enhances the bureau's power and prestige and with these its capacity to influence the course of events. Migue and Belanger have contended that agency officials also place a premium on the bureau's "discretionary budget," the excess of the bureau's funding above its necessary costs.[28] This "fat" in the budget can be employed for a variety of perquisites ranging from higher salaries and expanded staff to additional facilities or, perhaps, reduced effort.

We assume, for simplicity, that the bureau produces its output with only two inputs: labor (L) and waste emissions (E), such that:

$$Q = Q(L, E) \tag{11}$$

If, for example, output is held at some given level, then waste emissions can be reduced (for example, recycled) by the employment of additional labor. Moreover, the bureau is subject to a budget constraint:

$$B = wL + fE + cP \tag{12}$$

where B is the bureau's budget, w is the wage rate (given to the bureau), f is the effluent fee per unit of waste emissions, and c is the (constant) marginal cost of perquisites. Note that L is defined to include only the minimally necessary quantity of labor to provide a given output, and w is likewise defined as the lowest wage that will keep employees. Extra salary and labor are viewed as perquisites.

The budget-determination process remains a problem. Here, we shall simply assume that the bureau negotiates with the legislature for an appropriation which specifies both the bureau's budget and its level of output. For a given period, B and Q are predetermined by the legislative budget process; they are not decision variables for the bureau, and we shall therefore designate them as fixed: B_o and Q_o.

With output predetermined, the bureaucrat's maximization problem becomes that of maximizing perquisites (P) subject to the bureau's production and budget constraints. We form the Lagrangian:

$$M = P + \lambda_1[Q_o - Q(L, E)] + \lambda_2[B_o - wL - fE - cP], \quad (13)$$

where λ_1 and λ_2 are the Lagrangian multipliers. Maximizing Equation (13) and solving for the first-order conditions yields:

$$\frac{MP_E}{MP_L} = \frac{f}{w} \quad (14)$$

Equation (14) indicates that the bureau satisfies the same first-order condition as a cost-minimizing firm: it sets the ratio of marginal products of the inputs equal to the ratio of their (given) prices.

This result is essentially the same as that obtained from our earlier analysis of the Williamson model. Equation (14), like that for the managerial model of the maximizing firms, implies cost-minimization in only a limited sense: the bureau minimizes pollution abatement and other costs that do not generate perquisites. Effluent fees are effectively lost dollars; they provide no utility to the bureaucrat. By minimizing pollution-abatement costs, the bureau maximizes the remaining budget for the "procurement" of perquisites. Like cost-minimizing firms, a bureau behaving according to this model has an incentive to extend pollution-abatement activities to the point where marginal abatement cost equals the effluent fee.

An important qualification to this result introduces an indeterminacy similar to that in the satisficing models of firm behavior. Bureaucrats, like the employees of firms, are likely to have some preferences for the perquisite "ease of management." Just how pervasive such behavior is in public agencies is unclear; however, it could introduce some inefficiencies in the allocation of abatement quotas among polluters. Although, as the analysis suggests, bureaucrats are likely to have some incentive to economize on abatement costs, the discussion in the previous section on diffusion of innovations is also likely to apply to bureaus. Bureaus, like firms protected by managerial slack, do not need to respond as quickly to price changes and to the availability of new technologies as do competitive firms, since the survival of bureaus does not, in general, depend on an aggressively tight management.

Whatever the effectiveness of effluent fees in inducing pollution abatement from public agencies, a system of fees will have an impact on public output by increasing the bureau's cost of operation. Most models of bureaucracy imply unambiguously that the bureau's equilibrium output will exceed the socially desirable level.[29] The rationale for this result is straightforward. Bureaucrats are typically assumed to maximize an objective function containing, as arguments, their budgets or outputs (along with some other variables like perquisites); they trade off part of the "fat" in their budgets for increased size. That is, bureaucrats use some of their surplus to produce units of output for which marginal cost exceeds the marginal valuation to society.

By raising costs, a system of effluent fees will eat into the agency's fat. These higher costs will reduce the decisionmaker's capacity to generate excessive output, perquisites, and other items of value to the bureaucrat. From this perspective, fees for pollution may generate some beneficial side effects in terms of reducing the extent of excess production and of lowering the bureau's cost for any given level of output.

In an important new study, Wilson and coauthors have challenged the simple models of bureaucratic behavior for their failure to account for the immense variation in regulatory policies.[30] The Wilson volume is a collection of papers, each examining a public agency to determine how the behavior of these agencies can be related to a more general theory of the policies of regulation. These case studies support Wilson's thesis that public agencies will exhibit widely varying behavior depending on the context in which the bureau is formed, the economic and political environment in which the bureau operates, and the internal hierarchy of decision-making within the bureau. Particularly important to the behavior of a bureau is the perceived distribution of costs and benefits of different policies. Whenever the potential costs or benefits of a proposed policy are highly concentrated within a small segment of the population, this group can be expected to attempt to pressure the bureau into making the most favorable decisions for the group. When costs or benefits are more diffuse, transactions costs (and the free rider problem) reduce the expected gain to individuals from participation in special interest groups. Although the distribution of costs and benefits may affect bureaucratic decision-making, and _ceteris paribus_, ultimately affect final policies, Wilson is quick to emphasize the importance of the political climate in which a bureau was formed and operates. Legislative mandates can severely limit the scope of bureaucratic policies; instructions to bureaus can be very vague or very specific. The composition of employees is also important, particularly the mix of career bureaucrats, politicians, and professionals working in the bureau. Wilson suggests that we view public agencies as "coalitions of diverse participants who have somewhat different motives."[31] Yet whatever political pressures are expressed by special interest groups or from within the bureau itself, the policies the bureau follows may be directed by a chief administrator whose ideas do not coincide with, or reflect those of, special interest groups or coalitions within the bureau. In fact, the chief administrator might be appointed specifically to change the policies of the bureau, as was the case when Alfred Kahn was appointed to the Civil Aeronautics Board.

No simple predictions about the efficacy of effluent fees can be made on the basis of Wilson's broader perspective on the behavior of public agencies; on the contrary, the study vividly illustrates the difficulty of generalization. An examination of the particular circumstances underlying the behavior to be predicted is especially important for public agencies, where the motivations and pressures on decisionmakers are even more complex and varied than they are in the private sector. The precise optimality properties of effluent fees in a competitive world will most assuredly not hold in a world containing public agencies behaving according to the theories discussed in this section, but effluent fees may still have some compelling advantages over direct controls even for public agencies. Fees, for reasons spelled out in the analysis, are likely to provide systematic and continuing incentives for abatement. Moreover, the automatic nature of fees compared with the costs and political complexities of imposing direct and arbitrary controls may yet outweigh the problems of fees working imperfectly in the public sector.

14.4.4 Regulated Firms

The presence of regulated firms introduces some further complications both for the minimization of pollution abatement costs in the standards and charges approach and for the global optimality properties of a Pigouvian tax system. Regulated firms are not, in general, cost minimizers over all inputs or solely with respect to abatement costs. In exploring the behavior of regulated firms, we shall refer to private firms whose earnings are limited to some maximum "fair" rate of return to capital inputs. (For excellent comprehensive treatments of the analytics of the regulated firm, see Baumol and Klevorick, and Bailey.[32]) Our approach is to extend the standard Averch-Johnson (A-J) model to include effluents as a factor of production.[33]

The source of the difficulty is that the regulatory constraint distorts the relative prices of inputs to the firm. Therefore, cost minimization from the firm's perspective involves an excessive use of capital from the perspective of efficient resource allocation (the A-J effect). By extending its use of capital, the firm is able to enlarge the base upon which its constraint is determined. In consequence, the firm is not likely to extend abatement activities to the point where marginal abatement cost equals the effluent charge.

To show this result formally, we note that a regulated profit-maximizing firm maximizes:

$$\pi = R(K, L, E) - wL - fE - cK \qquad (15)$$

subject to:

$$\frac{R(K, L, E) - wL - fE}{K} = s$$

where: K = Capital (in value units)
L = Labor
E = Units of waste emissions
π = Profits
c = Competitive cost of capital
w = Wage rate
f = Effluent fee
s = Allowed rate of return
R(K,L,E) = Total revenue

We assume here that the allowable rate of return, s, exceeds the cost of capital and that the constraint binds so that we can write the constraint as an equality. Otherwise, the firm would simply behave as an unconstrained profit-maximizing monopolist. The constrained-maximization problem yields the Lagrangian:

$$M = R(L, K, E) - wL - fE - cK - \lambda[R(L, K, E) - wL - fE - sK] \quad (16)$$

Taking the requisite partial derivatives, setting them equal to zero, and solving for the firm's optimal use of inputs, generates the following three first-order conditions:

$$R_L = W \quad (17)$$

$$R_E = f \quad (18)$$

$$R_K = c - \frac{\lambda}{1-\lambda}(s-c) = \frac{c-\lambda s}{1-\lambda} \quad (19)$$

where R_i is the marginal revenue product of factor i.

On first glance at Equation (18), one might infer that the firm is a cost-minimizer with respect to pollution abatement. But this conclusion is incorrect.[34] Cost-minimization from society's perspective occurs only if the ratio of the marginal product to factor price (where the latter reflects true social opportunity costs) is the same for all factors. This relation obviously does not hold.

$$\frac{R_L}{w} = \frac{R_E}{f} = \frac{R_K}{c - \frac{\lambda}{1-\lambda}(s-c)} \quad (20)$$

If the regulatory constraint were not binding so that $\lambda = 0$, the profit-maximizing solution would imply cost minimization, since $R_K = c$. As it stands, however, the firm faces an implied cost of capital that is below the market price; this follows since $0 < \lambda < 1$, and by assumption, $s > c$. (For a formal proof, see Baumol and Klevorick.[35]) The result is that the firm does not employ the cost-minimizing combination of factor inputs from society's perspective.

As a hypothetical example, envision a regulated firm which employs excessive pollution-abatement equipment in the sense that the implied marginal cost of abatement exceeds the effluent fee. The rationale would be the expanded capital stock which would permit a higher absolute level of profits. In this instance, the firm would substitute capital for waste emissions to an extent that is unjustified by the true opportunity costs of the factors. Obviously, a set of effluent fees could not, under these circumstances, be expected to generate the least-cost set of pollution-abatement quotas among all polluters.

While these results are somewhat disturbing, there remains the important practical question of the actual magnitude of the A-J effect. The literature has not produced any compelling evidence of a widespread A-J effect.[36] In the absence of such evidence and in view of the consequent skepticism in the recent regulation literature concerning the A-J model, it is our sense that we should not regard the results in this section as constituting a compelling case against a system of fees. The extent of the complications resulting from the presence of regulated firms that are heavy polluters remains to be determined.

14.5 Conclusion

The analysis in this paper does not lead to any simple, sweeping conclusions, aside from the rather unsurprising result that the formal optimality properties of the Pigouvian tax, and also of the more limited charges and standards approach, do not generalize to all the various cases of "industry" structure. The recognition of this proposition really represents, however, the beginning rather than the end of the analysis. The basic and the hard question is that of how well a system of pricing incentives is likely to perform relative to the alternatives involving various forms of direct controls.

This is, of course, ultimately an empirical matter. At the same time, the analysis does offer some support for the use of pricing techniques. In particular, our examination of both private firms and public agencies with relatively complex objective functions suggests that fees or other market incentives are likely to set in motion efforts to reduce costs by engaging in abatement efforts; since fees result in a reduction in profits or perquisites of various sorts, managers or bureaucrats have a real incentive to seek out cost-saving abatement procedures. The importance of such incentives is underscored by an emerging empirical literature that indicates that existing programs of direct controls are generating enormous waste: abatement costs on the order of two to ten times as large as needed to attain the predetermined standards of environmental quality.[37] In particular, a recent study of CFC emissions estimates that, based on existing technology, a system of pricing incentives could result in a 40 percent savings in abatement costs compared to a likely system of direct controls.[38] The potential cost-savings from a more rational allocation of abatement quotas among polluters appear to be huge. Although our analysis suggests that fees can hardly be expected to generate the least-cost solution in any very precise sense, a substantial move in that direction could produce large gains. What remains to be determined is just how seriously the cost-saving properties of a system of market incentives are impaired by the "imperfections" in decision-making institutions as they relate to the control of polluting activities.

References

1. See, for example, W. Baumol and W. Oates, The Theory of Environmental Policy (Englewood Cliffs, N.J., Prentice-Hall, 1975).

2. For a comparison of the fee and marketable-permit approaches to the control of CFC emissions that favors marketable permits, see A. R. Palmer, W. E. Mooz, T. H. Quinn, and K. A. Wolf, Economic Implications of Regulating Chlorofluorocarbon Emissions from Nonaerosol Applications (Santa Monica, California, Rand, June 1980).

3. Federal Register, October 7, 1980, vol. 45, no. 196, pp. 66726-66734.

4. See chapter 3, section 3.2.1.1.

5. O.E. Williamson, "Managerial Discretion and Business Behavior," American Economic Review vol. 53 (December 1963).

6. W. Niskanen, Jr., Bureaucracy and Representative Government (Chicago, Ill., Aldine, 1977).

7. J.Q. Wilson, editor, The Politics of Regulation (New York, Basic Books, 1980).

8. H. Averch and L. Johnson, "Behavior of the Firm Under Regulatory Constraint," American Economic Review vol. 52 (December 1962).

9. Baumol and Oates, The Theory of Environmental Policy, pt. 1.

10. L. Lave and E. Seskin, Air Pollution and Human Health (Baltimore, Md., Johns Hopkins University Press for Resources for the Future, 1977.

11. Baumol and Oates, The Theory of Environmental Policy, chapter 7.

12. Ibid., chapter 8.

13. Ibid., chapter 10.

14. T. Teitenberg, "Taxation and the Control of Externalities: Comment," American Economic Review vol. 64 (June 1974); and S.R. Ackerman, "Effluent Charges: A Critique," Canadian Journal of Economics vol. 6 (November 1978).

15. J. Buchanan, "External Diseconomies, Corrective Taxes, and Market Structure," American Economic Review vol. 59 (March 1969) pp. 174-177.

16. Baumol and Oates, The Theory of Environmental Policy, chapter 6.

17. Council on Environmental Quality, Environmental Quality, The Third Annual Report of the Council on Environmental Quality (Washington, D.C., GPO, August 1972) pp. 269-301.

18. Lave and Seskin, Air Pollution, p. 230.

19. H. Leibenstein, "Allocative Efficiency vs. X-Efficiency," American Economic Review vol. 56 (June 1966).

20. R. Marris, The Economic Theory of Managerial Capitalism (London, Macmillan, 1964).

21. Williamson, Managerial Discretion."

22. K. Cohen and R. Cyert, Theory of the Firm (Englewood Cliffs, N.J., Prentice-Hall, 1962) chapter 16.

23. M.I. Kamien and N.L. Schwartz, "Market Structure and Innovation: A Survey," Journal of Economic Literature vol. 8 (March 1975).

24. See A. Chandler, Strategy and Structure (Cambridge, Mass., MIT Press, 1962); J.L. Bower, Managing the Resource Allocation Process (Cambridge, Mass., Harvard Business School, 1970).

25. W.E. Oates and D. Strassmann, "The Use of Effluent Fees to Regulate Public-Sector Sources of Pollution: An Application of the Niskanen Model," Journal of Environmental Economics and Management vol. 5 (September 1978).

26. Niskanen, Bureaucracy.

27. Wilson, The Politics of Regulation.

28. J. Migue and G. Belanger, "Toward a General Theory of Managerial Discretion," Public Choice vol. 17, 1974.

29. W. Orzechowski, "Economic Models of Bureaucracy: Survey, Extensions, and Evidence," in T. Borcherding, ed., Budgets and Bureaucrats (Durham, N.C., Duke University Press, 1977) pp. 229-259.

30. Wilson, The Politics of Regulation.

31. Ibid., p. 373.

32. W. Baumol and A. Klevorick, "Input Choices and Rate of Return Regulation: An Overview of the Discussion," Bell Journal of Economics and Management Science, vol. 1, no. 2 (Autumn 1970), pp. 162-190; and E. Bailey, Economic Theory of Regulatory Constraint (Lexington, Mass., Heath-Lexington, 1973).

33. Averch and Johnson, "Behavior of the Firm."

34. Baumol and Klevorick, "Input Choices," pp. 176-177.

35. Ibid., pp. 166-167.

36. D. Barron and R. Taggert, "A Model of Regulation Under Uncertainty and a Test of Regulatory Bias," Bell Journal of Economics vol. 8 (Spring 1977); H. Peterson, "An Empirical Test of Regulatory Effects," Bell Journal of Economics vol. 6 (Spring 1975); and R. Spann, "Rate of Return Regulation and Efficiency in Production: An Empirical Test of the Averch-Johnson Thesis," Bell Journal of Economics vol. 5 (Spring 1975).

37. S. Atkinson and D. Lewis, "A Cost-Effectiveness Analysis of Alternative Air Quality Control Strategies," Journal of Environmental Economics and Management vol. 1 (1974); T. Bingham, A. Miedema, with P. Cooley and J. Mathews, Final Report, Allocative and Distributive Effects of Alternative Air Quality Attainment Policies (Research Triangle Park, N.C., Research Triangle Institute, October 1974); and A. Kneese, S. Rolfe, and J. Harned, eds., Managing the Environment: International Economic Cooperation for Pollution Control (New York, Praeger, 1971) app. C.

38. Palmer and coauthors, Economic Implications, ch. IV.

Chapter 15

ENVIRONMENTAL STANDARDS AND THE MANAGEMENT OF CFC EMISSIONS

Dennis J. Snower

15.1 Introduction

This paper is concerned with the relation of environmental standards and the management of chlorofluorocarbon (CFC) emissions, viewed from two distinct perspectives. The first, covered in section 15.2, provides a set of practical guidelines for managing CFCs by showing how different social valuations of CFC concentrations may be matched with environmental and production standards. Viewed differently, that section contains criteria for determining the optimal production responses to a given set of CFC standards.

The second, considered in section 15.3, deals with socially desirable dynamic relations between CFC policies and macroeconomic policies; in particular, it examines how these policies should be affected by changes in resource supplies and international transfers of CFC emissions. After all, the problem of CFC emissions is an intertemporal one. In fact, this is one of the most prominent (and invidious) features of the problem: present CFC emissions affect social welfare in the future, and the implications of those intertemporal welfare effects for the formulation of environmental policy comprise the central focus here.

In sum, under the first perspective, environmental standards are given, and optimal relations between environmental and production stan-

---

Author's Note: I would like to express my gratitude to John Cumberland for his thought-provoking suggestions and to Derek Updegraff for his invaluable support in the empirical work of this chapter.

dards are then derived. Under the second perspective, dynamic and macroeconomic considerations are employed to develop implications on the way environmental standards should be set. In particular, I contend that (1) environmental standards should be pegged to levels of consumption and production; (2) environmental standards should be permitted to change over time; and (3) the optimal relationship between consumption, production, and standards should depend on factor supplies and international transfers of emissions.

15.2 Environmental Standards and Production Response

The obvious response to the problem of excessive CFC concentrations is to set an environmental standard for CFC emissions, that is, an upper bound on the flow of CFCs from production and consumption activities. This standard can be implemented through a wide variety of policy instruments--direct controls, taxes, pollution licenses, refundable deposits, and many others.

Characteristically, environmental economists and environmental policymakers have very different approaches to the task of setting environmental standards. The economists frequently have a predilection for maximizing a social welfare function subject to production opportunity constraints. The welfare function may contain pollutant concentrations, consumption and employment as arguments, while the constraints may depend on pollutant concentrations, factor supplies, and production technologies. What emerges from this welfare maximization exercise are the optimal pollutant emission flows together with the optimal levels of consumption, production, and employment. All these optimal levels may be adopted as "standards." The policymaker is supposed to use the policy instruments at his disposal to set the actual levels of pollutant emission, consumption, production, and employment at the levels of their respective standards.

This is not the way environmental policymakers approach the problem. They do not maximize a social welfare function because they do not attempt to formulate one. Instead, they simply impose standards in response to what they see as excessive pollution. The standards usually depend on the political and administrative pressures which policymakers face, together with their subjective evaluations of assorted benefit-cost analyses and news media reports. Commonly, two types of standards are used--environmental standards regulating the flow of pollutant emissions and production standards regulating the flow of output from selected industries.

The practice of setting environmental standards in an ad hoc way to deal with excessive pollution has received some support from economists, however. It has been argued (perhaps most prominently by Baumol and Oates) that accurate information on the social welfare effect of pollution is usually very difficult to acquire and that policymakers should therefore concentrate on implementing environmental standards efficiently rather than setting them optimally.[1] In this regard, the use of unit taxes has the desirable property of achieving any given vector of environmental standards and final outputs at minimum cost to society.[2]

However, environmental policymakers have not restricted themselves to setting environmental standards; frequently they have also used production standards to combat pollution problems. The banning of nonessential aerosol products in the United States is a good example of this. On the whole, economists have been rather puzzled by this practice. The use of production standards to reduce pollution has received no significant support in environmental economic theory. In section 15.2.1, however, we will consider a set of economic circumstances in which production standards are a much more convenient way of controlling pollution than environmental standards.

Yet even if it is admitted that both environmental and production standards have a role to play in the policymaker's arsenal of variables to be controlled, we are still left with the problem of choosing the levels at which to set these standards. Since policymakers are not inclined to make explicit relative valuations of pollution, consumption, employment, and so on, a welfare function cannot be constructed, and thus the economists' conventional approach cannot be pursued. Policymakers appear more likely to be able to choose between alternative sets of trajectories for pollution, consumption, and employment standards than to formulate relative valuations of these policy target variables.

Yet there are a number of problems involved in providing sets of trajectories from which policymakers may select. How can one ensure that the trajectories in each set are consistent with one another? For example, if the consumption and employment standards are observed in practice, what is to ensure that the pollution standard is attainable? For this purpose, it is necessary to have information on the production and pollution technologies of the economy. Such information may be difficult to acquire, but it is commonly much more accessible than the information needed to construct estimates of the relative social valuations of policy target variables.

An economy's production and pollution technologies together with its factor supplies yield its "production-pollution opportunity locus." This is the locus of all values of production and pollution emission such that it is impossible (1) to increase the production of one good without either decreasing the production of another good or raising an emission flow; and (2) to lower an emission flow without either raising another emission flow or decreasing the production of a good. A "technologically efficient" set of standards may be characterized as a point on this production-pollution opportunity locus. A technologically efficient trajectory of standards is a temporal sequence of points on a temporal sequence of production-pollution opportunity loci.

Naturally, once the policymaker has chosen a particular trajectory of standards, it is possible to infer properties of his implicit welfare function. After all, a welfare function may be understood simply as a device for choosing from a menu of production-pollution opportunities. Such inference may provide the policymaker with very useful information. For example, a trajectory which the policymaker has chosen may imply relative valuations of policy target variables with which the policymaker unambiguously disagrees. He then has an incentive to choose another trajectory of standards which is more in accord with relative valuations he can support. Alternatively, it may emerge that a trajectory choice may imply relative valuations which are temporally inconsistent. Here, too, the policymaker is induced to find another trajectory.

The main purpose of this section is to provide an empirically tractable framework within which this matching of trajectory choice and relative valuations can take place. We will construct a simple, illustrative example of this matching with regard to CFC emissions and the main outputs whose production or consumption generate these emissions. The model will be short run, covering a single year. Production and pollution technologies will be characterized by input-output coefficients.

There is another use to which this empirical framework can be put. It suggests a way of dealing with imperfect information on the costs and benefits of CFC emissions. Incomplete data on costs and benefits has given rise to a classic difficulty in the formulation of environmental policy. On the one hand, there is a general consensus that it is inadvisable to ignore the problem of CFC emissions until all the necessary information has been collected. On the other, specific policy proposals remain open to challenge by business, labor, or any other group that may wish to exploit the possibility of holding tenable but divergent views on the costs and benefits of CFC emissions.

This section suggests a way of examining whether it is possible to sidestep a full-fledged benefit-cost analysis and to find ranges within which costs and benefits can vary without significantly affecting the optimal levels of the standards. There appears to be a marked asymmetry in the quality of current information on the costs and benefits of CFC emissions. A full-fledged benefit-cost analysis would require, at minimum, (1) information on the technological relationships among production, consumption, and CFC emissions; (2) the relationship between these emissions and CFC concentrations in the stratosphere; (3) the relationship between these concentrations and the rates of ozone depletion and climate change; (4) the relationship between ozone depletion and climate change, on the one hand, and a variety of physiological and ecological impacts (principally, human health effects, agricultural effects, materials weathering, and several biological effects) on the other; (5) the welfare valuation of these impacts; and (6) the welfare valuation of consumption, production, and employment activities net of these impacts. It is noteworthy that information on item (1) is much more readily and comprehensively available than that on items (2) through (6).

Accordingly, the model to be presented here makes use of data on production and pollution technologies, and examines over what ranges the environmental and production standards are not sensitive to divergent estimates of items 2 through 6. Suppose that it can be shown that disagreements over estimates of these items fall within a range over which the standards are not sensitive. Then a strong case can be made for implementing these standards even though information is imperfect. Our model also suggests guidelines for collecting benefit-cost data. Clearly, emphasis should be placed on collecting data with respect to which the standards are potentially sensitive.

An attractive aspect of our model is the analytical simplicity with which such sensitivity analysis may be conducted. The model does not require us to examine how sensitive the standards are to variations in each of the physical science relationships (2) through (4) and each of the welfare-economic valuations (5) and (6). Instead, the model collapses all the information on (2) through (6) into two sets of parameters: one describing the net welfare effects of CFC emissions and the other describing the net welfare effects of consumption activities. The sensitivity analysis need only be undertaken with regard to these two parameters.

CFC emissions are generated by a number of different sectors. Policymakers who ignore the intersectoral origins of the CFC problem run the risk of overlooking something very important. The environmental economists' approach of maximizing a welfare function subject to a production-pollution opportunity locus yields an optimal set of environmental standards and production standards. If the free market mechanism, without government intervention, can optimally manage CFC emission, then the emissions and production which emerge under this mechanism must be identical to the environmental standards and the production standards respectively. On the other hand, if the market mechanism does not adequately perform this function, then the policymaker may find it necessary to impose these standards.

In this context it is important to note that the imposition of the environmental standards alone (whether through direct controls or taxes or other policy instruments) does not automatically guarantee that the production standards will be fulfilled. If the production response to the imposition of the environmental standards is suboptimal, then the environmental standards will be suboptimal as well. To manage CFC emissions in the optimal way, it is not always sufficient to impose CFC standards; production standards may be required as well. Thus, there is a need to consider the optimal relation between CFC standards and production standards.

In section 15.2.1 a case is made for the imposition of production standards. In section 15.2.2 an empirically implementable model of CFC generation is constructed, and the optimal relationship between CFC standards and production standards is examined theoretically in this context. Section 15.2.3 presents an illustrative empirical exercise with this model, in which environmental and production standards are matched with welfare valuations of pollution, consumption, production, and employment activities.

15.2.1 A Case for Production Standards

As noted, environmental policymakers have been known to impose production standards in order to deal with excessive pollution. These standards frequently take the form of bans on "nonessential" uses of pollutant-generating products, as has been done in the case of aerosol products emitting CFCs. Yet the nonessentiality of product uses is usually not defined in a rigorous way. Thus, the environmental policymakers may have considerable discretion in deciding which products to ban. Moreover, it is frequently not clear how the nonessentiality of a product use serves to justify the imposition of production standards. Why not use environmental standards instead?

In this section a definition of nonessentiality is provided and this definition is used to show why production standards may provide a convenient and efficient way of managing pollution. Let us evaluate nonessential product uses on the basis of the social benefits accruing from them. In figure 15-1, part A illustrates the benefit function for a particular good, say, aerosols. This function relates the quantity of the product used (Q) to the dollar value of total benefits it generates. The function has a kink; the marginal benefit for $Q < Q^*$ is discontinuously greater than the marginal benefit for $Q > Q^*$. The uses to which the good is put in segment $\overline{OQ^*}$ may be termed "essential;" the additional uses to which it is put may be termed "nonessential."

Suppose that the marginal cost curve intersects the marginal benefit curve at the discontinuity. Then the socially optimal level of production is Q^*. Shifts of the marginal cost curve have no effect on this optimal level, provided that the curve continues to pass through the discontinuity.

Now consider a firm which produces Q by means of a single factor, say, labor (L). In so doing, it generates an emission flow, ΔP. Its labor requirements and pollutant emissions are given by fixed coefficients: $L = a \cdot Q$ and $\Delta P = b \cdot Q$. The firm maximizes its profits subject to a

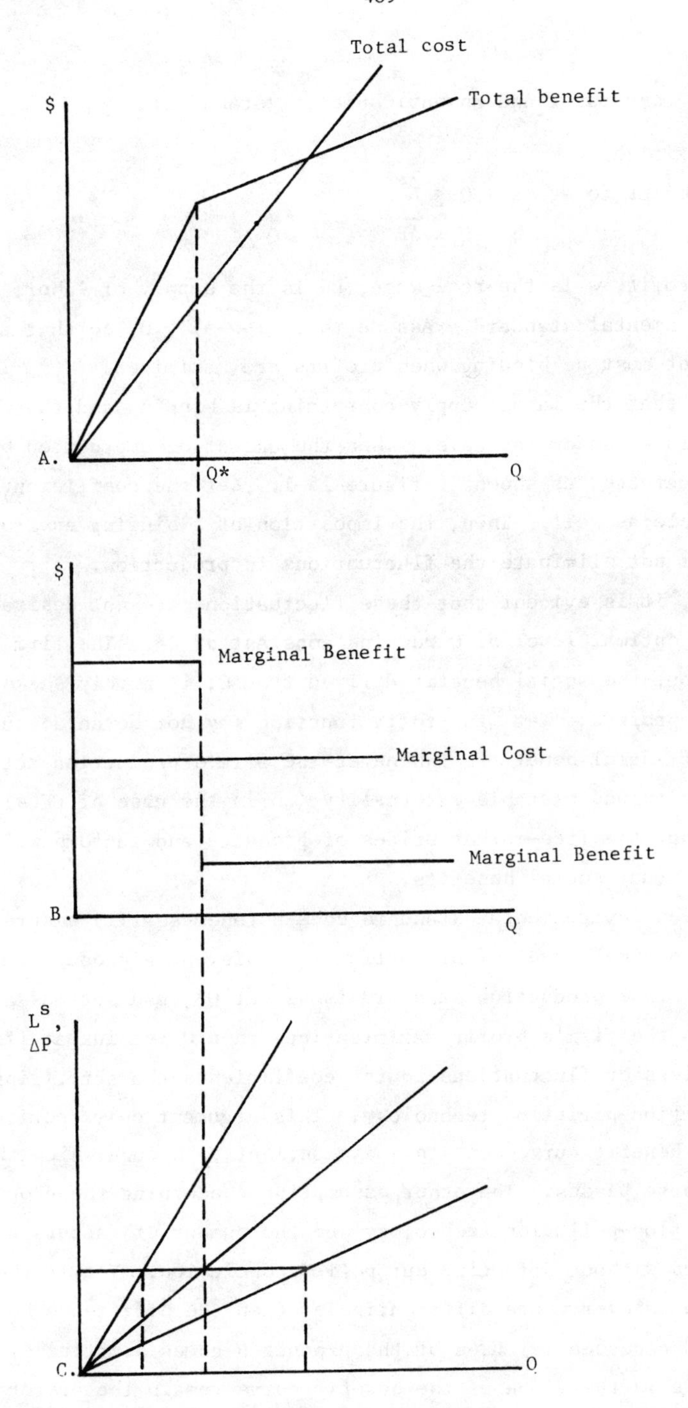

Figure 15-1. Production standards to ban nonessential uses of a product

labor supply constraint and an environmental standard:

$$\begin{aligned} &\text{Maximize} && \pi = Q - w \cdot L \\ &\text{Subject to} && a \cdot Q \leq L^s \\ & && b \cdot Q \leq \overline{\Delta P} \qquad Q \geq 0 \end{aligned}$$

where π is profit, w is the real wage, L^s is the supply of labor, and $\overline{\Delta P}$ is the environmental standard. Assume that $(1-w \cdot a) > 0$, so that at least one constraint must be binding when profits are maximized.

Suppose that the labor supply constraint is binding and that the coefficient a is a random variable. Then the amount of Q produced by the firm will fluctuate, as shown in figure 15-1. Let the coefficient b be a random variable as well. Then, the imposition of a binding environmental standard will not eliminate the fluctuations in production.

However, it is evident that these fluctuations are not desirable. The socially optimal level of Q remains constant at Q*. The firm is not concerned about the social benefit derived from Q; it merely seeks to maximize its profits. Yet its profit function may not be an adequate reflection of social benefits. Whenever the firm's production activity gives rise to an undepletable externality (as in the case of CFCs), it is inevitable that the free-market prices of products and factors will not reflect the actual social benefits.[3]

Whereas an environmental standard does not necessarily ensure that the socially optimal level of production is achieved, a production standard can do so. If the production standard is set at Q*, and acts as a binding constraint on the firm's profit maximization, then Q remains at its optimal level regardless of fluctuations in the coefficients characterizing the firm's production-pollution technology. This argument only requires that the marginal benefit curve contain a discontinuity through which the marginal cost curve passes. The other assumption concerning the properties of the production-pollution technology and the number of factors employed may be relaxed without affecting our policy conclusion. Yet if the total benefit curve is everywhere differentiable, then the difference between essential and nonessential uses of the product becomes a matter of degree. Sudden changes in the slope of the benefit curve remain the criterion whereby these uses may be distinguished.

An analogous argument may be made when the total cost curve is kinked while the total benefit curve is smooth. As long as the marginal benefit curve passes through the discontinuity of the marginal cost curve, the socially optimal level of production is unresponsive to shifts in the benefit curve or vertical shifts in the cost curve. In this case we may distinguish "expensive" from "inexpensive" uses of a product. If the private costs faced by a firm are not an accurate reflection of the social costs (in particular, if the social cost curve is kinked while the private cost curve is not) then production standards may be more effective than environmental standards in eliminating the expensive uses of a product.

15.2.2 Optimal Relationships Between Environmental and Production Standards

This section presents a simple model of interindustrial production and pollution activity, and in this context some properties of the optimal relations between environmental and production standards will be examined. The economy's production-pollution opportunity locus is characterized by the fixed-coefficients technologies expressed in the following equations:

$$Q \leqq A \cdot Q + D \qquad (1)$$

$$B \cdot Q \leqq f \qquad (2)$$

$$F \cdot Q + G \cdot (I-A) \cdot Q = E \cdot Q = \Delta P \qquad (3)$$

where, for simplicity, anthropogenic and natural treatment services have been omitted. Q is a vector of gross industrial output; D is the associated vector of final demands; A is an intermediate-good, input-output matrix; f is a vector of factor supplies; B is a matrix of factor-requirement, input-output coefficients; F is a matrix of production emission coefficients (that is, pollutant emission flow per unit of output produced); G is a matrix of final demand emission coefficients (that is, pollutant emission flow per unit of output used to satisfy final demand); $(I-A) \cdot Q$ is satisfied final demand; $E = F + G \cdot (I-A)$; and ΔP is a vector of emission flows. For the purposes of this section, the economy's production-pollution opportunities need only be specified for a single period, say, a year.

Let social welfare (W) be a linear function of satisfied final demands (consumption demands are immediate objects of welfare and investment demands enter the welfare function via a valuation of terminal capital stocks), factor services, and pollutant flows. (If the pollutants are long-lived, then pollutant stocks are relevant to social welfare. Yet since pollutant stocks inherited from last year are exogenously given, only this year's pollutant flow are objects of policy.)

$$W = \beta \cdot (I-A) \cdot Q - \gamma \cdot B \cdot Q - \delta \cdot \Delta P = \alpha \cdot Q - \delta \cdot \Delta P \qquad (4)$$

where β, γ, and δ are vectors of the marginal utilities of final demand and the marginal disutilities of providing factor services and emitting pollutants, respectively; α is a vector of the net marginal utilities of Q (including those pertaining to final demands and factor services but excluding the associated emission flows). The policymaker seeks to maximize the social welfare function Equation (4) subject to the constraints in Equations (1), (2), and (3). The endogenous variables are ΔP (the environmental standards) and Q (the production standards).

Within this framework, let us examine how different social valuations of the emission flows may be matched with environmental and production standards. For simplicity, we have grouped all CFCs into one category, so that we need only be concerned with the social valuations of a single emission flow. It is self-evident that as δ rises, the optimal emission flow, ΔP, falls. Yet what is the optimal production response associated with this diminished emission flow?

The answer to this question depends on the shape of the social welfare function and the shape of the production-pollution opportunity locus. Substituting the emission constraint Equation (3) into the social welfare function Equation (4), we obtain

$$W = \sum_i \alpha_i \cdot Q_i - \delta \cdot \sum_i e_i \cdot Q_i \qquad (5)$$

where Q_i, α_i, and e_i are elements of Q, α, and E, respectively. The remaining constraints (Equations (1) and (2)) describe the production opportunity locus, which is a concave hypersurface in Q space. As the social valuation of pollution is raised, the slope of the welfare function, Equation (5), changes and consequently the policymaker chooses a new point on the production opportunity locus. The new point stands for the new production standards associated with the new environmental standard. Thus, the responsiveness of the production standards to changes in the social valuation of pollution may be gauged by examining how the slope of the welfare function, Equation (5), responds to these changes.

The slope of the social welfare function with respect to two (arbitrarily chosen) outputs is

$$\phi = \frac{\partial Q_j}{\partial Q_i} = \frac{\alpha_i - \delta \cdot e_i}{\alpha_j - \delta \cdot e_j} \tag{6}$$

The responsiveness of this slope to a rise in the social valuation of pollution is

$$\frac{\partial \phi}{\partial \delta} = \frac{\alpha_j \cdot e_i - \alpha_i \cdot e_j}{(\alpha_j - \delta \cdot e_j)^2} \tag{7}$$

Thus, $$\frac{\partial \phi}{\partial \delta} \gtreqless 0 \quad \longleftrightarrow \quad \frac{e_i}{\alpha_i} \gtreqless \frac{e_j}{\alpha_j} \tag{8}$$

Equation (8) asserts that an increase in the social valuation of pollution raises (lowers) the slope of the social welfare function if the ratio of the emission flow to net benefits from Q_i exceeds (falls short of) the ratio of the emission flow to net benefits from Q_j. Let us apply Equation (8) to define EBI_{ij}, the "emission-benefit intensity" of good i relative to good j, as expressed in Equation (9):

$$EBI_{ij} = \frac{(e_i/\alpha_i)}{(e_j/\alpha_j)} \tag{9}$$

Figure 15-2 illustrates how the emission-benefit intensity affects the responsiveness of the optimal production standards to changes in the social valuation of pollution. The feasible region is given by the area ABCDEF. IF $EBI_{ij} > 1$, a rise in the social valuation of pollution increases the slope of the social welfare function and therefore causes the socially optimal production point to move in a counterclockwise direction along the boundary of the feasible region. Conversely, if $EBI_{ij} < 1$, a rise in the social valuation of pollution causes the slope of W to fall and thereby leads to a clockwise movement of the optimal production point along the boundary of the feasible region.

In sum: As the social valuation of pollution is raised, the socially optimal production target for good j rises (falls) relative to the socially

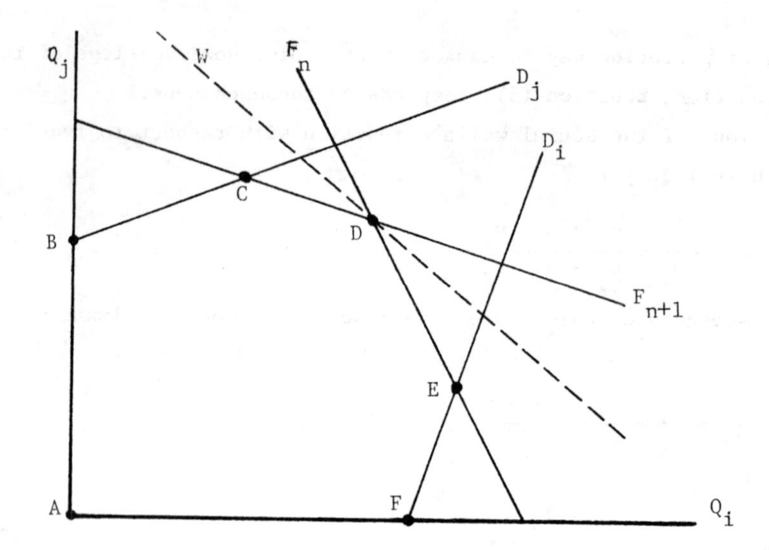

Figure 15-2. The choice of optimal production standards

optimal production target for good i, whenever the emission benefit intensity of good i relative to good j is greater than (less than) unity. The emission-benefit intensity may be seen as an empirically useful indicator of how optimal production and environmental standards are related.

15.2.3 An Empirical Application

The framework of the previous section may now be used to illustrate how environmental and production standards may be formulated empirically. As above, this illustration will only be concerned with short-term policy proposals. In particular, we will assume a one-year time horizon and derive standards for 1975. Based on the evidence in chapters 2, 3, and 7, F-11 and F-12 account for the bulk of CFC emissions, and we restrict our attention to these CFCs. Since their effects on the ozone layer are quite similar, we group them together in our formulation of the environmental standard. In other words, the standard is defined as an upper limit on the tonnage of F-11 and F-12 emitted per year.

The major outputs responsible for the emission of F-11 and F-12 may be divided into four categories (see chapter 3):

1. Foams (open-cell and closed-cell)

2. Aerosols (for example, antiperspirants, deodorants, hair care products, insecticides)

3. Vehicle air conditioners

4. Miscellaneous cooling equipment (for example, building cooling, food and beverage cooling, home refrigerators and freezers).

Thus, it is clear that the emission coefficients in our model must be defined with respect to emission flows from both production and consumption activities. Foams and aerosols are nondurable goods, and hence the corresponding emission coefficients describe emission flows per unit of output produced. The outputs of categories 3 and 4 are durable goods, and the corresponding emission coefficients must describe emission flows per unit of the stock of these goods. The appropriate stock figures for outputs of category 4 are difficult, if not impossible, to find and, thus, we will restrict our attention to the outputs of categories 1 through 3. Since F-11 and F-12 are grouped together, we implicitly assume that, for each output, the emission coefficients for F-11 and F-12 are the same.

In order to derive the feasible production region for the three outputs above, intermediate-good and factor input-output (I-O) coefficients are used. Yet, the appropriate I-O coefficients are not readily available for the three outputs. Thus, we place each output into a more aggregative economic sector for which the I-O coefficient may be gleaned from the Almon input-output model.[4] Foams are placed into the "plastic materials and resins" sector (SIC 2828; Almon sector 88); aerosols are placed into the "cleaning and toilet products" sector (SIC 2340; Almon sector 73); and vehicle air conditioners are placed into the "motor vehicles" sector (SIC 3714; Almon sector 145).[5] We assume that the proportion of each CFC-emitting output to its corresponding sectoral output remains constant through time. Then, the sectoral outputs may be used as proxies for the CFC-emitting outputs.

Another set of ingredients necessary for the derivation of the feasible production region comprise the demand external to the three sectors and the factor supplies available to these sectors. To keep our exercise simple, we assume that labor is the only factor in limited supply. The demand external to each sector consists of the final demand for the sectoral output and the intermediate demands for this output originating from all sectors except the three sectors above. The labor supply available to these sectors is taken to be the sum of the sectors' actual labor use.

For simplicity, we assume that labor is perfectly mobile among the three sectors, but that labor cannot move to or from the other sectors of the economy. The figures for the demands external to each sector and for the labor supply available to the three sectors were obtained from the Almon model.

There are three emission coefficients to be found: one for plastic materials and resins, one for cleaning and toilet products, and one for motor vehicles. The emission coefficient for plastic materials and resins is derived by dividing the 1975 U.S. emission of F-11 and F-12 from foams by the 1975 value of product shipments of plastic materials and resins. The former figure is obtained under the assumption that the United States was responsible for half of the 1975 worldwide emissions of F-11 and F-12, as given in the National Research Council report.[6] The latter figure, in constant 1972 purchaser prices, is obtained from the Almon model.[7] The emission coefficient for the cleaning and toilet products is derived analogously to the emission coefficient above.

The emission coefficient for motor vehicles is derived by dividing the 1975 U.S. emissions of F-11 and F-12 from vehicle air conditioners by the number of vehicles in existence in the United States in 1975. Although motor vehicles--cars, trucks, buses--may differ considerably from one another, we assume that the CFC-emitting properties of vehicle air conditioners differ negligibly from one another. Thus, the number of vehicles--rather than the stock of vehicles (in which vehicles of various types are weighted differently)--is relevant to the third emission coefficient. The number of vehicles registered in the United States in 1975 and the proportion of vehicles containing air conditioners are obtained from the Census Bureau.[8] The emission of F-11 and F-12 is obtained from the National Research Council.[9]

The socially optimal environmental standard for F-11 and F-12 and the corresponding production standards for the three outputs may be computed by solving the following linear program:

$$\text{Maximize} \quad W = \alpha_1 \cdot Q_1 + \alpha_2 \cdot Q_2 + \alpha_3 \cdot Q_3 - \delta \cdot P$$

$$\text{subject to } (1-a_{11}) \cdot Q_1 - a_{12} \cdot Q_2 - a_{13} \cdot Q_3 \leq \overline{D}_1$$

$$-a_{21} \cdot Q_1 + (1-a_{22}) \cdot Q_2 - a_{23} \cdot Q_3 \leq \overline{D}_2$$

$$-a_{31} \cdot Q_1 - a_{32} + (1-a_{33}) \cdot Q_3 \leq D_3$$

$$b_1 \cdot Q_1 + b_2 \cdot Q_3 \leq \overline{L}$$

$$e_1 \cdot Q_1 + e_2 \cdot Q_2 + e_3(Q_3 + K_{-1}) \leq \Delta P, \tag{10}$$

where K_{-1} is last year's (1974) stock of motor vehicles. For simplicity, we assume that the prices of Q_1, Q_2, and Q_3 reflect the marginal utilities of these goods net of marginal disutilities of the labor services required to produce them. Thus, we use the 1975 price indexes of these outputs (taken from the Almon model) to represent α_1, α_2, and α_3, respectively.

We examine how the socially optimal environmental standard and the production-response guidelines for the three outputs respond to changes in the social valuation of the emission stock (δ). In table 15-1, the standard and the guidelines are given for various values of δ. The corresponding unemployment figures, in thousands of persons (u), are given as well. Each unemployment figure represents the slack variable of the fourth constraint of program Equation (10). When $\delta = 0$, ΔP, Q_1, Q_2, and Q_3 are equal to their actual values in 1975. As the social valuation of the emission stock rises, the optimal environmental standard falls. Furthermore, we find an optimal order in which the three outputs are to be withdrawn from production. As δ rises from zero, Q_2 (cleaning and toilet products, as a proxy for aerosols) is the first to be withdrawn; Q_1 (plastic materials and resins, as a proxy for foams) are next to follow; and Q_3 (motor vehicles as a proxy for vehicle air conditioners) are last to be withdrawn. As we have seen in the previous section, this order depends on the relative emission-benefit intensities of the three sectors. Indeed, the emission-benefit intensity of Q_3 relative to Q_1 is greater than unity and the emission benefit intensity of Q_1 relative to Q_2 is greater than unity, as well.

In a richer empirical framework, including several technical processes whereby each output may be produced, outputs could be withdrawn from production partially before they are withdrawn completely. Nevertheless,

Table 15-1. Optimal Relations Between Environmental and Production Standards in Illustrative Example

δ	ΔP	Q_1	Q_2	Q_3	U
0.0	1108.517	5551.0	10153.0	55301.0	0.0
50.0	123.676	5551.0	0.0	55301.0	121.852
100.0	34.860	0.0	0.0	55301.0	210.668

the example above outlines the principle of our empirical exercise. Such an exercise may be quite helpful as a background to the formulation of environmental policy. By matching valuations of pollution with environmental and production standards, it may provide the policymaker with useful information about efficient policy alternatives.

Note that the policymaker need not decide on a precise social value of pollution prior to setting the environmental and production standards. Only ranges for such a valuation need be established. In the illustration above, a particular set of standards is optimal if the valuation of pollution falls between zero and 50, another set becomes optimal when the valuation falls between 50 and 100, and so on. Thus, it is not necessary to confront all the difficult problems in determining the social costs of pollution; confidence levels are sufficient.

The reason why the environmental and production standards are insensitive to changes in the valuation of pollution over the given ranges is that the economy's production technology has been characterized by fixed coefficients. For the purposes of short-run environmental planning, this does not appear to be an implausible assumption. Needless to say, the exercise above can also be conducted with regard to a production technology in which factor substitution or joint production occurs. Then the sensitivity of environmental and production standard to changes in the valuation of pollution becomes a matter of degree. In any case, the convenient characteristic of our empirical analysis is that it permits us to separate our description of the production technology, about which we know relatively much, from our description of the relation between emissions and concentrations, between concentrations and ozone depletion, and between ozone depletion and social costs, about which we know relatively little.

The methodology outlined here is meant to be an aid in clarifying policy choices rather than a formula for deriving the policies themselves.

15.3 Dynamic Standards for Management of CFC Emissions

The previous section noted the common practice of setting the levels of environmental standards independently of the levels of production and consumption. The current development of environmental standards for chlorofluorocarbons (CFCs) appears to conform to this practice. Accepting that practice as given, it was shown that a case could be made for production standards, as well as environmental standards; and a methodology was formulated for arriving at the appropriate production response to given standards. Here, however, we return to the underlying general point which began our discussion, which is that setting environmental standards independently of levels of production and consumption activities is, in general, socially suboptimal. We demonstrate that point explicitly in this section, expanding on it with these specific contentions:

1. Environmental standards for CFCs should be pegged to the levels of consumption and production activities.

2. Environmental standards for CFCs should be permitted to change through time. The optimal amount of CFC pollution cannot be expected to remain constant.

3. The optimal intertemporal relation between environmental standards for CFCs, on the one hand, and consumption and production activities, on the other, should depend on factor supplies and international transfers of CFC emissions.

These contentions are not merely matters of abstract economic theory. They have significant, far-reaching policy implications. If they are correct, we must face the immediate implication that CFC standards should be controlled in radically different ways from those used heretofore. Contention 1 implies that CFC standards should not be set independently of macroeconomic policy goals. Whenever there is a change in macroeconomic policy (for example, an income tax reduction which stimulates consumption expenditures or a reduction in government expenditures), there should be a corresponding change in the levels of CFC standards. It is socially undesirable that the government agency providing guidelines for setting CFC standards do so independently of the agencies which formulate budget or control the money supply. CFC guidelines should be part of a broad policy package which includes the macro-economic policy instruments.

This section illustrates how CFC standards may be related to production and consumption levels. (No attempt will be made, however, to relate values of macroeconomic instruments to these production and consumption levels.) For this purpose, an aggregative input-output model will be used, within which both the CFC standards and the consumption/production activities will be embodied. The reason for using an input-output model is that the control of CFC emissions is an intersectoral problem whose empirical treatment is most straightforward in an input-output context. The model is aggregative, because this section aims to focus on the relationship between CFC standards and macroeconomic entities. The extension of the model to many sectors will be sketched.

Contention 2 implies that CFC standards should be permitted to change through time. Policy efforts designed to find the optimal level of CFC emissions and then to maintain the CFC standard at this level are usually misguided. In general, there is no optimal level of CFC emissions; instead, there is an optimal time path for these emissions. Thus, CFC standards should not only be related to consumption and production at any given point in time, but they should also be related to CFC standards in the future. Present CFC standards should be formulated with future standards in mind. Consequently, policy instruments which cannot be changed easily with the passage of time are generally inappropriate for the control of CFCs.

The reason why it is generally undesirable to hold CFC standards stationary is to be found in the relationship between CFC emissions and the associated production/consumption activities. Whereas these activities generate a flow of CFCs into the environment, it is the concentration (stocks) of CFCs in the stratosphere which are relevant to social welfare, in terms of damaging radiation (DUV) and climate change (see chapter 2 for details). Thus, the problem of setting CFC standards is an intertemporal one, with present standards generally differing from future ones.

Contention 3 implies that it is senseless to seek the optimal intertemporal relation between CFC standards and production/consumption activities without reference to an economy's factor supplies or to CFC transfers from other countries.[10] For instance, a reduction in the availability of certain raw material resources or an increase in the use of aerosol sprays abroad alters the relationship above. These are potentially important issues. Despite the great fluctuations in resource supplies over the past decade, the interplay between resource availability and CFC pollution has received little attention. There also has been little international cooperation in the management of CFC emissions.

Although none of the three contentions above has thus far had a significant impact on practical policymaking, contention 1 and 2 have received a certain amount of general recognition in the literature on the optimal control of pollution.[11] However, this literature does not explicitly examine, as we do here, the differences among short-run, medium-run, and long-run goals of environmental macroeconomic policy, and these differences represent a simple way of describing temporal priorities. Yet contention 3 is probably the most novel contribution of this chapter. The impact of resource shortages and international pollution transfers on environmental and macroeconomic policies remains a rather neglected area in environmental economics.

In section 15.3.1 our environmental macroeconomic model is presented and a socially desirable relationship between CFC transfers and production/consumption activities is derived. Section 15.3.2 examines how this relation depends on factor supplies and international CFC transfers.

15.3.1 The Model

The model will first be described in a multisectoral context and then its macroeconomic analogue will be presented. The model contains these building blocks:

1. A description of the productive structure of the economy: how factor services and goods are used to produce goods which satisfy final demands

2. A description of emission and treatment activities: how the production and consumption of goods lead to the emission of CFCs and how factor services and goods can be used to reduce this emission

3. A list of CFC standards

4. A description of the relative social valuations of CFC concentrations and consumption levels.

Let F be an exogenously given vector of factor supplies, Q a vector of industrial outputs, T a vector of treatment services, and B_1 and B_2 factor technology matrices (containing coefficients of factors required per unit of output and per unit of treatment service, respectively). Then the derived demands for factors are constrained by the available supplies:

$$B_1 \cdot Q + B_2 \cdot T \leq F \qquad (11)$$

The treatment services here represent an analytically simple way of dealing with the possibilities for technical substitution whereby CFC emissions may be reduced. Technical substitution is analogous to treatment whenever the qualities of a product (in the Lancastrian sense) remain unchanged by such substitution. The additional economic resources required to incorporate substitutes for the CFCs in the appropriate outputs correspond to the inputs into treatment activities; the fall in emissions caused by technical substitution corresponds to the outputs of treatment activities. Substitutes for F-11 and F-12 can be found for a variety of products. Nitrogen, carbon dioxide, and nitrous oxide may be used as propellants in aerosol products that do not require fine sprays or exact control of discharge (for example, furniture polish, windshield deicer).

D (the vector of total final demands) is the sum of D^C (the vector of exogenously given satisfied consumption demands) and D^I (the vector of exogenously given investment demands). Let A_1 and A_2 be the intermediate-good technology matrices (containing coefficients of intermediate goods required per unit of output and per unit of treatment service, respectively). Then the supplies of goods and treatment services are constrained by the available final demands:

$$(I - A_1) \cdot Q - A_2 \cdot T \leq D^C + D^I = D \qquad (12)$$

The assumptions that factor supplies and product demands are exogenously given are not as restrictive as they may appear at first. Suppose that we would have assumed, instead, that F and D are linear functions of Q and T (these are the "income effects" in the factor supply and product demand functions) as well as of an unspecified number of exogenous variables. Provided that the income effects are not strong enough to change the sign of the gradient (in (Q, T) space) of any constraint in Equation (11) or (12), the qualitative conclusions of our analysis are unaffected.

Let E_1, E_2 and E_3 be emission matrices (containing coefficients of CFCs entering the stratosphere per unit of output, per unit of treatment service, and per unit of final demand, respectively), δ a matrix of reaction constants governing nature's treatment of CFCs; P, a vector of CFC stocks; and $\dot{P}_s$, a vector of CFC standards (that is, upper bounds on permissible emission flows). Actual CFC emission flows are constrained by the CFC standards:

$$E_1 \cdot Q - (I-E_2) \cdot T + E_3 \cdot D - \delta \cdot P \leqq \dot{P}_s \qquad (13)$$

Treatment services are scaled in such a way that one unit of treatment service corresponds to one unit of CFC cleansed. Natural treatment services are usually divided into "active" and "inactive" types. Active processes (for example, the removal of CFCs in the stratosphere) entail the destruction of ozone, whereas inactive processes (for example, trapping of CFCs in polar ice) do not. Natural removal rates are usually represented as the product of a reaction constant (for example, the photo-dissociation rate) and the concentration of the reaction species. If D includes durable goods, then it is necessary to include lagged values of these elements of D in the right-hand of Equation (13). Vehicle and building air conditioners and home refrigerators are examples of consumer durables which generate CFCs. (For simplicity, this complication has been omitted here.)

During any given year, the policymaker's preference function (which may be called, euphemistically, the instantaneous "social welfare" function) contains the consumption levels and the CFC concentration (namely, stocks) as arguments. To avoid the difficult problem of valuing the terminal stocks of CFC concentrations, we let the policymaker's time horizon run from the present into the infinite future. The policymaker's rate of time preference is assumed to be constant as r (with $r > 0$).

Thus, the policy problem is to maximize

$$W = \int_0^\infty e^{-rt} \cdot U(D^C, P)\, dt \qquad (14)$$

Subject to the constraints of Equations (11) through (13), where U is the instantaneous social welfare function. The endogenous variables are contained in the vectors Q, T, D, $\dot{P}_s$, and P. D^I is a vector of first differences of the produced factor supplies contained in F, and $\dot{P}$ is dP/dt.

Now consider a simple macroeconomic analog of this policy problem. There is a single produced good (Q) and a single treatment service (T), each of which generate a homogeneous flow of CFCs. The stock of CFCs is P. There is a single nonproduced scarce factor of production, the raw material resource R. Consumption (D) is the only final demand. Then the constraints of the policy problem here are (15), (16) and (17), corresponding to (11), (12) and (13), respectively:

$$b_1 \cdot Q + b_2 \cdot T \leqq R \tag{15}$$

$$(1-a_1) \cdot Q - a_2 \cdot T = D \tag{16}$$

$$e_1 \cdot Q - (1-e_2) \cdot T + e_3 \cdot D - \delta \cdot P \leqq \dot{P}_s \tag{17}$$

where R represents a flow of resource services per instant of time; $(1-a_1)$, $(1-e_2) > 0$; Q, T, D, and $P \geqq 0$; and all the variables and constants of these constraints are scalars. Equation (16) holds as an equality since D is satisfied consumption demand. Subject to these constraints, the policy seeks to maximize social welfare (W), which depends positively on D and inversely on P. It is evident that Equation (17) holds as an equality in the optimal solution: to set the CFC standard above the actual flow of CFCs cannot contribute to social welfare. Constraint (15) must hold as an equality as well. If there were an excess supply of resource services, it would be possible to increase both Q and T in such a way that the CFC flow remains constant. Yet this implies a social welfare gain; hence unemployed resource services cannot exist at the social optimum.

The equality constraints in Equations (15) and (16) yield the following values of Q and T:

$$Q = \frac{a_2 \cdot R + b_2 \cdot D}{a_2 \cdot b_1 + (1-a_1) \cdot b_2} \tag{18}$$

$$T = \frac{(1-a_1) \cdot R - b_1 \cdot D}{a_2 \cdot b_1 + (1-a_1) \cdot b_2} \tag{19}$$

Substituting these values into the equality constraint (17) we obtain the following transformation function:

$$\dot{P} = \gamma_1 \cdot R + \gamma_2 \cdot D - \delta P \tag{20}$$

where $\dot{P}$ stands for both the actual and the prescribed flow of CFCs and

$$\gamma_1 = \frac{e_1 \cdot a_2 - (1-e_2) \cdot (1-a_1)}{a_2 \cdot b_1 + (1-a_1) \cdot b_2}$$

$$\gamma_2 = \frac{e_1 \cdot b_2 + (1-e_2) \cdot b_1}{a_2 \cdot b_1 + (1-a_1) \cdot b_2} + e_3$$

Since all the input-output coefficients are positive and $(1-a_1)$, $(1-e_2) > 0$, γ_2 is positive.

To interpret the sign of γ_1, let $\dot{P}_Q = e_1 \cdot Q$ be the flow of CFCs generated by the production of Q and $D_Q = (1-a_1) \cdot Q$ be the net demand satisfied by the production of Q. Moreover, let $\dot{P}_T = -(1-e_1) \cdot T$ be the net CFC flow cleansed by T and $D_T = -a_2 \cdot T$ be the intermediate demand generated by T. Then

$$\gamma_1 \gtreqless 0 \Longleftrightarrow \dot{P}_Q/D_Q \gtreqless (\dot{P}_T/D_T) \tag{21}$$

Let us consider the following conceptual experiment. Suppose that the resource supply increases and that, as a result, Q and T rise in such a way that the incremental increase in net emissions generated by Q is equal to the increase in net emissions cleansed by T. Then, if $\gamma_1 < 0$, the production activity satisfies more demand for Q than the treatment activity generates. Thus, it is possible for consumption to rise. In other words, the treatment activity requires less than the incremental increase in production to cleanse the incremental emission flow which the incremental production increase generated. Consequently, an increase in the supply of resources permits consumption to rise with the emission flow unchanged. We take this to be the usual case, and thus let $\gamma_1 < 0$.

To fix ideas, assume a Cobb-Douglas utility function.[12] The policymaker's problem is to maximize

$$W = \int_0^\infty e^{-rt} \cdot D^\alpha (\overline{P}-P)^\beta \, dt \tag{22}$$

subject to the transformation function of Equation (20) where $\overline{P}$ is the maximum socially tolerable rate of emissions and the parameters $0 < \alpha$,

$\beta < 1$. In this policy problem, P is the state variable and D and $\dot{P}_s = \dot{P}$ may be adopted as the control variables. However, given the level of R, $\dot{P}$ is uniquely related to D and P. Thus D alone will suffice as control variable. Define the level of environmental quality, ε, to be $(\overline{P}-P)$. Then the necessary conditions for an optimal solution of the policy problem are:

$$\frac{\dot{D}}{D} = \frac{\delta \cdot (1+\beta) + r}{(1-\alpha)} + \left(\frac{\gamma_2 \cdot \beta}{\alpha}\right)\frac{D}{\varepsilon} + \left(\frac{\beta}{1-\alpha}\right) \cdot \left[\delta \cdot \overline{P} - \gamma_1 \cdot R\right] \cdot \frac{1}{\varepsilon} \quad (23)$$

$$\frac{\dot{\varepsilon}}{\varepsilon} = - \gamma_2 \cdot \frac{D}{\varepsilon} + \left[\delta \cdot \overline{P} - \gamma_1 \cdot R\right] \cdot \frac{1}{\varepsilon} - \delta \quad (24)$$

Consequently, at the stationary state, in which $(\dot{D}/D) = 0$ and $(\dot{\varepsilon}/\varepsilon) = 0$,

$$D = D^* = \frac{\delta + r}{d \cdot (1-\alpha)} \cdot \left[\delta\ \overline{P} - \gamma_1 \cdot R\right] > 0 \quad (25)$$

$$\varepsilon = \varepsilon^* = \frac{\gamma_2 \cdot \beta}{d \cdot \alpha \cdot (1-\alpha)} \cdot \left[\delta \cdot \overline{P} - \gamma_1 \cdot R\right] > 0 \quad (26)$$

where $d = \left(\frac{\gamma_2 \cdot \beta \cdot \delta}{\alpha}\right) + \gamma_2 \cdot \left(\frac{\delta \cdot (1+\beta) + r}{1 - \alpha}\right)$.

At the initial point in time, $t = 0$, the level of environmental quality is given: $\varepsilon = \varepsilon(0)$. The policymaker inherits this from the past. By setting D, the policymaker can control the level of environmental quality in the future. Each value of $D = D(0)$, when inserted in equations (23) and (24), yields a different time path for D and ε. It is easy to show (for example, see Arrow and Kurz[13]) that for any given value of $\varepsilon(0)$, there exists a unique value of D(0) such that $(D,\varepsilon) \rightarrow (D^*, \varepsilon^*)$ as $t \rightarrow \infty$, and the resultant time path for D and ε satisfies sufficient conditions for social optimality. This time path lies on either of the two branches of the upward-sloping stable path, labeled SP in part A of figure 15-3. In figure 15-3, part B, this path is translated into one relating the level of consumption to the stock of CFCs.

In our model, macroeconomic policies are those which influence D and environmental standards for CFCs control $\dot{\varepsilon}$; D and $\dot{\varepsilon}$ should be set in such a way that the economy remains on the stable path. This policy prescription illustrates our Contentions 1 and 2. It is not sensible to devise macroeconomic policies independently of environmental standards and, unless $(D,\varepsilon) = (D^*, \varepsilon^*)$, both consumption and environmental quality should be made to change through time.

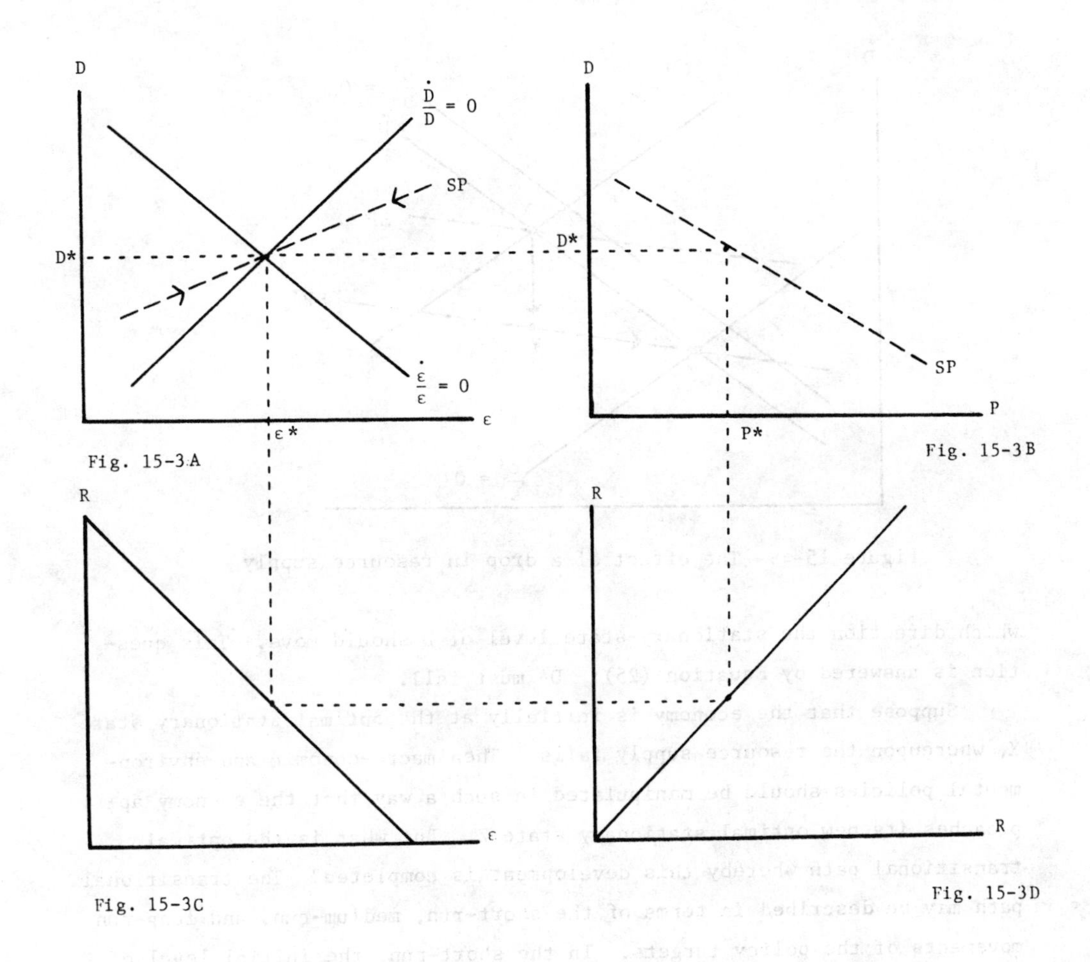

Figure 15-3. The optimal time paths of consumption, environmental quality, and CFC stocks

15.3.2 Resource Shortages and International CFC Transfers

To motivate contention 3, consider a drop in the supply of resource services, R (for example, a fall in the flow of petroleum supplied by the members of the OPEC cartel). From Equations (23) and (24) it is evident that the $(\dot{D}/D) = 0$ function shifts upward in ε - D space and the $(\dot{\varepsilon}/\varepsilon) = 0$ function shifts downward in this space, as illustrated in figure 15-4. Thus, the stationary-state level of ε must fall, but it is not clear in

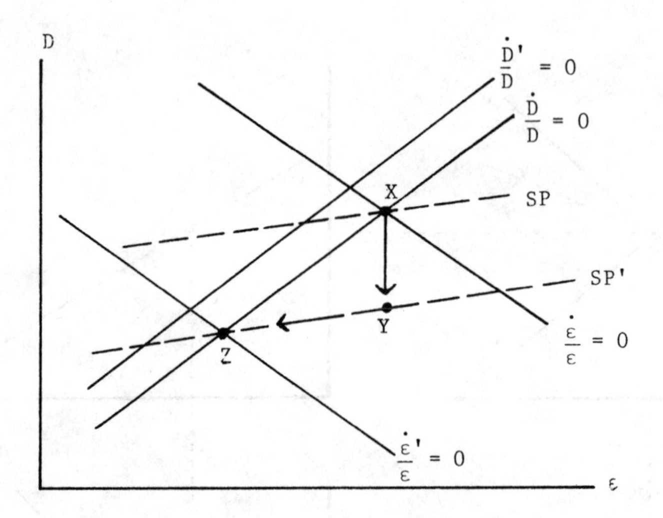

Figure 15-4. The effect of a drop in resource supply

which direction the stationary-state level of D should move. This question is answered by Equation (25): D* must fall.

Suppose that the economy is initially at the optimal stationary state X, whereupon the resource supply falls. Then macroeconomic and environmental policies should be manipulated in such a way that the economy approaches its new optimal stationary state Z. But what is the optimal transitional path whereby this development is completed? The transitional path may be described in terms of the short-run, medium-run, and long-run movements of the policy targets. In the short-run, the initial level of environmental quality ($\varepsilon(0)$) - and therefore also the initial level of CFCs concentrations ($P(0)$) - remains unchanged, while the level of D is free to vary. In the medium run, both D and ε (together with P) are free to vary, but the new optimal stationary state is not attained. Lastly, in the long run, the entire transition from the initial to the final stationary state is undertaken.

In the short run there is a vertical shift from the initial stationary state to a point on the new stable path. Yet the position of the new stable path cannot be deduced unambiguously from a qualitative examination of the shifts in the $(\dot{D}/D) = 0$ and $(\dot{\varepsilon}/\varepsilon) = 0$ functions. There are three possibilities: the stable path shifts downward (as illustrated in figure

15-4) and thus calls for a short-run fall in D, or it shifts upward and occasions a short-run rise in D, or it remains unchanged, in which case there is also no change in D.

The medium-run and long-run movements in D and ε are unambiguous, however. The stable path is upward-sloping; hence, both D and ε must fall steadily in the medium run. As the level of D falls, resource services are transferred out of the consumption-good sector and into the treatment-service sector. Consequently, the level of environmental quality must decline by smaller and smaller amounts with the passage of time. In the long-run, as noted, both consumption and the level of environmental quality decrease.

In sum, a fall in resource supplies calls for contractionary macroeconomic policies and less stringent environmental standards. The appearance of resource shortages makes it desirable for policymakers to pursue less ambitious environmental goals.

The socially desirable effect of international CFC transfers on macroeconomic-environmental policies is quite analogous. It is well known that CFC emissions are a global problem. One country's emissions affect another country's welfare. To take these intercountry influences into account, let us amend the transformation function as follows, modifying Equations (20), (23), (24), (25), and (26), respectively:

$$P = \gamma_1 \cdot R + \gamma_2 \cdot D - \delta \cdot P + I \tag{27}$$

where I is the (exogenously given) net flow of CFC emissions from abroad. As a result, the necessary conditions for social optimality become

$$\frac{\dot{D}}{D} = \frac{\delta \cdot (1+\beta) + r}{(1-\alpha)} + \left(\frac{\gamma_2 \cdot \beta}{\alpha}\right) \cdot \frac{D}{\varepsilon} + \left(\frac{\beta}{1-\alpha}\right) \cdot \left[\delta \cdot \overline{P} - \gamma_1 \cdot R - I\right] \cdot \frac{1}{\varepsilon} \tag{28}$$

$$\frac{\dot{\varepsilon}}{\varepsilon} = -\gamma_2 \cdot \frac{D}{\varepsilon} + \left[\delta \cdot \overline{P} - \gamma_1 \cdot R - I\right] \cdot \frac{1}{\varepsilon} - \delta \tag{29}$$

and the stationary state may be characterized as

$$D = D^* = \frac{\delta + r}{\alpha \cdot (1-\alpha)} \cdot \left[\delta \cdot \overline{P} - \gamma_1 \cdot R - I\right] \tag{30}$$

$$\varepsilon = \varepsilon^* = \frac{\gamma_2 \cdot \beta}{d \cdot \alpha \cdot (1-\alpha)} \left[\delta \cdot \overline{P} - \gamma_1 \cdot R - I\right] \tag{31}$$

A rise in I shifts the $(\dot{D}/D) = 0$ function upwards and the $(\dot{\varepsilon}/\varepsilon) = 0$ function downward in ε - D space. D* and ε* both fall. Once I becomes sufficiently large, the $(\dot{D}/D) = 0$ and $(\dot{\varepsilon}/\varepsilon) = 0$ functions no longer intersect in the positive quadrant of ε - D space and then the sufficient conditions for social optimality can no longer be satisfied. Yet as long as I does not fall into this range, the optimal effect of international CFC transfers on D and ε are qualitatively the same as the effect of resource shortages. An increased flow of CFC emissions from abroad requires the pursuit of less ambitious environmental policies at home.

The primary message of this section is that standards for CFC emissions should not be set in isolation from other economic phenomena. These standards are relevant to a number of important industries (for example, the automobile and the plastics and construction industries) and thus they may be expected to have significant macroeconomic impacts. Similarly, macroeconomic policies may be expected to have a nonnegligible influence on CFC emissions. Thus, CFC standards should be formulated conjointly with macroeconomic policies, not only in a static, but also a dynamic sense. Moreover, both sets of policies should be responsive to fluctuations in resource supplies and international emission transfers.

References

1. W. J. Baumol, and W. E. Oates, "The Use of Standards and Prices for Protection of the Environment," Swedish Journal of Economics (March 1971) pp. 42-54; and W. J. Baumol and W. E. Oates, The Theory of Environmental Policy (Englewood Cliffs, Prentice Hall, 1975).
2. Ibid.
3. Baumol and Oates, The Theory of Environmental Policy.
4. C. Almon, M. Buckler, L. Horowitz, and T. Reinbold, 1985: Extended Forecasts of the American Economy (Lexington, Mass., D.C. Heath).
5. Since the publication of Almon and coauthors, Extended Forecasts, the Almon model has been expanded to 200 sectors. The sector numbers here are given in terms of the 200-sector aggregation scheme.
6. National Research Council, Halocarbons: Environmental Effects of Chlorofluoromethane Release (Washington, D.C., National Academy of Sciences, 1976).

7. Almon and others, Extended Forecasts.

8. United States Department of Commerce, Bureau of the Census, Statistical Abstract of the United States, 1975 (Washington, D.C., 1975).

9. National Research Council, Halocarbons.

10. Other economic phenomena, such as technological progress, are relevant as well; yet these lie beyond the scope of this paper. With regard to technological progress, see D. Snower, "Dynamic Environmental Targets and Technological Progress," International Economic Review (1982).

11. R. d'Arge, and K. Kogiku, "Economic Growth and the Environment," Review of Economic Studies vol. 40 (January 1973) pp. 61-77; B Forster, "Optimal Consumption Planning in a Polluted Environment," Economic Record vol. 49 (December 1973) pp. 534-545; E. Keeler, M. Spence, and R. Zeckhauser, "The Optimal Control of Pollution," Journal of Economic Theory vol. 4 (February 1972) pp. 19-34; K. Goran-Maler, Environmental Economics: A Theoretical Inquiry (Baltimore, Md., John Hopkins University Press for Resources for the Future, 1974); C. Plourde, "A Model of Waste Accumulation and Disposal," Canadian Journal of Economics vol. 5 (February 1972) pp. 119-125; and V. K. Smith, "Dynamics of Waste Accumulation: Disposal Versus Recycling," Quarterly Journal of Economics vol. 86 (November 1972) pp. 600-616.

12. Yet our qualitative conclusions hold for the entire family of utility functions U(D,P), $U_D > 0$, $U_{DD} < 0$; $U_P < 0$, $U_{PP} < 0$; and $U_{DP} \leq 0$.

13. Kenneth J. Arrow, and Mordecai Kurz, "Optimal Growth with Irreversible Investment in a Ramsay Model," Econometrica vol. 38 (March 1970) pp. 331-344.